AF693980

Coral Reefs of the World

Volume 7

Series editors
Bernhard Riegl, Dania Beach, Florida, USA
Richard E. Dodge, Dania, Florida, USA

More information about this series at http://www.springer.com/series/7539

Dawit Tesfamichael • Daniel Pauly
Editors

The Red Sea Ecosystem and Fisheries

Springer

Editors
Dawit Tesfamichael
Fisheries Centre
University of British Columbia
Vancouver, BC, Canada

Department of Marine Sciences
University of Asmara
Asmara, Eritrea

Daniel Pauly
Fisheries Centre
University of British Columbia
Vancouver, BC, Canada

Additional material to this book can be downloaded from http://extras.springer.com.

ISSN 2213-719X ISSN 2213-7203 (electronic)
Coral Reefs of the World
ISBN 978-94-024-1349-6 ISBN 978-94-017-7435-2 (eBook)
DOI 10.1007/978-94-017-7435-2

Springer Dordrecht Heidelberg New York London

Softcover reprint of the hardcover 1st edition 2016

Printed on acid-free paper

Springer Science+Business Media B.V. Dordrecht is part of Springer Science+Business Media (www.springer.com)

Foreword

In the summer of 1984, after nearly a month spent at sea surveying reefs of the Saudi Arabian coast, a tiny wooden boat pulled up beside our sailing catamaran. The three men on board, Egyptians on a long excursion from their home port in the Gulf of Suez, had come to barter lobsters for freshwater. Several things made the encounter memorable. They had no shelter, limited provisions and were far from land in a rickety and barely seaworthy vessel. But they seemed perfectly at ease and left with broad smiles and laughter as we waved them off four lobsters lighter and a jerrycan of water heavier. The other noteworthy thing was that they were the first fishermen we had seen in weeks.

Despite their great richness and abundance of life, the Red Sea reefs we dived seemed scarcely to have been touched by fishing. Here and there we came across the remains of seasonal fishing camps on offshore islands or would see a wiry figure standing at the reef edge with hook and line. But there was little in the way of commercial fishing outside a handful of ports, most of them in the southern Red Sea. Tallying the counts of sharks and turtles our team made at hundreds of sites along the length of Saudi Arabia's coast, I was able to detect a weak signal of fishing in reduced abundance near coastal towns. But for the most part, the reefs were intact, brimming full of heavy-bodied groupers and toothsome snappers and emperors.

I didn't know it then, but others had their sights on these fish at the time. Fisheries development officers from the Food and Agriculture Organisation, United Nations Development Program and others were busy persuading governments of the region to invest more in fishing. They saw the Red Sea as an untapped source of fisheries wealth, and reading their reports now there is the sense that fish left in the water were seen as a flagrant waste of opportunity; people in Red Sea nations were sitting around a blue gold mine, they argued. So ventures were established to right this wrong, ports developed, engines added to boats and trawlers brought in. Some fisheries prospered, while others were rapidly expanded only to expire shortly after, like that for sharks along the Egyptian coast. Few have since lived up to the expectations of their proponents. The Red Sea may be rich, but it turns out not to be very productive, which makes it difficult to balance profit with long-term sustainability.

This timely book reconstructs the history of fishing in the Red Sea from the mid-twentieth century to the present. It takes us from the benign neglect of low-impact artisanal fishing, through an era of fisheries expansion and intensification in what could be called the optimistic 1960s and 1970s, to the present era of declining landings despite sustained fishing effort. Throughout this half century of time, there has been little in the way of management or control of fishing, which bodes ill for their long-term prosperity.

The nature of Red Sea fisheries is still changing, with countries like Saudi Arabia bringing in South-East Asians to man their fisheries, unwittingly introducing destructive fishing practices with them. In other places, migrant construction workers fish to supplement their incomes leading to intensive exploitation along the shoreline. In Egypt tourism has supplanted commercial fishing, and stocks are in good health, while conflict has kept nations like Sudan from developing their fisheries. This book documents this diversity of fishing history in unprecedented detail, country by country, bringing a grand and much-needed perspective to a hitherto little-known region. It concludes there is much to be gained from good management of the Red Sea's fragile and valuable resources and much to be lost if such management is not introduced.

Marine Conservation
York, UK

Callum Roberts

Preface and Acknowledgements

The Red Sea is one of the hotspots for coral reef ecology. The unique physical and biological characteristics of the Red Sea enabled it to host reef communities characterized by high rates of endemism for its diverse groups and the northernmost coral reefs in the world. However, there are many anthropogenic impacts on the ecosystems of the Red Sea, as it is a major shipping route, and its coasts are becoming lined with resorts, harbours and urban developments. Still, the most direct human impact of humans with the Red Sea ecosystems is fishing. Fishers' experience of the sea, knowledge and traditions depend on and are informed, more than anything else, by their interaction with the ecosystem through fishing. Hence, examining the fisheries of the Red Sea becomes important to understand this ecosystem.

The seeds of this book were planted early during the PhD program of Dawit Tesfamichael, who, under the supervision of Professors Daniel Pauly and Tony J. Pitcher, started building an ecosystem model (presented here as Chap. 9) in order to understand the dynamics of the Red Sea ecosystem and the impacts of fisheries thereon. One of the required inputs to this model were long time series of catches, which were initially assumed to be straightforwardly available. Reality stepped in, however: not only did the countries bordering the Red Sea lack many of the required long time series of fishery statistics, but also what little there was not reliable. This also applied to the fishery statistics disseminated by the Food and Agriculture Organization (FAO) on behalf of the countries bordering the Red Sea countries, which lacked data for the small-scale fisheries and for discards, as well as details on catch composition, and which did not account for the differences between domestic and foreign catches. Thus, the catches of Red Sea fisheries had to be 'reconstructed by countries, sectors and species (groups) composition'. (Re)estimating catches, without which fisheries research cannot be done, was going back to the basics of the discipline. This, however, usually does not get enough emphasis. This is changing now, due to an initiative taken by the *Sea Around Us* at the University of British Columbia, to establish better fishery statistics in the world, of which this book is one of many products.

Reconstructing the catches of the Red Sea Large Marine Ecosystem (LME) from 1950 to 2010 demanded the collaboration of local fisheries administrations and research personnel from the countries bordering the Red Sea. The main collaborators are included as co-authors, the reason why this book is edited rather than authored by us. However, many more persons than our co-authors need to be acknowledged here. Notably, we are very grateful for the support and cooperation we received from the fisheries administration of the Red Sea countries; without their involvement and their willingness to share their data, this work would not have been possible. Dawit Tesfamichael also received the help of fishers and managers who shared their knowledge and expertise during field interviews in Eritrea, Sudan and Yemen, and he thanks them for their time and insights.

There are several individuals in each country that should be acknowledged individually. However, as their contributions were country specific, we have opted to thank them in each country's chapter. Here, we would like to mention and express our gratitude to Tony Pitcher for his insights and co-supervising Dawit Tesfamichael's PhD programme, UBC librarian Sally Taylor for helping us locate old records, the FAO Library for locating rare reports, Dirk Zeller and Kyrstn Zylich for reviewing individual chapters and their underlying data, Evelyn Liu for drafting all our figures and Christopher Hoornaert for preparing our maps.

The research leading to this book was funded by *Sea Around Us*, a scientific collaboration between the University of British Columbia and The Pew Charitable Trusts. Additional support was provided by Eritrea's Coastal, Marine & Island Biodiversity Conservation Project (ECMIB) and the Khaled bin Sultan Living Oceans Foundation, all of which are gratefully acknowledged.

Note, finally, that the Fisheries Centre of the University of British Columbia was renamed, in mid-2015, the "Institute for the Oceans and Fisheries", and given a broader mission. Here, however, we maintain the name current while the work leading to this book was performed.

September 2015

Dawit Tesfamichael
Daniel Pauly

Contents

Contributors

Abdalla Nassir Elawad Animal Resources Research Corporation, Ministry of Animal Resources and Fisheries, Khartoum, Sudan

Rhona Govender Fisheries Centre, University of British Columbia, Vancouver, BC, Canada

Sahar Fahmy Mehanna National Institute of Oceanography and Fisheries, Ministry of State for Scientific Research, Suez, Egypt

Sammy Mohamud Research Division, Ministry of Fisheries, Massawa, Eritrea

Daniel Pauly Fisheries Centre, University of British Columbia, Vancouver, BC, Canada

Peter Rossing Fisheries Centre, University of British Columbia, Vancouver, BC, Canada

Hesham Saeed Marine Research and Resource Center, Hodeidah, Yemen

Dawit Tesfamichael Fisheries Centre, University of British Columbia, Vancouver, BC, Canada

Department of Marine Sciences, University of Asmara, Asmara, Eritrea

About the Editors

Dawit Tesfamichael is a fisheries researcher whose passion for marine science started in the Red Sea. He is fascinated by ecosystems and the interactions with humans. His research has been focused on the functioning of ecosystems, with emphasis on fisheries. He has published on the fisheries of the Red Sea in scientific journals and other outlets.

Daniel Pauly is a Professor of Fisheries at the Fisheries Centre of the University of British Columbia, Vancouver, Canada. His main areas of research are tropical fishery systems and global analyses of fisheries. His award-winning work is documented in over 500 scientific articles and book chapters, and he has authored or co-authored about 30 books and reports.

Introduction to the Red Sea

1

Dawit Tesfamichael and Daniel Pauly

Abstract

The Red Sea, characterized by a number of unique oceanographic and biological features, is a hotspot for coral reef ecology. It also provided humans for millennia, from the earliest record of human consumption of seafood to its current role as an important fishing ground for the seven countries along its shores. Contemporary fisheries need monitoring and management, and catch data are crucial to both. However, reliable time-series of catch data are lacking for most Red Sea fisheries. Here, the catches of Red Sea fisheries are 'reconstructed' from 1950 to 2010 by country (i.e., Egypt, Sudan, Eritrea, Yemen, Saudi Arabia, Jordan and Israel) and sector (artisanal, subsistence, industrial and recreational), and in terms of their species composition. Historical documents, published and unpublished reports and other grey literature, databases, field surveys, anecdotal information, interviews, and information on processed seafood products were used as sources.

When reliable data were available for a number of years, they were used as anchor points, and missing years were interpolated, based on assumptions of continuity, and given the best knowledge of the fisheries available. The reconstructed catches (which also include discards) were compared to the statistics submitted by the above-mentioned countries to the Food and Agricultural Organization (FAO) of the United Nations. Overall, the total Red Sea catch was low (around 50,000 $t \cdot year^{-1}$) until 1960, increased to a peak (around 177,000 t) in 1993, and is declining since. Overall, it was 1.5 times higher than the catch officially submitted to FAO by the countries bordering the Red Sea. Artisanal fisheries generally contributed about half of the total Red Sea catch, while the composition of the catch was extremely varied, with no single species or even family dominating. In addition to the national catch reconstructions, the local (Arabic) names of common commercial fishes, an ecosystem model and a time series of the effort are also presented. The resulting catch trends provide crucial historical records and important guidance for the development of future fisheries management policies aiming at resource conservation and sustaining the livelihoods of the coastal communities. Extra material for this chapter is available from http://extras.springer.com/.

Keywords

Catch time series • Mis-reported catch • Catch composition • Ecosystem based management • Red Sea

Electronic supplementary material The online version of this chapter (doi:10.1007/978-94-017-7435-2_1) contains supplementary material, which is available to authorized users.

D. Tesfamichael (✉)
Fisheries Centre, University of British Columbia,
Vancouver, BC, Canada

Department of Marine Sciences, University of Asmara,
Asmara, Eritrea
e-mail: dawittes@gmail.com

D. Pauly
Fisheries Centre, University of British Columbia,
Vancouver, BC, Canada

D. Tesfamichael, D. Pauly (eds.), *The Red Sea Ecosystem and Fisheries*, Coral Reefs of the World 7,
DOI 10.1007/978-94-017-7435-2_1

Introduction

The Red Sea is an elongated narrow sea between Northeastern Africa and the Arabian Peninsula, ranging from 30°N to 12°30'N and from 32°E to 43°E with a straight length of about 2,000 km and an average width of 208 km (Fig. 1.1). The maximum width is 354 km in the southern part (Morcos 1970), and the total area is 4.51×10^5 km^2. The Red Sea is connected to the Indian Ocean in the south through the narrow strait of Bab al Mandab, the door of fortune. Bal al Mandab, which is only 29 km wide, has a sill 137 m below sea level, which limits the circulation of water between the Red Sea and the Gulf of Aden. The Red Sea is also connected to the Mediterranean Sea through the Suez Canal since its opening in 1869. The average depth of the Red Sea is 491 m, with a maximum of 2,850 m. In the north, the Red Sea is divided into the Gulfs of Suez and Aqaba. The Gulf of Suez is generally wide, shallow and muddy, while the Gulf of Aqaba is narrow and deep.

Geological Evolution

The Red Sea was formed by plate tectonics, i.e., by the African and Arabian plates drifting apart, and is part of a larger tear that includes the Dead Sea and the East African rift systems. Geologically, the Red Sea is a young ocean that is still growing or spreading (Braithwaite 1987). The zone was already structurally weak during the Pan-African orogeny 600 Ma. The separation of the Arabian and African plates is believed to have started in the Tertiary period, between the Eocene and Oligocene periods; it accelerated during the late Oligocene, with intense magmatic activity and the development of a continental rift (Makris and Rihm 1991). The Red

Fig. 1.1 The Red Sea (17,640 km^2) and the surrounding countries, including their Exclusive Economic Zones (EEZs) and shelf areas

Sea depression is believed to have been flooded by the Mediterranean as a result of extensive sinking in the early Miocene (Girdler and Southren 1987). Since its inception, the Red Sea went through a series of connections and disconnections with the Mediterranean in the North and the Indian Ocean in the south. At the end of Miocene, upheaval of land occurred and the Red Sea was disconnected from the Mediterranean to become a separated salty lake. At the beginning of the Pliocene, the Red Sea was reconnected with the Mediterranean, and for the first time, it was also connected with the Indian Ocean, but at the end of the Pliocene, the northern connection with the Mediterranean was closed off due to crustal plate movement. The connection with the Indian Ocean was closed off during the Pleistocene, when the Red Sea again became an isolated sea. At the end of the Pleistocene, a glacial period, its connection with the Indian Ocean was re-established, whereas the connection with the Mediterranean remained closed until it was artificially opened via the Suez Canal in 1869 (Goren 1986; Getahun 1998). The Red Sea being young and still expanding is used as a case study to understand and explain plate tectonics, mid ocean ridges and formation of oceans.

Physical Oceanography

The Red Sea area is generally arid, rainfall is very sparse with annual average ranging from 1 to 180 mm (Edwards 1987). Evaporation, with an annual average of 2 m (Morcos 1970), largely exceeds precipitation, and the deficiency is made up by the flow of water from the Indian Ocean through Bab al Mandab. In winter, warmer and less saline water flows into the Red Sea in the surface layer, while cooler and saltier water flows into the Gulf of Aden in the lower layer. In summer, there are three layers of water flow in the strait. In addition to the two winter flows, warm water flows on the surface from the Red Sea to the Gulf of Aden (Smeed 2004). The exchange with the Gulf of Aden is a major driving mechanism of the Red Sea ecosystem functioning (Triantafyllou et al. 2014). Sea and air temperatures are high in the Red Sea with mean annual sea surface temperature of 28 °C. Additionally, the Red Sea is undergoing an intense and rapid increase in temperature, which is attributed to climate change. The average temperature for the period 1994–2007 was 0.7 °C higher than the period 1985–1993. The increase in 1994 was the strongest shift in the last 160 years (Raitsos et al. 2011).

Another remarkable characteristic of the Red Sea is its high salinity, about 35 psu on average at the surface; readings as high as 40.5 psu are also reported. The high salinity of the Red Sea is due to the combination of its geological history and its location in dry and hot environment. Though originally the Red Sea depression was flooded with Mediterranean water, it soon started to become more saline due to high evaporation. Later during the glacial period, the Red Sea was an isolated salty lake with salinity higher than the present by 10 psu. The highly saline water was diluted by water from Indian Ocean when the Red Sea was reconnected with the Indian Ocean (Thunell et al. 1988; Rohling 1994). However, it is still more saline than the Indian Ocean water due to high evaporation (Morcos 1970). The salinity in the Red Sea increases from south to north. In the south (12.5°N), through which water from the Indian Ocean flows to the Red Sea, the salinity is around 36.5 psu, similar to the Gulf of Aden. It increases to 38 psu at 17°N, 39 psu at 22°N and 40 psu at 26°N, the gulfs of Suez and Aqaba (Edwards 1987).

Biological Oceanography and Origin of the Biota

The Red Sea is not very productive, mainly due to lack of nutrient-rich terrestrial run off; also, there is almost no upwelling to lift nutrient-rich deep water to the surface where photosynthesis can occur. Moreover, for most of the year, the vertical mixing of water is prevented by a permanent thermocline as the temperature of the sub-surface water is always lower than the warm surface temperature. The thermocline is deeper in winter than summer (Edwards 1987). In the northern part, the deep waters are renewed by cold dense water from the surface which is cooled by cold winds (Sofianos and Johns 2015). The deep waters have higher nutrient contents causing the flourishing of green algae especially in the Gulf of Aqaba. Generally, the southern part of the Red Sea is more productive than the northern part due to the flow of nutrient rich water from the Indian Ocean, the main nutrient input, and the re-suspension of nutrients from the bottom sediments by turbulent mixing over shelf areas (Sheppard et al. 1992). The average primary productivity for the Red Sea large marine ecosystem (LME) based on SeaWiFS database was calculated to be 150–300 $gC \cdot m^{-2} \cdot year^{-1}$, which is considered moderately productive at a global scale (McGinley 2008). The shallow Gulf of Suez is also productive and supports many exploited fish populations. The Red Sea receive about 6 t of dust per year (Jish Prakash et al. 2015); the dust particles bring nutrients to the Red Sea; however, this contribution has never been quantified.

The connections of the Red Sea with the Mediterranean in the north and the Indian Ocean in the south account for the species that colonized it at different times. Though the Red Sea was first populated by Mediterranean species, its current biota resembles that of the Indian Ocean. When the Red Sea was disconnected from Mediterranean and for the first time connected with the Indian Ocean in the beginning of the Pliocene period (about 5–6 million years ago), it was populated by Indian Ocean fauna. Later during the glacial period of the Pleistocene, the level of the world's oceans was low. The Red Sea was isolated with high level of salinity (about 50 psu at the surface) and low temperature (about 2 °C lower than the present) (Thunell et al. 1988). This resulted in the extinction of many species. Later, when the Red Sea was reconnected with the Indian Ocean at the end of the glacial period, 10,000–12,000 years ago, an opportunity was created for Indian Ocean species to re-populate it (Goren 1986). After the opening of the Suez Canal in 1986, organisms migrated from the Red Sea to the Mediterranean Sea, less in the other direction. These 'Lessepsian migrations' (Por 2012) are now becoming more frequent, due to rapid warming of the Eastern Mediterranean (Keskin and Pauly 2014).

As a result of its connection to the speciose Indo-Pacific fauna, the Red Sea currently has very high fish diversity, with more than 1,400 species of fishes reported in FishBase (www.fishbase.org). It is also characterized by high degree of endemism, due to the closures alluded to above, with estimates of endemic fish species reaching 10–17 % (Ormond and Edwards 1987). Because the Red Sea has very low nutrient input, as explained above, species that can survive its extreme environments have very good chance to dominate, as there are fewer competitors. One example is the phytoplankton *Trichodesmium erythraeum*, a blue-green alga (cynobacterium) that can overcome nitrate depletion by fixing atmospheric nitrogen dissolved in the water. In calm waters, its filaments float to the sea surface of the Red Sea and form a rather reddish scum, the likely origin of the name 'Red Sea' (and incidentally, of Eritrea's as well).

On the shores of coastal lagoons and sheltered bays mangroves are common. The most common species is *Avicennia marina*. *Bruguiera gymnorhiza* and *Ceriops tagal* also occur, though they are less common. The shallow waters of the lagoons and bays are home to seagrass beds. About 500 species of algae are reported from the Red Sea. Most algae in the north and central part are macroscopic, non-calcareous, brown, green and red algae. In the south, large brown algae such as *Sargassum* dominate (Walker 1987).

Five sea turtle species are reported from the Red Sea: hawksbill, green, olive ridley, loggerhead and leatherback. Hawksbill and green turtles are the most common, and are reported to nest along Red Sea beaches (Frazier et al. 1987). There is no active hunting for sea turtles in the Red Sea at the moment, but they are accidentally caught in fishing nets. The rich seagrass beds support dugongs, which are reported from Gulf of Suez in the north to Eritrea's Dahlak Archipelago in the south (Preen 1989). There used to be active fishing for turtles and dugongs. The reports of cetaceans from the Red Sea are sparse. Seven species of dolphins are commonly reported, as well as occasional spotting of killer whale and false killer whale. Frazier et al. (1987) suggested that the narrow strait of Bab al Mandab and the low productivity in the Red Sea as reasons for the scarcity of cetaceans. As far as seabirds are concerned, the enclosed nature of the Red Sea acts as a barrier for pelagic fishes on which many seabirds feed. As a result pelagic seabirds, such as shearwaters and petrels, are poorly represented. Because of its elongated shape, the Red Sea has high coast to sea area ratio and its seabird fauna is dominated by coastal species (Evans 1987).

Human Settlements

According to archeological evidence, human settlement on the Red Sea coast started millennia ago (Horton 1987) and the Red Sea has the oldest records of early Middle Stone Age artefacts (about 125 kyr ago) of human use of marine resources, in the form of giant clam and other shell middens (Walter et al. 2000). The Red Sea was also used as an important trade route between the Indian Ocean and the Mediterranean. However, in contrast with the rest of the world, where most of the population lives in a narrow strip of land along the coast (Edgren 1993), the population density on the Red Sea coast is still very low, except for very few major ports and cities. This is mainly due to the arid and hot climate, which resulted in most of the settlements being farther inland, in milder climate and where freshwater is less scarce. Until recently, this greatly limited the degree of coastal shoreline alteration, pollution and resource extraction. Thus, many Red Sea communities still depend on harvesting marine resources for subsistence using traditional methods of shell collection and fishing (Fig. 1.2).

However, in the last few decades, the wider availability of technology coupled with cheap oil, at least for the oil producing countries, is changing the demography of the Red Sea coast. The major port cities are metropolitan hubs, with diverse economic activities. For some countries, e.g., Saudi Arabia, fishing has become less important compared to other economic activities, resulting in the importation of foreign workers to do the fishing activities as most Saudis left the trade (see the Saudi Arabia chapter). Egypt has a strong recreational and tourism industry, and its coast is quite populated, creating pressure on the coastal ecosystems. Air conditioners and desalination plants are making life easier. A typical example is the Saudi Arabia coast where vibrant cities, such as Jeddah, have grown fast and new cities (e.g., Yanbu) are developing. In such cities, coastal reclamation and dredging are becoming common for residential, commercial and industrial purposes. Pollution is prevalent around urban areas and ports, and lack of sewage treatment is a serious problem throughout the Red Sea, as is the pollution from oil refineries. Overall the impact of human activities is growing (Frihy et al. 1996).

Research Expeditions

One of the earliest scientific expeditions to the Red Sea is the Danish *Arabia Felix* of 1761–1767, which spent October

Fig. 1.2 Fishing boats in the Red Sea; (*left*): *huris* in Eritrea (Photo: Steffan Howe); (*right*): *sambuk* in Saudi Arabia (Photo: Julia Spaet)

1762–August 1763 in the Red Sea area. It included the Swedish naturalist Peter Forsskål, a student of Linnaeus, who made an extensive collection of plants and animals, and particularly fish. His report was published posthumously by Carsten Niebuhr, the sole survivor (Forsskål 1775). There were many fragmented accounts of expeditions, most of them unsuccessful, to the Red Sea in the eighteenth and nineteenth centuries. One important and outstanding work in describing the Red Sea ecosystem and its organisms is that of Carl Benjamin Klunzinger, a German medical doctor who worked as a quarantine inspector in the Egyptian Red Sea port of Quseir from 1863 to 1869 and 1872 to 1875. His descriptions include coral fauna, fish, crustaceans, hemichordates and also meteorological (Klunzinger 1870, 1872), and cultural observations (Klunzinger 1878). An Austrian research vessel, the *Pola*, conducted an expedition in 1895–1896 to the northern Red Sea (Luksch 1898) and 1897–1898 to the south (Luksch 1900), including the first oceanographic studies and sampling of deep sea life down to 2,000 m (Head 1987a). The specimens from the expedition are kept in the Natural History Museum in Vienna (Stagl et al. 1996).

More recent expeditions include the *John Murray* expedition carried out using the Egyptian research vessel *Mabahiss* 1933–1934 (Rice 1986; Tesfamichael 2005), which collected oceanographic and biological samples throughout the Red Sea and the Arabian Sea (Norman 1939). From 1959 to 1964, the International Indian Ocean Expedition brought some vessels to sample the Red Sea, whose oceanography was compiled by Morcos (1970). An Israeli expedition to the southern Red Sea in 1962 and 1965 (Ben-Tuvia 1968), and the Israeli Marine Biological Station at Eilat, which was opened in 1968, also contributed to the knowledge of the Red Sea. At present, a lot of initiatives are taken by the countries bordering the Red Sea and new information is collected.

Coral Reef Ecosystems

The Red Sea is one of the hotspots for coral reef ecology in the world (Roberts et al. 2002). Although it covers only 0.12 % of the global ocean, it accounts for 6.2 % of global coral reefs (Wilkinson 2008) (Fig. 1.3). In coral bio-geography, the Red Sea is considered part of the Indo-Pacific region and contains the highest diversity of reef communities outside of the Southeast Asian 'coral triangle' (DeVantier et al. 2000). There are 333 reported coral species (Dubinsky and Stambler 2011), of which many are found in other Indo-Pacific locations. The Red Sea also has high level of endemism, estimated at about 10 % (DeVantier et al. 2000; Sheppard 2000).

The high and relatively stable temperature of the Red Sea favours the formation of coral reefs, which are well developed in its northern part, starting from the tip of Sinai Peninsula (Sheppard et al. 1992). The longest continuous fringing reef in the Red Sea extends from Gubal, at the mouth of the Gulf of Suez, to Halaib, at the Egyptian border with Sudan (Pilcher and Alsuhaibany 2000). In the south, the reefs are more patchy, as the turbid waters of the shallow shelf prevent the formation of extensive reefs. Saudi Arabia, at 6,660 km^2, has the largest area of coral reefs in the Red Sea (Bruckner et al. 2011b). There are diverse coral reef structures along the Red Sea coast: fringing (both along the mainland and around islands), platform patch, barrier and ridge reefs (Bruckner et al. 2011a; Al-Sofyani et al. 2014). The ridge reefs are characteristics of the Red Sea. Sanganeb Atoll, located in Sudan near the border with Egypt, is the only atoll in the Red Sea; it surges from a depth of 800 m to form a structure that has been recognized as regionally important for conservation, and proposed to UNESCO as a World Heritage Site in the 1980s (Pilcher and Alsuhaibany 2000). Some bays, locally called *sharms*, create conducive

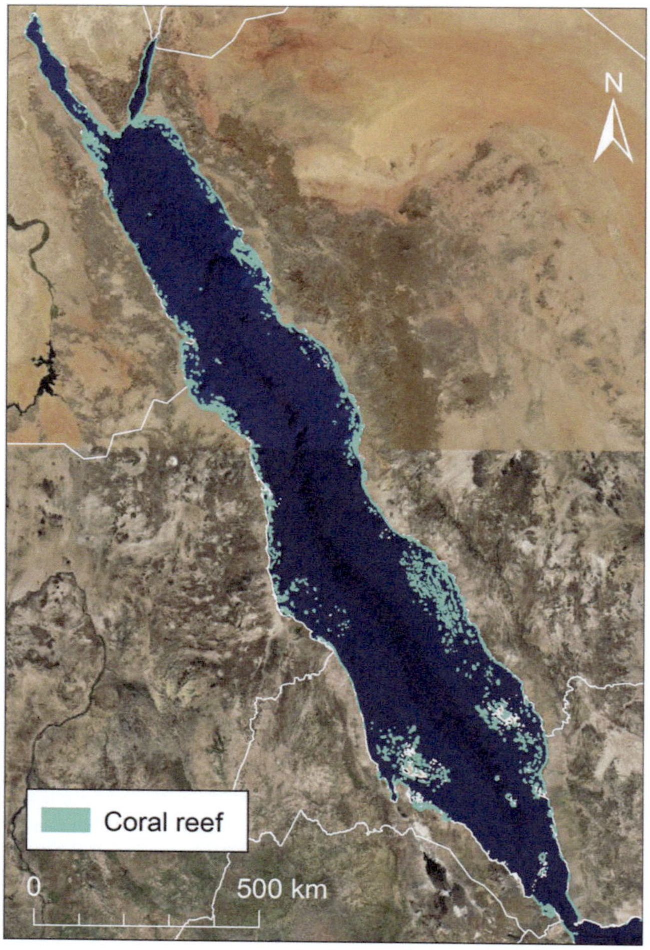

Fig. 1.3 Distribution of coral reefs in the Red Sea (Image base layer credits: Esri, DigitalGlobe, GeoEye, i-cubed, USDA, USGS, AEX, Getmapping, Aerogrid, IGN, IGP, swisstopo, and the GIS user community)

environment for coral development and are unique to the Red Sea (Bruckner et al. 2011a). However, some *sharms* near river mouths are turbid and do not have coral reefs. The reefs in the Gulf of Suez are some of the northernmost reefs in the western Indo-Pacific and have a different community and geographical structures compared to other reefs in the Red Sea (Riegl et al. 2012). The Red Sea also has deep water corals found down to 760 m (Qurban et al. 2014).

The coral reefs of the Red Sea were some of the earliest to be described by researchers from Europe, i.e., Peter Forsskål, Christian Gottfried Ehrenberg, Eduard Rüppell and Carl Benjamin Klunzinger starting in the 1700s, but contemporary information is scarce. Moreover, more than 50 % of the published research on coral reefs of the Red Sea is from the Gulf of Aqaba, which accounts for less than 2 % of the total Red Sea area (Berumen et al. 2013). However, even for the Gulf of Aqaba, there are a lot of issues not well studied and which need attention (Loya et al. 2014). The quality and detail of coral reef research in the Red Sea need to improve. In the international coral reef monitoring standard, the quality of data used for monitoring the Red Sea coral reefs is at the 'low' level (Wilkinson 2008).

The coral reefs of the Red Sea are important habitats for fish and invertebrates (Fig. 1.4). Generally, the populations of invertebrates are healthy except for localized depletions of giant clams in Egypt, other molluscs in Sudan, and lobsters and sea cucumbers throughout most of the Red Sea (Kotb et al. 2008). As far as fish abundance and distribution are concerned, diverse types of fishes inhabit the coral reefs of the Red Sea and there are still areas where large predators, which are the main targets of fishers, are available (Kotb et al. 2008). Sharks are one of the most heavily exploited groups (Tesfamichael and Pitcher 2006; Tesfamichael et al. 2014; Spaet and Berumen 2015). Environmental gradients affect the distribution of fish populations in the Red Sea (Nanninga et al. 2014). In a study conducted in Saudi coast, the density of herbivorous fishes was found to be higher in areas with less live coral cover, due to bleaching, than in reefs with higher live cover (Khalil et al. 2013). In another study from Saudi Arabia, top predators such as jacks, snappers and groupers dominated off shore reef communities resulting in an inverted (top-heavy) biomass pyramids, while in shore communities were dominated by lower trophic-level fish (bottom-heavy) pyramid, which is an evidence of trophic cascade (Kattan 2014). Spawning aggregations have been observed in the Red Sea coral reefs (Gladstone 1996).

Coral reefs recycle their nutrients, which enables them to maintain a high productivity, much like an oasis in a desert. They attract fishers mainly subsistence and small-scale artisanal operators. Globally, coral reefs support small scale fishing activities that provide basic needs to about 500 million people (Wilkinson 2008). In the Red Sea, almost all the commercially important fishes for the handline fishery are found in the reef areas (Barrania 1979; Vine and Vine 1980). Fishing pressure is increasing in the Red Sea and is affecting the coral reef ecosystem, especially spawning and nursery sites (Gladstone 1996; Kotb et al. 2008). In addition to fishing for consumption, ornamental fish are collected in most of the Red Sea countries. Another important economic activity based on the coral reefs of the Red Sea is tourism, mainly diving and snorkeling (Hawkins and Roberts 1994; Kotb et al. 2008). The diverse coral reef ecosystems coupled with clear and warm waters attract many tourists. At present, tourism is developed mainly in the north: in Egypt and Israel. At 250,000 dives per year, the coral reefs off Eilat, Israel, are among the most frequently visited places by recreational divers in the world (Zakai and Chadwick-Furman 2002). Tourism has not developed well in other countries of the Red Sea, mainly because of political instability, but also local customs, which do not encourage tourism activities.

Fig. 1.4 A coral reef ecosystem in Eritrea (Photo: Yohannes Tecklemariam)

The current status of the coral reefs of the Red Sea is debatable. Some, relative to other areas, claim it is in good condition with live coral cover of 30–50 % (Kotb et al. 2008), while others assert that the coral cover showed significant decline and suggest that the claim the Red Sea reefs are healthy suffer from shifting environmental baselines (Price et al. 2014). Using data from more than two decades, Riegl et al. (2012) showed that dividing the Red Sea into three latitudinal faunistic zones, as sometimes done by researchers, was not apparent and coral size had decreased, recruitment had remained stable, and size distribution had not changed significantly, but mean coral size had decreased, due to a decline of large corals. The richest spots were found in Farasan Islands (18–23 spp.) and the Northern Islands of Egypt at Gubal Saghir. Overall, the health and coral cover increases significantly towards the north (Price et al. 2014).

There are multiple threats to the coral reefs of the Red Sea. The main damage has been due to coastal developments for urban and industrial centers, which include land-filling, dredging, port activities, oil spill, sewage and pollution (Kotb et al. 2008). The magnitude of the impacts of coastal construction has increased significantly in the last 30 years, beach oil has declined, but shore debris have increased (Price et al. 2014). The impacts are higher in areas where population size has increased, e.g., Jeddah and Yanbu in Saudi Arabia (Kotb et al. 2008). Other direct human activity affecting coral reefs in the Red Sea are diving and snorkeling, which raise sediments and also break reef structures. This is especially true along the Egyptian and Israeli coasts where there is high intensity of diving (Hawkins and Roberts 1994; Zakai and Chadwick-Furman 2002). In the shallow reefs of Eilat, Israel, where reef degradation has occurred, algal cover was up to 72 %, an indication of damage to the reef, while in the nearby areas in Aqaba, the turf cover was only 6 % (Bahartan et al. 2010). Reefs dominated by algae had higher densities of herbivorous fish (Khalil et al. 2013).

The Red Sea coral reefs suffered from coral bleaching in 1998. The damage was worse in the southern part, but some signs of recovery were seen especially in the central and northern Red Sea. In addition, extreme low tides in 2007 caused coral bleaching and mortality (Kotb et al. 2008). Outbreaks of crown-of-thorns starfish devastating coral reefs have been reported at different places and times along the Red Sea (Wilkinson 2008). The impact of climate change has also been observed in the region (Baker et al. 2004; Raitsos et al. 2011; Riegl et al. 2012). The steady increase in sea surface temperature (SST) has been the key factor in coral reef skeletal growth by 30 % since 1998, rather than the increased acidity of the water (Cantin et al. 2010). Moreover, when and where temperatures decreased, there were signs of recovery indicating the resilience of Red Sea coral reef to changing temperatures (Baker et al. 2004). Using satellite derived sea surface and ground based air temperature, Raitsos et al. (2011) showed that the Red Sea is going through an intense warming period starting the mid-1990s, with an abrupt increase of 0.7 °C since 1994. The Red Sea reef dwellers are adapted to very warm environments; however, they can be vulnerable to further and rapid warming. Thus, understanding abrupt temperature change becomes an important issue, as ecosystems have a better chance to adapt in a slowly rather than in a rapidly changing environment (Raitsos et al. 2011). On the other hand, the adaptation of the Red Sea coral reefs to high temperature causes them to have high bleaching threshold and thus the Red Sea may become a refuge for corals exposed to climate change (Riegl and Piller 2003; Fine et al. 2013).

Considerable progress has been made in the understanding and management of the coral reefs of the Red Sea both by the individual countries and the Regional Organization for the Conservation of the Environment of the Red Sea and Gulf of Aden (PERSGA) (Kotb et al. 2008). PERSGA has been actively involved in assessing the status of Red Sea resources, current issues, needs for additional actions and constraints. Every country bordering the Red Sea has either proposed or approved some form of Marine Protected Areas (MPAs). Some of the important coral reef ecosystems in the Red Sea could be good candidates for MPAs and conservation including the Farasan Islands, Dhalak Islands and Sanganeb Atoll (Gladstone 2000).

Fisheries

The Red Sea has multiple uses, the major one being a shipping route from the Indian Ocean to Europe. Recently, interest in the tourism industry has been increasing, notably in Egypt, which has undergone extraordinarily rapid expansion from a handful of hotels in the 1980s to many hundreds today. As far as resource extraction is concerned, however, fishing is still the most important sector in Red Sea. The Red Sea has a long history (and prehistory) of resource exploitation by humans. Archaeological studies of middle stone age middens from the Eritrean Red Sea coast indicate that humans were eating giant clams and other molluscs about 125,000 years ago, possibly the most ancient such practice on record in the world (Walter et al. 2000). The artisanal fisheries have traditionally operated in relative harmony with the ecosystem because of low population; non-destructive traditional fishing technology; and poor communication and infrastructure. However, depletion of resources have been observed in areas frequently visited by fishers (Tesfamichael 2001) or resources that are specifically targeted by fishers such as sea cucumbers (Tewelde and Woldai 2007; Kotb et al. 2008) and sharks (Tesfamichael 2012; Spaet and Berumen 2015). Recently, more advanced and destructive methods are being used. Currently, fishing operations in the Red Sea range from foot fishers catching fish mainly for their own consumption, to very large trawlers with freezing facilities.

The fisheries in the Red Sea are typical tropical fisheries, multi-gear and multi-species. Most fishing is performed from wooden boats ranging from 5 to 18 meters, locally called *'Sambuk'* and *'Huris'*. Sambuks are larger, and have inboard engines; Huris are smaller and use outboard engines. Both Sambuks and Huris use similar fishing gears, mostly handlining and gillnet (Fig. 1.5). The main difference in the operation of Sambuk and Huri are length of the fishing trip, crew size and capacity (Tesfamichael and Pitcher 2006). Most of the countries do not have fisheries regulations (e.g., quota, total allowable catch) or if they have they are ineffectively enforced, thus the fisheries are practically open-access fisheries. Currently there is no regional fisheries management organization (RFMO) for the Red Sea. An initiative is underway to establish one under the umbrella of FAO.

Total annual potential landings from the Red Sea were once estimated at 360,000 $t \cdot year^{-1}$ (Gulland 1971), but this value needs further scrutiny. Though the Red Sea accounts for 0.12 % of the total world ocean area, its contribution to the world catch is only 0.07 % (Head 1987b). Nevertheless, it is important to the countries in the region. Fishing produces a cheap source of animal protein and provides livelihood for the communities on the coast. Since the countries on the Red Sea coast are generally less industrialized, fisheries can provide multiple livelihoods.

Of the seven countries that border the Red Sea, Jordan and Israel have too small coastlines to support any major fishery. Of the other countries, Egypt and Yemen have well established fisheries and have been utilizing their resources for a long time. Egyptian and Yemen fishermen also fish in other countries' waters. Sudan is the country which utilize its fishery resources the least, besides Jordan and Israel. Saudi

Fig. 1.5 Fishing activities in the Red Sea, Saudi Arabia; (*left*): handlining (Photo: Julia Spaet); (*right*): gillnetting (Photo: Mohamed Gabr)

Arabia has recently established an industrial fishery, in addition to the artisanal fishery that has been active for many years. Eritrea had a strong small pelagic fishery in the past, then the fishery was dormant until it resumed after the country's independence in 1991.

Fishery Data and Assessment

A key part of documenting a fishery is reporting its catches. Given the catch level of a fishery, inferences can be drawn on the intensity of the pressure it exerts, and the approximate number of people involved in, and/or dependant on that fishery. Also, from additional information on the catch composition, inferences can be drawn on the technology that is deployed, the trade linkages that a fishing community has with its neighbours, its income from fishing, etc. In fact, reliable catch data are the most straightforward source of information for a variety of disciplines, ranging from history and maritime anthropology to fisheries economics (Pauly 2006).

For fisheries scientists, the value of catch data is even greater. Indeed, catch data are crucial to their main task, which is to perform fish stock assessments in support of fisheries management. Herein, the key feature of stock assessments is to evaluate the status or level of fishing activity in relation to the productivity of the ecosystem, so that fish from a given stock can be caught in such a manner that the various components of the ecosystem and its regeneration potential are not compromised. If these conditions are met, the ecosystem will sustain fishing for a long time. To accomplish this task, there are two different subtasks to be considered: first establishing the potential of the ecosystem and second establishing where the fishery is relative to that potential. Many assessment tools have been developed to estimate the biological potential of a fishery system and use them as benchmarks for the level of exploitation. Maximum sustainable yield (MSY), and the ratio between the estimated original (un-fished) biomass and the current biomass are two of the many metrics used globally to establish levels beyond which extraction is advised not to go (Beverton and Holt 1993; Hilborn and Walters 1992). Of course, there are criticisms of those approaches, the assumptions they use and their applicability to different ecosystems, and they even share part of the blame for the decline of many fisheries (Larkin 1977; Punt and Smith 2001). However, until better alternatives are available to replace the traditional stock assessment tools, they will be used, despite their limitations. Moreover, while new approaches are being developed, many fisheries in the world do not even have estimates of those metrics and/or are not managed at all.

Overall, reliable catch data, jointly with the methods to estimate the biomass of fish and their productivity, are crucial components of effective assessment and management of fisheries. Time series of total catch, preferably by species, is thus the most basic and important information that can be gathered about a fishery (Caddy and Gulland 1983; Pauly and Zeller 2003). It is even more useful when coupled with fishing effort data. Notably, catch and effort data can help with preliminary assessment of the status of populations upon which fisheries depend. However, this should be done with caution (Harley et al. 2001), because catch per unit of effort (CPUE), although an indicator of fish biomass, is not always proportional to abundance. CPUE can remain more or less stable while abundance is declining, a phenomenon called 'hyperstability', observed on schooling pelagic fish and spawning aggregations (Hilborn and Walters 1992; Pitcher 1995; Sadovy and Domeier 2005). On the other hand, CPUE can decline more than the actual decline of abundance, which is called 'hyperdepletion' (Hilborn and Walters 1992). This can occur, for example, when only a portion of the population is vulnerable to the fishery (Walters and Bonfil 1999; Kleiber and Maunder 2008). However, for many fisheries, CPUE is the best type of information available for assessment, and not using it is short-sighted.

The Rationale for Catch Reconstructions

There are many ways catch data can be collected. The most common are log books filled in by fishers, the records of observers onboard fishing vessels and data collection at the landing sites and on markets (e.g., auction and exports). For Red Sea countries, many of these methods are very difficult to implement. Most of the local (artisanal) fishers are illiterate. The communities are predominantly based on oral traditions, thus logbooks are out of question. The majority of the boats are small, and on-board observers are impractical to deploy. Data recording at landing sites, although still arduous, is the most practical for routine catch and effort data collection. The challenge here is that the number of landing sites along the coast is quite large, and some of them are not even known to the fisheries administrations. Setting up proper data collection systems is thus not straightforward, given the complexity of fisheries and fish marketing.

There are many possible fates of a fish following its encounter with fishing gear (Fig. 1.6). The actual effect of a fishery in an ecosystem should be measured by the amount of fish killed (rather than fish landed). The actual measure of fishing mortality can be concealed by lack of data on the mortality of the fish at the different parts of Fig. 1.6. For example, for some Red Sea countries, more than half of the fish catch does not go through fish markets, where official recording occurs (Chakraborty 1983). If only the data from landing sites is used to calculate the fishing mortality, this will underestimate its actual magnitude. Thus, proper planning and systematic collection procedures are needed

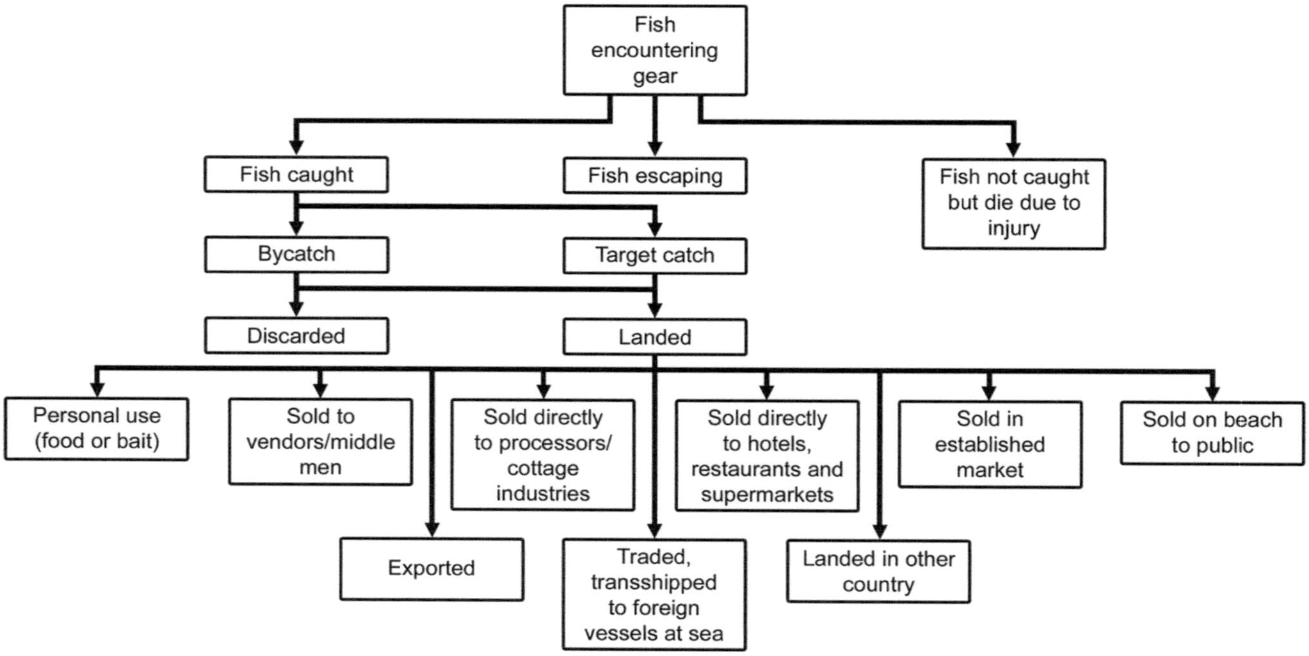

Fig. 1.6 Possible fates of fish following an encounter with a fishing gear, based on Mohammed (2003)

(Gulland 1975; Sparre 2000) and for the Red Sea, it is urgent (Tesfamichael 2012). Systematic data collection requires resources, and thus developed countries usually have better catch and related statistics than developing countries (Alder et al. 2010), while the latter also have to contend with a generally higher biodiversity, which makes the catch highly diverse, and hence comprehensive catch statistics difficult to produce (Pauly and Watson 2008). Note, however that even in developed countries with better statistics, overfishing is rampant, e.g., in the North Atlantic (see e.g., Christensen et al. 2003).

The Food and Agricultural Organization (FAO) of the United Nations assembles annual catch data submitted by member countries and harmonizes and disseminates them since 1950 (Garibaldi 2012; Pauly and Froese 2012), and Garibaldi (2012) gives a comprehensive description of the FAO database and its evolution. Because it consists of continuous, long time series and is easy to access, the FAO database is used extensively for research and policy at regional or international scales. Thus, 600 articles in peer-reviewed journals cited the FAO database in the last 15 years, notably because its standardized data makes comparisons straightforward (Garibaldi 2012).

FAO's mandate is very broad, and when it comes to fishery data, it can only compile and distribute what is submitted to it. This is the main bottleneck to the quality of the data. Countries do not necessarily have the incentive to submit reliable data, except as a moral obligation to contribute to a global system. Thus, it is not uncommon for countries to send incorrect or incomplete fishery data (Pauly and Froese 2012), and FAO does not have a legal or procedural mandate to refuse such data. Even more problematic, the technical reports produced by FAO staff or consultants are not reflected in the database. Thus, the global estimates of discards documented in successive *Technical Papers* and other FAO documents were never included in the FAO statistics (Zeller and Pauly 2005).

Another example, applying specifically to the Red Sea, is that most of the early fishery data for the Red Sea comes from national or regional projects executed by FAO, especially the project 'Development of fisheries in areas of the Red Sea and Gulf of Aden', which ran from the late 1970s to the mid-1980s under the United Nations Development Programme (UNDP) and FAO. Among other things, the projects surveyed the fisheries and estimated national catches (Chakraborty 1984), but they were not incorporated into the FAO catch database. Moreover, while the countries around the Red Sea are all members of FAO, and hence they send their fishery data to FAO, many suffer from political and institutional instability, which affects their fishery agencies, and thus there are gaps and inconsistencies in the data supplied to FAO.

FAO's mandate, while broad, does not include detailed analysis and review of the data supplied by member countries, which thus remain limited in their reliability and usefulness. It is assessed by FAO itself that the catch data it receives from over half of its developing country members, and one quarter of developed country members are unreliable (Garibaldi 2012). The following are the major constraints with the fishery statistics in the FAO database, and affect all countries, and not only those around the Red Sea:

1. The FAO database reports global marine catches spatially only to the extent that they are allocated to 19 giant 'statistical areas'. In the cases of Red Sea catches, this is area 51, the 'Western Indian Ocean', extending from the tip of the Gulf of Suez in the North to the Antarctic Convergence in the South, and from Sri Lanka in the East to South Africa in the West;
2. The level of taxonomic aggregation of the catch is usually very high, and a large part of the catch is reported as 'miscellaneous' or 'unidentified species', which masks qualitative changes occurring within the ecosystem;
3. FAO's member countries often send in catch data (usually emanating from a Department of Fisheries or similar institution) through their Ministry of Trade, or some central statistics office or other government agency not directly connected with fisheries, where they are often over-aggregated and/or otherwise modified before being sent off;
4. Some countries may have political reasons to misreport their catch, including over-reporting of catches as China did to FAO for at least two decades (Watson and Pauly 2001) and, gravest of all:
5. When data for certain fisheries are not available (because the fisheries in question were not monitored), no estimates for the missing catch data are submitted. Subsequently, absent catch data for a given year become an annual catch of precisely '0' tonne (Pitcher et al. 2002). Thus, the FAO database does not account for illegal, unreported and unregulated (IUU) catch (Alverson et al. 1994; Kelleher 2004), nor does it suggest where gaps in its coverage may occur.

FAO has taken initiatives to improve the content of its catch database, and indeed, it has improved over time. Also, there is a university-based research project, the *Sea Around Us* (www.seaaroundus.org), which aims to improve the quality of global marine fishery data. Being non-governmental, *Sea Around Us* is not limited by formal procedures. Hence, country catch reports can be critically examined, and when fisheries were omitted, their catch can be estimated using the best available information. In effect, the major issues with the FAO database can be overcome through reconstructing historical catch time series (Pauly 1998; Pauly and Zeller 2003; Pauly and Froese 2012). Reconstructed time series of catch (and effort) data from the past are not merely useful for historical purposes. Rather, they provide a basis for overcoming the shifting baseline syndrome (Pauly 1995), i.e., for improved assessment of past and current impacts of fishing on marine ecosystems, and for ecological restoration (Scott Baker and Clapham 2004; Pitcher 2005). The lessons learned from catch reconstruction in different circumstances of the fisheries can be informative, similar to 'scenarios' in adaptive management of resources (Walters 1986).

Catch reconstruction involves quantifying the catch of each fishery known to have existed, based either on 'hard' catch data, or when such data are not available, on the 'shadow' that the fishery – a social activity - throws on the society in which it is embedded. This shadow may consist of household fish consumption figures, number and income of fishers, export figures, etc… (Pauly 1998). Estimates from catch reconstruction, while approximate, will generally be closer to reality than the misreported catches, e.g., the precise estimate of zero in the official databases alluded to in the above (Pitcher et al. 2002; Zeller et al. 2007).

The main objective of this book is to present reconstructed catches of the Red Sea fisheries from 1950, the year FAO started to publish annual statistical reports on the fisheries of the world, up to 2010. Included here are all the Red Sea countries: Egypt, Sudan, Eritrea, Yemen, Saudi Arabia, Jordan and Israel and all the fishing sectors of these countries. The major outputs are a time series of standardized fishery catches for the Red Sea, by sector and species or other groupings. We do not claim these catch reconstructions data to be final. Rather, we see them as the start of an iteration, and as a basis to kick start the discussion on how to improve fishery data for the Red Sea, and ultimately, the management of its fisheries resources.

Sources and Catch Reconstruction Procedures

The main procedure in catch reconstruction is digging into different sources reporting the catches of the countries, critically analyzing them, and organizing them to a common standard, which can be used for comparison and carrying out analysis for the assessment of the resources (Mohammed 2003; Tesfamichael and Pauly 2011). The sources used here include peer-reviewed published papers, grey literature (mainly government, consultant, and FAO reports), and national databases, complemented by field trips by the first author to Egypt, Sudan, Eritrea, and Yemen from December 2006 to September 2007. The information collected was enriched by the insights of local experts and colleagues who provided data through personal communications. The catch reconstruction for the whole Red Sea was first compiled in the form of individual country reports, co-authored by country experts: Egypt (Tesfamichael and Mehanna 2012), Sudan (Tesfamichael and Elawad 2012), Eritrea (Tesfamichael and Mohamud 2012), Yemen (Tesfamichael et al. 2012b), Saudi Arabia (Tesfamichael and Rossing 2012), and Jordan and Israel (Tesfamichael et al. 2012a), which give country-specific details (see also www.seaaroundus.org/eez/). Here, a summary of the general methodology and the procedure to establish one coherent data set for the whole Red Sea are described.

Sources

The earliest data sources for the Red Sea countries were technical reports of the assessments of the fishery resources for planning the development of the fishing industry, starting in the decades following WWII. The 1950s was also a period where several of these countries became independent and started to run their national economies, and food security became a critical issue. These assessments/surveys were made by foreign experts (except for Egypt and Israel), usually recruited through the FAO. The earliest sources available were for Saudi Arabia (El-Saby and Farina 1954), Sudan (Kristjonsson 1956), Eritrea (Ben-Yami 1964), Egypt (Al-Kholy and El-Hawary 1970) and Yemen (Lisac 1971; Losse 1973). Other early assessments were performed through bilateral arrangements or consultants hired directly by the countries (e.g. see Ben-Yami 1964; Atkins 1965; Grofit 1971 for Eritrea). In the 1970s and 1980s, in part because of the Cold War and ensuing East-West competition, development aid was pouring into the Red Sea countries. A fraction of these funds were assigned to fisheries development projects, which led to an improvement in documented knowledge about the fisheries (catches, catch composition, gear, etc). A regional project for the Red Sea area, 'Development of fisheries in areas of the Red Sea and Gulf of Aden', was carried out from the end of the 1970s until the mid-1980s and led to an improvement of the quality (comprehensiveness and taxonomic resolution) of fishery catch data. Additional sources were also used, notably tax offices and export records. For example, the catch of the Eritrean beach seine small pelagic fishery was reconstructed from export figures for fish meal, which was the output of the fishery (Ben-Yami 1964).

Organized databases and/or annual fishery statistical reports are a relatively new development for the Red Sea countries. The oldest database is that of Egypt, which starts in 1979, while Saudi Arabia started publishing its annual fishery statistics in the 1980s. Eritrea has had annual reports since its independence in 1991, but its fishery database started only in 1996. Sporadic annual reports are available for Yemen and a database system is being established. Sudan does not have any fishery data reporting system yet; however, daily catch data are collected at the main fishing market of Port Sudan, which are stored, but not issued as annual reports. All these sources were accessed for the catch reconstruction of the respective countries.

Once the sources were accessed, their contents were analyzed for their spatial, temporal and sectoral coverage. Some reports were written only for a certain section of the countries or only a specific sector of the fisheries. Then, the sources were critically examined with regards to the method(s) and assumptions used in collecting their data. For some years, data were available from different sources, some simply regurgitating previously reported data. In such cases, an effort was made to locate the original reports. When there were multiple independent sources, the ones which have detailed explanations of the methodology and comprehensive coverage were selected. In a few cases, the information from one source was used to correct data from another.

Interviews

Field interviews (Fig. 1.7) were conducted, in Sudan, Eritrea and Yemen, by the first author and assistants with fishers ranging from 15 to 82 years of age, and with fishing village elders and the employees of fisheries administrations (Tesfamichael et al. 2014). The main goal of the interviews was to assess long-term change in fisheries productivity by

Fig. 1.7 A researcher interviewing a Yemeni fisher while his son watches (Photo: Dawit Tesfamichael)

accessing fishers' memories, which provided two major inputs to the catch reconstructions. First, the interviews were very useful in filling data gaps. For some periods there were no records at all, so interviewees were asked to explain what occurred during those periods, i.e., whether the catches were higher or lower than, or about equal, to the adjacent periods with records. The other type of information supplied by the interviews was the unreported catch, i.e., the catch missed by official records. For many artisanal fisheries in the Red Sea, this included the catch given freely to some members of the community and the catch landed at remote landing places, away from data collectors. Regarding the former, there is a strong tradition, shared by the maritime cultures of Red Sea countries, that part of the catch is expected to be given freely to family, friends and people who need assistance (e.g., the elderly, disabled, and widows). The amount given freely is called *'kusar'* and is a form of food security social network. Not to give *'kusar'* leads to loss of prestige, which may have serious consequences, e.g., with regards to market transactions and eventual marriages. The amount was about half of the total catch in the 1950s and 1960s; however, as the catches started to decrease and the fish accrued market value, the proportion of the catch devoted to *kusar* started to decrease.

Another input from the interviews was explanations of discrepancies among reports. The insights from older fishers and people who have been involved in the management of fisheries helped resolve ambiguities in reports and/or records. Although they did not give specific quantitative values, their ability to give comparative qualitative information helped to base the assumptions used in quantifying the catch. In the absence of any other source, anecdotal information can be a good starting point (Pauly 1995) and quantitative data can be inferred from qualitative information, given some anchoring (Tesfamichael and Pitcher 2007). In addition to acquiring information through interviews, the effectiveness of a fishery management scheme and compliance of fishers is higher when fishers are involved in information gathering and management processes.

Missing Data

For the years data were missing, interpolations or extrapolations were made to fill in the data gaps. These were made on the basis of explicitly stated assumptions, given the best knowledge of the fisheries available at the time. Population size and per capita consumption were frequently used as a proxy for inferring catches. In a few instances, information from one country was used for another country with a similar fishery, particularly in the case of catch composition data.

Comparison and Compilation

Using the different sources and procedures, the catches of each country were reconstructed by sector, and the catch compositions were inferred. Then, the reconstructed catches were compared to the catch data reported to FAO (see http://www.fao.org/fishery/statistics/software/fishstat/en) by the respective country. The FAO data are used as a reference for comparison because they are a good source of time series catch data for the Red Sea countries and are used by many organizations (local and foreign) for analysis and planning. Thus, the part of the reconstructed catch of a given taxon that was accounted for in the FAO data was assigned as 'reported' catch in our analysis and results. When the reported catch of a taxon was higher than what is reported for that taxon in the FAO database, the difference was assigned to the 'unreported' catch. In contrast, when the FAO catch for a taxon was higher than the reconstructed catch, it was assigned as 'over-reported' catch. As will be seen in the country chapters, reported and unreported catches are identified separately in the catch reconstructions. Note that if there was a part of the catch that was not reported (e.g., catches were sold outside landing sites where catch data recordings are carried out and we managed to get an estimate of the amount), then that part of the catch is referred as 'unreported' catch in our computations. This should not be confused with the reported and unreported catches of the results as compared to the FAO data. Once the catches were reconstructed for each country, by sector, and the catch composition calculated, they added up to represent the catches of the Red Sea as a whole, i.e., as a Large Marine Ecosystem (see also www.seaaroundus.org/lme/33.aspx).

Summary Results and Discussion

The total reconstructed catch from the Red Sea from 1950 to 2010 was 6,333,000 t, 1.5 times higher than what is reported to FAO by the surrounding countries for the same period. The total catch was low (around 50,000 t $\cdot$ year^{-1}) until 1960, when it started its first increase until a decline in the early 1970s (Fig. 1.8), due to the war between Israel and Egypt. The catch increased again from the mid-1970s, until it reached a peak of 177,000 t in 1993. This phase is characterised by massive boat motorization and the introduction of industrial fishing by several Red Sea countries. This increased the effort and also allowed the expansion of the fisheries to areas they did not access previously. The total catch remained high, with some fluctuations, until the mid-2000s when it started to decline. This decline is here interpreted as a sign of resource depletion, especially in Yemen (Tesfamichael et al. 2012b).

The reconstructed catch was higher than the FAO data, except for the last few years. An obvious reason why the reconstructed catch is generally higher is that we included discards, subsistence and recreational fisheries, which are not usually included in FAO data for the Red Sea. The higher FAO catch in the last few years of our analysis was caused by double counting of some fishery catches in the FAO database. This is due mainly to Egypt fishing outside its EEZ in the waters of Sudan, Eritrea and Yemen, and reporting all their catches as Egyptian, while Sudan, Eritrea and Yemen report some of these same catches to FAO as well, as they are taken within their EEZ. One can argue this catch should be reported by area, i.e., by the EEZ it was taken from, or by the country that has taken it. Here, in view of the current emphasis on ecosystem-based fisheries management, we focused on the area, i.e., the EEZ from which the catch originates, as it provides the spatial context for fisheries management. For completeness, we also indicate, the country fishing in the database. Presently, there is no a regional fishery management agency for the Red Sea LME, and whatever management there is extends only to national schemes, pertaining to single EEZs. By country, Yemen has the highest percentages of the Red Sea catch (36 %), followed by Egypt (28 %), Saudi Arabia (23 %), Eritrea (11 %) and Sudan (2 %), while Jordan and Israel contribute less than 0.2 % each.

The artisanal fisheries accounted for 49 % of the total catch from 1950 to 2010 (Fig. 1.9). Their contribution was dominant throughout the whole period, unlike the industrial sector (22 %), which is important only in the later part of the period covered here. This has major economic and social implications, as artisanal fisheries employ a higher number of fishers per tonne of catch (Pauly 2006), which translates to higher employment and livelihood in the communities. The discards (near exclusively from industrial fishing), which are usually ignored in official reports, represented 16 % of the total catch. The subsistence catch was 12 %, while the recreational fishery (1 %), which started only recently, is still negligible. Egypt is the country with the most developed recreational fishery and even in that country, recreational catches are low.

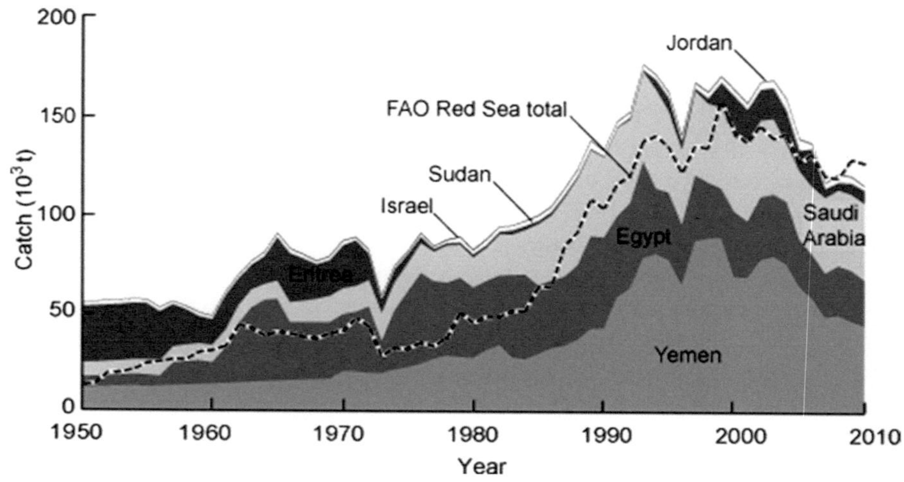

Fig. 1.8 Reconstructed catch of the countries bordering the Red Sea from 1950 to 2010 and its comparison to the data reported to FAO

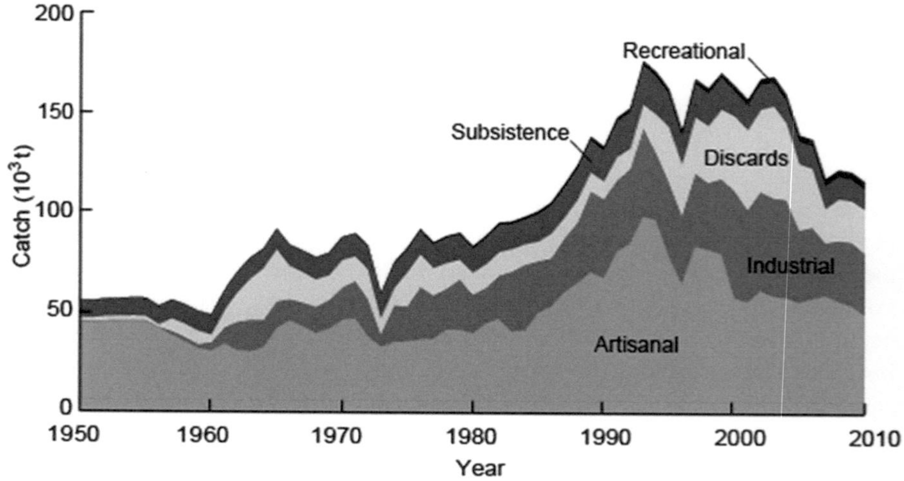

Fig. 1.9 Reconstructed catch of the Red Sea fisheries by sector from 1950 to 2010

Comparing the reconstructed catch with the FAO data in terms of taxonomy, only 42 % of the reconstructed catch was accounted in the FAO data, i.e., the reported catch (Fig. 1.10). The remaining 58 % was not accounted for at all. This included 43 % unreported, but landed catch and 15 % discarded bycatch catch, which is also not reported.

A total of 209 taxa or taxonomic groups were identified as contributing to Red Sea catches, in addition to a group 'others' encompassing the minor taxa that were not represented separately. The taxa contributing most to the catch was Indian mackerel (*Rastrelliger kanagurta;* 8 %), Spanish mackerel (*Scomberomorus commerson*; 7 %), and jacks (Carangidae; 7 %). Emperors (Lethrinidae) and ponyfishes (Leiognathidae) each accounted for 5 % of the total catch, the former prized fishes, the latter the dominant taxon in the discarded catch of industrial trawlers. These percentages suggest that there is no a single taxon that is overly dominant in the Red Sea fisheries, a reflection of their multi-species nature, and one of the main challenges in managing the Red Sea fisheries. The major taxonomic groups of the total catch composition are presented in (Fig. 1.11). Only a few taxa are included here for better visual effect of the figure; supplementary tables with extensive taxonomic composition are presented in the electronic supplementary materials (ESM) http://extras.springer.com/ and the spatial distribution of the catch is given at www.seaaroundus.org.

In the following, a brief per-country account is given, starting with Egypt and moving counter-clockwise along the Red Sea coast as the different chapters are introduced. For Egypt (Chap. 2), the reconstructed catch is higher than the fisheries catch statistics that Egypt submits to FAO from the beginning of 1960s until the beginning of 1990s, but the reverse occurs after the mid-1990s. This discrepancy is due to the fact that Egypt fishes outside its own waters (e.g., in Eritrean waters starting early 1990s and these catches are not included in the reconstruction, as the focus of the reconstruction is to quantify the amount fished in the waters of various countries (also clearly identifying the fishing country) and

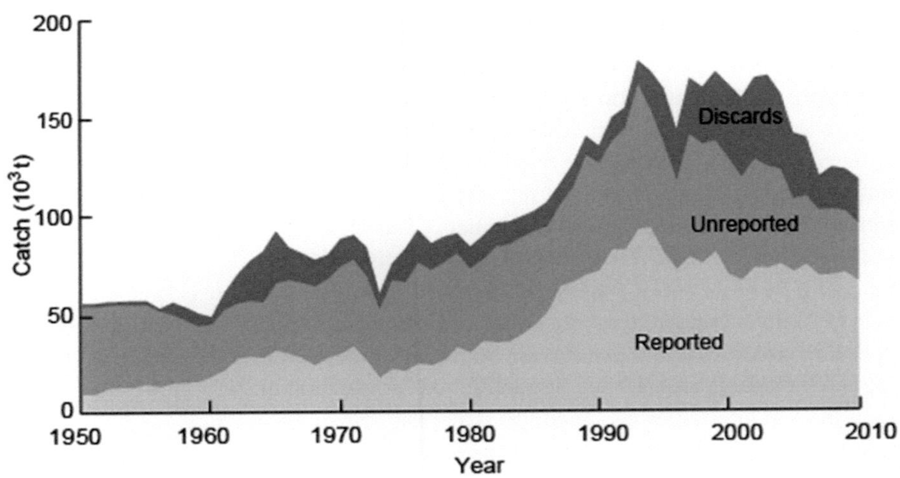

Fig. 1.10 Reconstructed catch of the Red Sea fisheries by components from 1950 to 2010

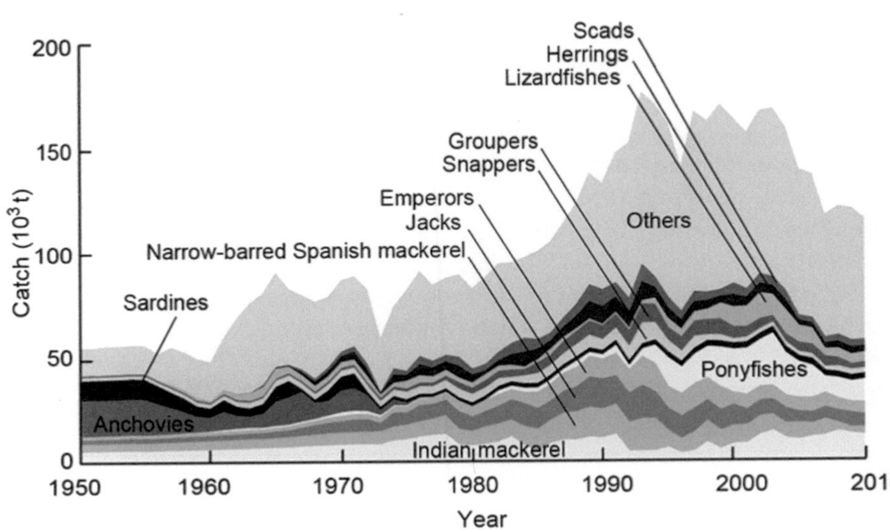

Fig. 1.11 The major taxonomic composition of the total reconstructed catch of the Red Sea from 1950 to 2010

not where they were landed. The catch of Egyptian vessels from Eritrean waters is reported in the reconstruction of Eritrea.

The Sudanese data (Chap. 3) submitted to FAO does not include the catches of shells (trochus and mother-of-pearl), which were very important before 1980s. Generally, there is no large difference between the reconstructed data and the data submitted to FAO for Sudan. The sudden spike of Sudanese catch reported to FAO in 1983, on the other hand, is likely due to a reporting error, as there was no major change in the fisheries likely to cause such a sudden jump for only 1 year. The higher catches reported to FAO after the 1990s are also suspicious, as they contradict locally available data.

For Eritrea (Chap. 4), Yemen (Chap. 5) and Saudi Arabia (Chap. 6), the reconstructed catches are higher than those reported to FAO, due to the latter not including various fisheries and omitting discards. The major discrepancies between the reconstructed data and the data submitted to FAO for Eritrea are in the early decades (1950s and 1960s) and later, after 2000. Between those periods the fishery was largely inactive, hence catches were low. For Yemen in the Red Sea, the reconstructed catch is higher than the reported catch, the difference being more consistent for Yemen than for any other country. There is a clear difference between the reconstructed and reported catch for Saudi Arabia in the Red Sea until the mid-1980s. After the mid-1980s, trawlers were introduced into the Saudi fishery, and hence the differences between the two data sets consist mainly of discards. The reconstructed catches of Jordan and Israel (Chaps. 7 and 8) are negligible compared to those of the other countries, which is understandable given their minuscule footholds in the inner Gulf of Aqaba. They also exhibited less fluctuation than the FAO data. For each country (Chapter), the ecosystem and fisheries are described and the reconstructed catch presented from 1950 to 2010. The details of the sources and procedures of the reconstruction are put at the end. Supplementary tables and results are available at (ESM) http://extras.springer.com/ and the spatial distribution of the catch is given at www.seaaroundus.org.

In addition to the catch reconstruction for each country bordering the Red Sea, an ecosystem model of the Red Sea Large Marine Ecosystem (LME) is presented (Chap. 9) using Ecopath with Ecosim (EwE), where the interactions of the organisms in the ecosystem and the impact of fisheries are quantified. The model can be used to simulate policy scenarios and predict the outcomes, which is an important tool for informed management of the ecosystem. Last but not least, a list is provided of common commercial fishes caught by the fisheries in the Red Sea and their corresponding local names (Chap. 10). The names include valid scientific names, common English names, local (Arabic) names written in both Arabic script and Roman characters. We believe this will help researchers, resource users and managers. Jointly, the information presented here can help in better understanding the Red Sea and provide a basis for the management schemes that the future will require (Tesfamichael 2012).

Acknowledgements The work documented here was mostly supported by the *Sea Around Us*, itself funded by the Pew Charitable Trusts. There were many people who helped us in each country that should be acknowledged individually. However, as their contributions were country-specific, we have opted to thank them in each country's chapter. Here, we would like to thank Gwilym Rowlands for Fig. 1.3 and Callum Roberts for reviewing the chapter.

References

Alder J, Cullis-Suzuki S, Karpouzi V, Kaschner K, Mondoux S, Swartz W, Trujillo P, Watson R, Pauly D (2010) Aggregate performance in managing marine ecosystems of 53 maritime countries. Mar Policy 34(3):468–476

Al-Kholy AA, El-Hawary MM (1970) Some penaeids of the Red Sea. Bull Inst Oceanogr Fish Cairo 1:339–377

Al-Sofyani A, Marimuthu N, Wilson JJ (2014) A rapid assessment of Scleractinian and non-Scleractinian coral growth forms along the Saudi Arabian coast, Red Sea. J Ocean Univ China 13(2):243–248

Alverson DL, Freeberg MH, Murawski SA, Pope JG (1994) A global assessment of fisheries bycatch and discards. FAO Fish Tech Pap No. 339, Rome, 233 p

Atkins WSP (1965) Development of fishing industry in Ethiopia. W. S. Atkins and Partners, Surrey/Addis Ababa, 282 p

Bahartan K, Zibdah M, Ahmed Y, Israel A, Brickner I, Abelson A (2010) Macroalgae in the coral reefs of Eilat (Gulf of Aqaba, Red Sea) as a possible indicator of reef degradation. Mar Pollut Bull 60(5):759–764

Baker AC, Starger CJ, McClanahan TR, Glynn PW (2004) Coral reefs: corals' adaptive response to climate change. Nature 430(7001):741

Barrania AA (1979) The exploratory socio-economic survey of the Sudanese Red Sea fisheries. FAO/UNDP Project for Development of Fisheries in Areas of the Red Sea and Gulf of Aden, Rome, 36 p

Ben-Tuvia A (1968) Report on the fisheries investigation of the Israel South Red Sea expeditions, 1962. Sea Fish Res Stn Haifa Bull 52:21–55

Ben-Yami M (1964) Report on the fisheries in Ethiopia. Ministry of Foreign Affairs, Department for International Cooperation, Jerusalem, 106 p

Berumen ML, Hoey AS, Bass WH, Bouwmeester J, Catania D, Cochran JEM, Khalil MT, Miyake S, Mughal MR, Spaet JLY, Saenz-Agudelo P (2013) The status of coral reef ecology research in the Red Sea. Coral Reefs 32(3):737–748

Beverton RJH, Holt SJ (1993) On the dynamics of exploited fish population. Chapman & Hall, London

Braithwaite CJR (1987) Geology and palaeogeography of the Red Sea region. In: Edwards AJ, Head SM (eds) Red Sea. Pergamon Press, Oxford, pp 22–44

Bruckner A, Rowlands G, Riegl B, Purkis S, Williams A, Renaud P (2011a) Khaled bin Sultan living oceans foundation Atlas of Saudi Arabian Red Sea Marine Habitats. Panoramic Press, Phoenix

Bruckner AW, Faisal M, Alnazry HH (2011b) A paradigm shift for fisheries management to enhance recovery, resilience, and sustainability of coral reef ecosystems in the Red Sea. Sustainable fisheries: multi-level approaches to a global problem. American Fisheries Society, Maryland, pp 85–111

Caddy JF, Gulland JA (1983) Historical patterns of fish stocks. Mar Policy 7(4):267–278

Cantin NE, Cohen AL, Karnauskas KB, Tarrant AM, McCorkle DC (2010) Ocean warming slows coral growth in the central Red Sea. Science 329(5989):322–325

Chakraborty D (1983) Marine fisheries statistics in Sudan – an expanded plan of development. FAO, Rome, 39 p

Chakraborty D (1984) Fish landings on the Red Sea and Gulf of Aden coast of the member countries: a preliminary estimate. Development of fisheries in areas of the Red Sea and Gulf of Aden, United Nations Development Programme (UNDP) and Food and Agriculture Organization (FAO) of the United Nations, Suez, 29 p

Christensen V, Guénette S, Heymans JJ, Walters CJ, Watson R, Zeller D, Pauly D (2003) Hundred-year decline of North Atlantic predatory fishes. Fish Fish 4(1):1–24

DeVantier L, Tourak E, Al-Shaikh K, De'ath G (2000) Coral communities of the central-northern Saudi Arabian Red Sea. Fauna Arab 18:23–66

Dubinsky Z, Stambler N (eds) (2011) Coral reefs: an ecosystem in transition. Springer, New York, 551

Edgren G (1993) Expected economic and demographic developments in coastal zones world wide. In: Beukenkamp P, Günther R, Klein R, Misdorp D, Sadacharan LDV (eds) World coast '93. Noordwijk, Coastal Zone Management Centre, The Netherlands National Institute for Coastal and Marine Management, pp 367–370

Edwards FJ (1987) Climate and oceanography. In: Edwards AJ, Head SM (eds) Red Sea. Pergamon Press, Oxford, pp 45–69

El-Saby MK, Farina L (1954) Report to the government of Saudi Arabia on possibility for development of marine fisheries. FAO, Rome, 29 p

Evans PGH (1987) Sea birds of the Red Sea. In: Edwards AJ, Head SM (eds) Red Sea. Pergamon Press, Oxford, pp 315–338

Fine M, Gildor H, Genin A (2013) A coral reef refuge in the Red Sea. Global Change Biol 19(12):3640–3647

Forsskål P (1775) Descriptiones animalium, avium, amphibiorum, piscium, insectorum, vermium; quae in itinere orientali observavit. Hauniae, Copenhagen

Frazier J, Bertram GC, Evans PGH (1987) Turtles and marine mammals. In: Edwards AJ, Head SM (eds) Red Sea. Pergamon Press, Oxford, pp 288–314

Frihy OE, Fanos AM, Khafagy AA, Abu Aesha KA (1996) Human impacts on the coastal zone of Hurghada, northern Red Sea, Egypt. Geo-Mar Lett 16(4):324–329

Garibaldi L (2012) The FAO global capture production database: a six-decade effort to catch the trend. Mar Policy 36(3):760–768

Getahun A (1998) The Red Sea as an extension of the Indian Ocean. In: Sherman K, Okemwa EN, Ntiba MJ (eds) Large marine ecosystems of the Indian Ocean: assessment, sustainability, and management. Blackwell Science, Cambridge, MA, pp 277–281

Girdler RW, Southren TC (1987) Structure and evolution of the northern Red Sea. Nature 330(6150):716–721

Gladstone W (1996) Unique annual aggregation of longnose parrotfish (*Hipposcarus harid*) at Farasan Island (Saudi Arabia, Red Sea). Copeia 2:483–485

Gladstone W (2000) The ecological and social basis for management of a Red Sea marine-protected area. Ocean Coast Manag 43(12):1015–1032

Goren M (1986) A suggested model for the recolonization process of the Red Sea at the post glacial period. In: Uyeno T, Arai R, Taniuchi T, Matsuura K (eds) International confererence on Indo-Pacific Fishes, 29 Jul–3 Aug 1985, 29 Jul–3 Aug 1985, Tokyo

Grofit E (1971) The Red-Sea fisheries of Ethiopia (1966–1969). Ministry of Foreign Affairs, Department for International Cooperation, Jerusalem, 32 p

Gulland JA (ed) (1971) The fish resources of the ocean. Fishing News Books, Surrey

Gulland JA (1975) Manual of methods for fisheries resource survey and appraisal. Part 5. Objectives and basic methods. FAO Fish Tech Pap No. 145, Rome, 134 p

Harley SJ, Myers RA, Dunn A (2001) Is catch-per-unit-effort proportional to abundance? Can J Fish Aquat Sci 58:1760–1772

Hawkins JP, Roberts CM (1994) The growth of coastal tourism in the Red Sea: present and future effects on coral reefs. Ambio 23(8):503–508

Head SM (1987a) Introduction. In: Edwards AJ, Head SM (eds) Red Sea. Pergamon Press, Oxford, pp 1–21

Head SM (1987b) Red Sea fisheries. In: Edwards AJ, Head SM (eds) Red Sea: key environments. Pergamon Press, Oxford, pp 363–382

Hilborn R, Walters CJ (1992) Quantitative fisheries stock assessment. Choice, dynamics and uncertainty. Chapman and Hall, New York

Horton M (1987) The human settlement of the Red Sea. In: Edwards AJ, Head SM (eds) Red Sea. Pergamon Press, Oxford, pp 339–362

Jish Prakash P, Stenchikov G, Kalenderski S, Osipov S, Bangalath H (2015) The impact of dust storms on the Arabian Peninsula and the Red Sea. Atmos Chem Phys 15(1):199–222

Kattan A (2014) Baselines and comparison of coral reef fish assemblages in the Central Red Sea, King Abdullah University of Science and Technology, Saudi Arabia. Available at http://hdl.handle.net/10754/336498

Kelleher K (2004) Discards in the world's marine fisheries: an update. FAO Fish Tech Pap No. 470, Rome, 134 p

Keskin C, Pauly D (2014) Changes in the 'Mean Temperature of the Catch': application of a new concept to the North-eastern Aegean Sea. Acta Adriat 55:213–218

Khalil M, Cochran JM, Berumen M (2013) The abundance of herbivorous fish on an inshore Red Sea reef following a mass coral bleaching event. Environ Biol Fishes 96(9):1065–1072

Kleiber P, Maunder MN (2008) Inherent bias in using aggregate CPUE to characterize abundance of fish species assemblages. Fish Res 93(1–2):140–145

Klunzinger CB (1870) Eine zoologische Excursion auf ein Korallenriff des rothen Meeres. Verhandlungen der zoologisch-botanischen Gesellschaft Wien 20:389–394

Klunzinger CB (1872) Zoologische Excursionen auf ein Korallenriff des Rothen Meeres bei Kossér. Zeitschrift der Gesellschaft für Erdkunde zu Berlin 7:20–55

Klunzinger CB (1878) Upper Egypt: its people and its products. A descriptive account of the manners, customs, superstitions, and occupations of the people of the Nile Valley, the desert, and the Red Sea coast, with sketches of the natural history and geology, vol ixv, figs. Blackie and Son, London, pp 1–408

Kotb MMA, Hanafy MH, Rirache H, Matsumura S, Al-Sofyani AA, Ahmed AG, Bawazir G, Al-Horani FA (2008) Status of coral reefs in the Red Sea and Gulf of Aden Region. In: Wilkinson C (ed) Status of coral reefs of the world: 2008. Global Coral Reef Monitoring Network and Reef and Rainforest Research Centre, Townsville, pp 67–78

Kristjonsson H (1956) A brief survey of the Sudanese Red Sea fisheries – report to the government. FAO report no. 510, FAO, Rome

Larkin PA (1977) An epitaph for the concept of maximum sustained yield. Trans Am Fish Soc 106(1):1–11

Lisac H (1971) Report to travel to Yemen Arabic and the People's Republic of Southern Yemen (16 Jan–1 Feb 1972) FAO Fish. Travel Aide Mem, Rome

Losse G (1973) Marine resources of the Yemen Arabic Republic. A report prepared for the UNEP/FAO Food and Nutrition programme, Rome

Loya Y, Genin A, el-Zibdeh M, Naumann MS, Wild C (2014) Reviewing the status of coral reef ecology of the Red Sea: key topics and relevant research. Coral Reefs 33(4):1179–1180

Luksch J (1898) Expedition S. M. Schiff "*Pola*" in das Rothe Meer, Nördliche Hälfte (Oct 1895–May 1896). Wissenschaftliche Ergebnisse. VI. Physikalische Untersuchungen. Denkschrifte der Akademie der Wissenschaften (Mathematische-Naturwissenschaftliche Klasse) vol 65, pp 351–422. Wien

Luksch J (1900) Expedition S. M. Schiff "*Pola*" in das Rothe Meer, Südliche Hälfte (Sept 1897–Apr 1898). Wissenschaftliche Ergebnisse. XVIII. Physikalische Untersuchungen. Denkschrifte der Akademie der Wissenschaften (Mathematische-Naturwissenschaftliche Klasse) vol 69, pp 337–398. Wien

Makris J, Rihm R (1991) Shear-controlled evolution of the Red Sea: pull apart model. Tectonophysics 198(2–4):441–466

McGinley M (2008) Red Sea large marine ecosystem. The Encyclopedia of Earth. Available at http://www.eoearth.org/view/article/155677/

Mohammed E (2003) Reconstructing fisheries catches and fishing effort for the southeastern Caribbean (1940–2001): general methodology. In: Zeller D, Booth S, Mohammed E, Pauly D (eds) From Mexico to Brazil: Central Atlantic fisheries catch trends and ecosystem models. Fisheries Centre Research Reports, vol 11(6). Fisheries Centre, University of British Columbia, Vancouver, pp 11–20

Morcos SA (1970) Physical and chemical oceanography of the Red Sea. Oceanogr Mar Biol 8:73–202

Nanninga GB, Saenz-Agudelo P, Manica A, Berumen ML (2014) Environmental gradients predict the genetic population structure of a coral reef fish in the Red Sea. Mol Ecol 23(3):591–602

Norman J (1939) Fishes. In: History. BMN (ed) The John Murray expedition 1933–1934: Scientific reports VII, London, pp 1–116

Ormond R, Edwards A (1987) Red Sea fishes. In: Edwards AJ, Head SM (eds) Red Sea. Pergamon Press, Oxford, pp 251–287

Pauly D (1995) Anecdotes and the shifting baseline syndrome of fisheries. Trends Ecol Evol 10(10):430

Pauly D (1998) Rationale for reconstructing catch time series. EC Fish Coop Bull 11(2):4–10

Pauly D (2006) Major trends in small-scale marine fisheries, with emphasis on developing countries, and some implications for the social sciences. Mar Stud 4(2):7–22

Pauly D, Froese R (2012) Comments on FAO's state of fisheries and aquaculture, or 'SOFIA 2010'. Mar Policy 36(3):746–752

Pauly D, Watson R (2008) Adjusting for context in evaluating national fisheries statistics reporting systems. In: Alder J, Pauly D (eds) A comparative assessment of biodiversity, fisheries and aquaculture in 53 countries' exclusive economic zones. Fisheries Centre Research Reports, vol 16(7). Fisheries Centre, University of British Columbia, Vancouver, pp 57–61

Pauly D, Zeller D (2003) The global fisheries crisis as a rationale for improving the FAO's database of fisheries statistics. In: Zeller D, Booth S, Mohammed E, Pauly D (eds) From Mexico to Brazil: Central Atlantic fisheries catch trends and ecosystem models. Fisheries Centre Research Reports, vol 11(6). Fisheries Centre, University of British Columbia, Vancouver, pp 1–9

Pilcher N, Alsuhaibany A (2000) Regional status of coral reefs in the Red Sea and the Gulf of Aden. In: Wilknson C (ed) Status of coral reefs of the world: 2000. Australian Institute of Marine Science, Dampier, pp 35–64

Pitcher TJ (1995) The impact of pelagic fish behaviour on fisheries. Scientia Marina 59(3–4):295–306

Pitcher TJ (2005) Back-to-future: a fresh policy initiative for fisheries and a restoration ecology for ocean ecosystems. Philos Trans R Soc 360(1453):107–121

Pitcher TJ, Watson R, Forrest R, Valtysson HP, Guenette S (2002) Estimating illegal and unreported catches from marine ecosystems: a basis for change. Fish Fish 3(4):317–339

Por FD (2012) Lessepsian migration: the influx of Red Sea biota into the Mediterranean by way of the Suez Canal, Ecological Studies 23. Springer, Berlin

Preen A (1989) Dugongs, vol 1, the status and conservation of Dugongs in the Arabian Region, MEPA coastal and marine management series. MEPA, Jeddah, 200 p

Price ARG, Ghazi SJ, Tkaczynski PJ, Venkatachalam AJ, Santillan A, Pancho T, Metcalfe R, Saunders J (2014) Shifting environmental baselines in the Red Sea. Mar Pollut Bull 78(1–2):96–101

Punt AE, Smith ADM (2001) The gospel of maximum sustainable yield in fisheries management: birth, crucifixion and reincarnation. In: Reynolds JD, Mace GM, Redford KR, Robinson JR (eds) Conservation of exploited species. Cambridge University Press, Cambridge, UK, pp 41–66

Qurban MA, Krishnakumar PK, Joydas TV, Manikandan KP, Ashraf TTM, Quadri SI, Wafar M, Qasem A, Cairns SD (2014) In-situ observation of deep water corals in the northern Red Sea waters of Saudi Arabia. Deep-Sea Res I Oceanogr Res Pap 89:35–43

Raitsos DE, Hoteit I, Prihartato PK, Chronis T, Triantafyllou G, Abualnaja Y (2011) Abrupt warming of the Red Sea. Geophys Res Lett 38(14), L14601

Rice AL (ed) (1986) Deep-sea challenge: the John Murray/Mabahiss expedition to the Indian Ocean 1933–34. UNESCO, Paris

Riegl B, Piller WE (2003) Possible refugia for reefs in times of environmental stress. Int J Earth Sci 92(4):520–531

Riegl BM, Bruckner AW, Rowlands GP, Purkis SJ, Renaud P (2012) Red Sea Coral reef trajectories over 2 decades suggest increasing community homogenization and decline in coral size. PLoS One 7(5), e38396

Roberts CM, McClean CJ, Veron JEN, Hawkins JP, Allen GR, McAllister DE, Mittermeier CG, Schueler FW, Spalding M, Wells F, Vynne C, Werner TB (2002) Marine biodiversity hotspots and conservation priorities for tropical reefs. Science 295(5558):1280–1284

Rohling EJ (1994) Glacial conditions in the Red Sea. Paleoceanography 9(5):653–660

Sadovy Y, Domeier M (2005) Are aggregation-fisheries sustainable? Reef fish fisheries as a case study. Coral Reefs 24(2):254–262

Scott Baker C, Clapham PJ (2004) Modelling the past and future of whales and whaling. Trends Ecol Evol 19(7):365–371

Sheppard CRC (2000) The Red Sea. In: Sheppard CRC (ed) Seas at the millennium: an environmental evaluation. Pergamon, Amsterdam, pp 35–45

Sheppard C, Price A, Roberts C (1992) Marine ecology of the Arabian region: patterns and processes in extreme tropical environments. Academic, London

Smeed DA (2004) Exchange through the Bab el Mandab. Deep-Sea Res II Top Stud Oceanogr 51(4–5):455–474

Sofianos S, Johns W (2015) Water mass formation, overturning circulation, and the exchange of the Red Sea with the adjacent basins. In: Rasul NMA, Stewart ICF (eds) The Red Sea. Springer earth system sciences. Springer, Berlin/Heidelberg, pp 343–353

Spaet JLY, Berumen ML (2015) Fish market surveys indicate unsustainable elasmobranch fisheries in the Saudi Arabian Red Sea. Fish Res 161:356–364

Sparre PJ (2000) Manual on sample-based data collection for fisheries assessment: examples from Viet Nam. FAO, Rome, 171 p

Stagl V, Sattmann H, Dworschak PC (1996) The material of the "*Pola*" Red Sea expeditions (1895–1898) in the collections of the natural history Museum in Vienna. In: Uiblein F, Ott J, Stachowitsch M (eds) Deep-sea and extreme shallow-water habitats: affinities and adaptations. Biosystematics and ecology series, vol 11. Austrian Academy of Science, Vienna, pp 29–41

Tesfamichael D (2001) Characterising the Eritrean artisanal fisheries using catch and effort data analysis. MSc thesis, Wageningen University, Fish Culture and Fisheries Group, Wageningen, 42 p

Tesfamichael D (2005) Reconstructing the John Murray/Mabahiss expedition. The Sea Around Us Project newsletter 29:4

Tesfamichael D (2012) Assessment of the Red Sea ecosystem with emphasis on fisheries. PhD thesis, University of British Columbia, Resource Management and Environmental Studies, Vancouver, 241 p

Tesfamichael D, Elawad AN (2012) Reconstructing Red Sea fisheries of Sudan: foreign aid and fisheries. In: Tesfamichael D, Pauly D (eds) Catch reconstruction for the Red Sea large marine ecosystem by countries (1950–2010). Fisheries Centre Research Reports, vol 20(1). Fisheries Centre, University of British Columbia, Vancouver, pp 51–70

Tesfamichael D, Mehanna SF (2012) Reconstructing Red Sea fisheries of Egypt: heavy investment and fisheries. In: Tesfamichael D, Pauly D (eds) Catch reconstruction for the Red Sea large marine ecosystem by countries (1950–2010). Fisheries Centre Research Reports, vol 20(1). Fisheries Centre, University of British Columbia, Vancouver, pp 23–50

Tesfamichael D, Mohamud S (2012) Reconstructing Red Sea fisheries of Eritrea: a case study of the relationship between political stability and fisheries development. In: Tesfamichael D, Pauly D (eds) Catch reconstruction for the Red Sea large marine ecosystem by countries (1950–2010). Fisheries Centre Research Reports, vol 20(1). Fisheries Centre, University of British Columbia, Vancouver, pp 71–104

Tesfamichael D, Pauly D (2011) Learning from the past for future policy: approaches to time-series catch data reconstruction. West Indian Ocean J Mar Sci 10(2):99–106

Tesfamichael D, Pitcher TJ (2006) Multidisciplinary evaluation of the sustainability of Red Sea fisheries using rapfish. Fish Res 78(2–3):227–235

Tesfamichael D, Pitcher TJ (2007) Estimating the unreported catch of Eritrean Red Sea fisheries. Afr J Mar Sci 29(1):55–63

Tesfamichael D, Rossing P (2012) Reconstructing Red Sea fisheries catches of Saudi Arabia: national wealth and fisheries transformation. In: Tesfamichael D, Pauly D (eds) Catch reconstruction for the Red Sea large marine ecosystem by countries (1950–2010). Fisheries Centre Research Reports, Vol 20(1), Fisheries Centre, University of British Columbia, Vancouver, pp 153-178

Tesfamichael D, Govender R, Pauly D (2012a) Preliminary reconstruction of fisheries catches of Jordan and Israel in the inner Gulf of Aqaba, Red Sea, 1950–2010. In: Tesfamichael D, Pauly D (eds) Catch reconstruction for the Red Sea large marine ecosystem by countries (1950–2010). Fisheries Centre Research Reports, Vol 20(1), Fisheries Centre, University of British Columbia, Vancouver, pp 179–204

Tesfamichael D, Rossing P, Saeed H (2012b) The marine fisheries of Yemen with emphasis on the Red Sea and cooperatives. In: Tesfamichael D, Pauly D (eds) Catch reconstruction for the Red Sea large marine ecosystem by countries (1950–2010). Fisheries Centre Research Reports, Vol 20(1), Fisheries Centre, University of British Columbia, Vancouver, pp 105–152

Tesfamichael D, Pitcher TJ, Pauly D (2014) Assessing changes in fisheries using fishers' knowledge to generate long time series of catch rates: a case study from the Red Sea. Ecol Soc 19(1):18

Tewelde M, Woldai M (2007) Socioeconomic study of the sea cucumbers fisheries in Eritrea. Eritrea's Coastal Marine & Island Biodiversity Conservation Project (ECMIB), Massawa, 35 p

Thunell RC, Locke SM, Williams DF (1988) Glacio-eustatic sea-level control on Red Sea salinity. Nature 334(6183):601–604

Triantafyllou G, Yao F, Petihakis G, Tsiaras KP, Raitsos DE, Hoteit I (2014) Exploring the Red Sea seasonal ecosystem functioning using a three-dimensional biophysical model. J Geophys Res Oceans 119(3):1791–1811

Vine PJ, Vine MP (1980) Ecology of Sudanese coral reefs with particular reference to reef morphology and distribution of fishes. In: Symposium on coastal and marine environment of the Red Sea, Gulf of Aden and Tropical Western Indian Ocean, vol 2, Khartoum, pp 87–140

Walker DI (1987) Benthic algae. In: Edwards AJ, Head SM (eds) Red Sea. Pergamon Press, Oxford, pp 152–168

Walter RC, Buffler RT, Bruggemann JH, Guillaume MMM, Berhe SM, Negassi B, Libsekal Y, Cheng H, Edwards RL, von Cosel R, Neraudeau D, Gagnon M (2000) Early human occupation of the Red Sea coast of Eritrea during the last interglacial. Nature 405(6782):65–69

Walters C (1986) Adaptive management of renewable resources. McMillan Publisher Co, New York, 374 p

Walters CJ, Bonfil R (1999) Multispecies spatial assessment models for the British Columbia groundfish trawl fishery. Can J Fish Aquat Sci 56(4):601–628

Watson R, Pauly D (2001) Systematic distortions in world fisheries catch trends. Nature 414(6863):534–536

Wilkinson C (ed) (2008) Status of coral reefs of the world: 2008. Global Coral Reef Monitoring Network, Townsville

Zakai D, Chadwick-Furman NE (2002) Impacts of intensive recreational diving on reef corals at Eilat, northern Red Sea. Biol Conserv 105(2):179–187

Zeller D, Pauly D (2005) Good news, bad news: global fisheries discards are declining, so are total catches. Fish Fish 6(2):156–159

Zeller D, Booth S, Davis G, Pauly D (2007) Re-estimation of small-scale fisheries catches for U.S. flag island areas in the Western Pacific: the last 50 years. US Fish Bull 105(2):266–277

Egypt

2

Dawit Tesfamichael and Sahar Fahmy Mehanna

Abstract

The Egyptian Red Sea ecosystem and the fisheries catches in its Exclusive Economic Zone (EEZ) are presented from 1950 to 2010. Five fisheries sectors are identified and their catch reconstructed by taxonomic group. Published papers, gray literature, reports, databases and on-site observations were used as sources. Where data gaps were identified, they were accommodated with assumptions based on the best available knowledge, which are clearly stated and can be substituted by different ones given better information. The result showed that purse seining is the dominant fishery, followed by trawling, subsistence, artisanal and recreational fisheries. The total catch of Egypt in its Red Sea EEZ was around 6000 $t \cdot year^{-1}$ in the early 1950s, which rapidly increased in 1960 and remained at a high level except for a sharp decline in 1973 due to the Israel-Arab war. The peak catch of about 50,000 t was obtained in 1993; catches then declined to about 25,000 $t \cdot year^{-1}$ by the end of 2000s. A total of 42 taxonomic groups were identified in the catches, in addition to many which could not be individually identified and were categorized as 'others'. Horse mackerel, scads and other jacks, and herring jointly accounted for 34 % of the total catch. The estimated total catch was compared with data Egypt submitted to the Food and Agriculture Organization (FAO) and clear differences were observed. While the reconstructed total catch is, overall, 1.1 times what Egypt submitted to FAO, this relatively close match masks a much stronger dominance of the reconstructed over the official catches from 1950 to the mid-1990s, followed by a period of high official catches, which includes fish caught outside Egypt's Red Sea EEZ.

Keywords

Time series • Catch • Catch composition • Mis-reported catch • Over investment • Foreign waters fishing

Introduction

Egypt has access both to the Red Sea and Mediterranean Sea, besides expansive inland waters supplying fish, i.e., the Nile and its canals, coastal and delta lakes and the man-made Lake Nasser. The importance of aquaculture in Egypt is increasing, accounting more than half of total fish supply in some years. During the 1980s, marine fisheries constituted about 22 % of total fish catch, while lakes formed about 50 %; the Nile yielded 11 % and aquaculture production was

D. Tesfamichael (✉)
Fisheries Centre, University of British Columbia, Vancouver, BC, Canada

Department of Marine Sciences, University of Asmara, Asmara, Eritrea
e-mail: dawittes@gmail.com

S.F. Mehanna
National Institute of Oceanography and Fisheries, Ministry of State for Scientific Research, Suez, Egypt

D. Tesfamichael, D. Pauly (eds.), *The Red Sea Ecosystem and Fisheries*, Coral Reefs of the World 7,
DOI 10.1007/978-94-017-7435-2_2

Fig. 2.1 Egyptian Exclusive Economic Zone (EEZ) and shelf area to 200 m depth in the Red Sea

about 17 %. In the 1990s, the percentages were 20 % for marine fisheries, 32.7 % for lakes, 13.3 % for the Nile and 34 % for aquaculture (Mehanna 2007). Based on annual fishery statistics data for Egypt, in 2009, the contributions of the Red Sea and Mediterranean Sea were 4.4 % and 7 % of the total fish catch, respectively, while lakes contributed 15.7 % and the Nile 8 %; the lion's share – 65 % – was from aquaculture (GAFRD 2012).

Egypt's coast along the Red Sea extends from the border with Israel in the Gulf of Aqaba to Sudan in the south (Fig. 2.1). In terms of fisheries, the Gulf of Suez is the most important part of the Egyptian Red Sea. The continental shelf of the Gulf of Suez covers 8400 km^2, which is about the same as the rest of the Egyptian shelf in the Red Sea (Sanders and Morgan 1989). Egypt is divided into 27 governorates, out of which three border the Red Sea: Suez, Red Sea and South Sinai. Fishing is done mainly from Suez and Red Sea proper, while South Sinai is known for its coastal tourism.

Due to the lack of good and extensive fisheries data recording systems, the extent of Egyptian fisheries in the Red Sea and its effects are not well known. Most of the research in Egypt is on various biological aspects of fishes (e.g., Bebars et al. 1993; Yassien 2003; El-Ganainy 2005; EL-Ganainy and Sabra 2008; Sabrah and El-Ganainy 2009; Amin 2011, 2012a, b), usually published in local journals with limited distribution. With internet access getting easier, accessibility of the journals is improving. Publications looking at fisheries level are rare (e.g., Ibrahim et al. 1998; Mahmoud et al. 2009). In this chapter, we present the Egyptian catch in the Red Sea by gear and taxonomic composition from 1950 to 2010. First, we briefly describe the coral reef ecosystem of Egypt in the Red Sea, then each major fishery is introduced, followed by the catches of the fisheries. The sources and methods of the estimation of catches and their composition are presented at the end.

Coral Reef Ecosystems

Egypt has coral reef structures along its coast from the Gulf of Suez to the border with Sudan and the Sinai Peninsula. The reefs in the Gulf of Suez are some of the northernmost reefs in the western Indo-Pacific and have a different community and geographical structures compared to other reefs in the Red Sea (Riegl et al. 2012). Extensive fringing reefs have developed inside the Gulfs of Suez and Aqaba and extend from Guhai in the north all the way to Ras Hedarba in the south at the border with Sudan (Kotb et al. 2008). These fringing reefs are not continuous, but are interrupted by seasonal creek and rivers that flow to the Red Sea creating soft bottom lagoons, local called '*sharms*', which are inimical to the development of corals. A total of 209 hard and 16 soft coral species are identified in the Egyptian Red Sea and live coral cover can be as much as 48 % in some areas (Kotb et al. 2008). Overall, coral cover is higher in northern part close to the entrance to the Gulf of Suez and in the south around Foul Bay and close to the border with Sudan. Species richness of corals is higher in islands in the north close to the entrance of the Gulf of Suez (Riegl et al. 2012). The most abundant coral reef fish are butterflyfishes. Distant second are parrotfishes and groupers are third (Kotb et al. 2008).

There are many stressors and threats to the coral reef ecosystem of Egypt in the Red Sea. They range from large scale impacts of climate change to specific areas affected by local developments. The impact of climate change was clearly seen in the coral bleaching and the extreme low tide that occurred in 2007, which exposed reef flats and resulted in extensive coral mortality down to 20 m depth at 'Rocky Island' in the southern Egyptian Red Sea. In addition, Egyptian corals have been affected by outbreaks of crown-of-thorn starfish (Kotb et al. 2008). The main direct human impact on Egyptian reefs is a rapid increase of the tourism industry (Hawkins and Roberts 1994). Egypt has one of the most developed and expansive tourism for diving, snorkelling and sun bathing along the Red Sea (Fig. 2.2). The impacts of tourism are many: ecosystem alterations from divers and snorkelers (e.g., stepping and breaking corals), anchorage by boats, boat groundings and developments resulting in landfills, dredging and sewage outlets. For example, as there are few natural beaches in the Red Sea coast of Egypt artificial beaches were created, leading to increasing turbidity and sedimentation (Kotb et al. 2008).

The coastal communities depend heavily on the marine ecosystem for food and employment, and cannot easily switch to other livelihoods. High dependency and low flexibility result in overfishing and were strongly and negatively correlated to conservation attitudes among the fishers. For example, only 11.4 % fishers interviews were aware that there were MPAs in their coasts (Marshall et al. 2010). This could be partly due to the top-down management schemes, which compromises the success of conservation measures because they do not account for the socio-economic realities of the local communities.

Other factors affecting coral reefs in Egyptian Red Sea are fishing for marine curios (Hawkins and Roberts 1994), the ornamental fish trade (officially banned in the last few years) and destructive fishing, mainly by new entrants to the fishery lacking the traditional knowledge (Kotb et al. 2008). Empirical researches revealed the cumulative impact of the different stressors on the coral reef ecosystem to be serious: a large proportion of reefs exhibited damages from activities

Fig. 2.2 Coastal tourism in Egyptian waters is one of the most developed in the Red Sea (Photo: Sahar F. Mehanna)

such as anchorage (Riegl and Velimirov 1991; Jameson et al. 1999), species depletion especially sea cucumbers and giant clams (Kotb et al. 2008). Riegl et al. (2012) found, based on long time series data, a decline in coral size and clear signs of decreasing coral growth rate. The Egyptian authorities have introduced MPAs as a management tool; however, their effectiveness is debatable (Gladstone et al. 2003).

Fisheries

Egypt has one of the most developed fisheries in the Red Sea (Tesfamichael and Pitcher 2006). Egypt's industrial fishery, which predominantly consists of trawling and purse seining, was the earliest to develop in the Red Sea. Egypt's industrial fishery has distant-water fishery, which operate in areas beyond national jurisdiction all over the Red Sea and outside the Red Sea, including in high seas/international waters as far as the eastern Atlantic (Feidi 1976). Because the number of industrial boats was growing too rapidly, the Egyptian government put a moratorium on the entry of new boats to the fishery (Mehanna and El-Gammal 2007). The main landing site in the Red Sea is Ataka, in the Gulf of Suez; Hurgada and El-Tor are major landing sites as well. Other, less important landing sites include Salakhana, Berenis, Quseir and Shalateen (Fig. 2.1). Ataka has a detailed data recording system: for every vessel, boxes are weighed and species recorded for all types of fisheries, even traditional ones. In Egypt, the management of fisheries lies under the jurisdiction of the General Authority for Fish Resources Development (GAFRD), which is mandated to collect landings data. A relatively good data recording system was initiated in 1979 by the "Project for the development of fisheries in areas of the Red Sea and Gulf of Aden", but the quality of data from remote areas remains debatable. Another institute involved in fisheries is the National Institute of Oceanography and Fisheries (NIOF), whose task is mainly to conduct research.

The Egyptian fisheries can be divided into four major sectors: industrial (i.e., large-scale commercial, mainly purse seine and trawl gears), artisanal (i.e., small-scale commercial, also called 'semi-industrial'), subsistence (labelled 'traditional' in Egypt) and recreational. The GAFRD database divides the Egyptian fisheries in to the first three categories, i.e., the recreational fishery does not appear in the database. Sometimes, the fishery records are divided geographically for administrative purposes. In this chapter, only the division by gear type is used, because it is an important criterion which can be used in designing management tools. Purse seine and trawl are categorized here as industrial fisheries. Semi-industrial fishing (here referred to as artisanal) is done by motorized boats of smaller size than industrial vessels. The 'traditional' fishery, categorized here as subsistence fishery uses the least technologically advanced boats. Some of its practitioners have no boats, and fish from land. Both semi-industrial and traditional fisheries are mainly reef-associated fisheries.

The main fishing gears deployed by Egyptian fishers in the Red Sea are bottom trawl and purse seine (the industrial fishery), handlines, longlines and gillnets (artisanal) and a variety of gears used by the traditional subsistence fishery. The industrial fisheries operate mainly in the Gulf of Suez and its adjacent areas, and Foul Bay, which borders Sudan. Semi-industrial fisheries are active around Ataka, Salakhana, Sakkala and El-Tor. Boats in Foul Bay, unlike the other areas, do not operate only one gear; the same boat can be involved in trawling, purse seining and handlining (Sanders et al. 1984b). Most of the catch from the Red Sea is consumed within Egypt, a small proportion is exported to Saudi Arabia and Jordan. Egypt imports additional sea food from neighbouring and other countries.

Industrial Fishery

Purse Seine Fishery

The purse seine fishery has been active in Egypt for a long time. The earliest operation began using sailing boats. The main operation, using motorized purse seiners, started in the Gulf of Suez in 1960 with five 150 hp vessels and in Hurgada in 1964 with four vessels. The number of purse seiners increased gradually, with the aim of identifying the level of fishing effort generating optimum catches (Rafail 1970, 1972). The vessels use lamps (gas or kerosene) to attract the fish; once the fish aggregate, they are caught by the purse seine and hauled to the deck of the vessel. The purse-seiners are operated at night using lighted dinghies. The operation starts in October and finishes at the end of May, and it ceases for about 10 days every month during the full moon. About 65 % of the catch comes in the first 3 months, October to December (Sanders et al. 1984a). At the beginning of each season, the fishing trip takes 2–5 days because most fishing is undertaken relatively close to the landing site of Ataka at Suez City. Later in the season, fishing trips take longer, as they have to venture further afield (Mehanna and El-Gammal 2007). The three distinct areas for purse seining along the Egyptian Red Sea coast are the Gulf of Suez, Hurgada area and Foul Bay (Rafail 1970).

Usually, about 82 purse seiners operate in the Gulf of Suez and 30 outside, mainly in Foul Bay. The vessels range between 12.5 and 30 m. They are powered by engines of 150–700 hp, with the majority in the 400–700 hp range. The nets' length range between 200 and 300 m and their depth range from 50 to 80 m. The nets are hauled manually. The crew numbers range between 25 and 40, with no changes from the past (Barrania and El Shennawi 1979; El-Gammal and Mehanna 2002).

Purse seiners target small pelagic schooling fishes, which are also caught by artisanal trammel and gillnets in Salakhana and El-Tor (starting in the 1970s). In the past, only 10 % of the catch was consumed in Suez City, the rest was shipped to Cairo (Barrania and El Shennawi 1979); now, the majority of catch is sold in the Suez fish markets. The catch is consumed mainly fresh and some part salted. There are events where salted fish are consumed, e.g., Sham Al-Naseem Feast or Spring Festival. This feast is an Egyptian national holiday marking the beginning of spring; Egyptians eat salted fish, lettuce, and onions on this day.

Trawl Fishery

Egypt has one of the earliest industrialized trawl fisheries in the Red Sea, which started in 1921, and increased after the Second World War. Egyptian investment in the industrial fisheries is unparalleled in the Red Sea. Egyptian trawlers first targeted shrimp using otter trawlers, which were allowed to operate only from September to June, because the rest of the year is the spawning period for many fishes. The best catches of shrimp were obtained from October to mid-January. A small proportion of shrimp was caught by seiners in shallow waters (Al-Kholy and El-Hawary 1970). The fishing grounds are generally divided into the Gulf of Suez, areas adjacent to the Gulf of Suez and Foul Bay. At the beginning of the season, the vessels operate in the northern part of the gulf and their trips take about 5 days. Later in the season, they move further south and the trips take longer. Trips from Foul Bay can be up 20 days, out of which 6 days are to travel to and from the fishing ground. The main base for this trawl fishery is Ataka, although Sakkala is also used when vessels operate outside the gulf. Note that none of the trawlers operate exclusively outside the gulf. The vessels that operate in Foul Bay, which represents about 10 % of the three trawling grounds, are used for purse seining as well (El-Gammal and Mehanna 1999).

In the mid-2000s, the number of vessels operating in the Gulf of Suez was 78 vessels, while about 100 vessels operated outside the gulf and outside Egyptian waters. The vessel length varied between 20 and 30 m (Fig. 2.3), and each was powered by a main engine of 200–1200 hp (mostly 400–600 hp). All vessels had mechanized winches, and some of them echo-sounders. The trawl net they employed was of the Mediterranean type, and its length ranged between 20 and 30 m with an average mesh size of 1.5 cm in the cod-end, and a sweep length between 200 and 250 m. The fishing trip was about 5–10 days and the number of crew varied from 10 to 15. The trawl fishery is seasonal, generally from October to May. The number of fishing days during the first 3 months of the fishing season constitutes 42 % of the total effort (Mehanna and El-Gammal 2007).

Artisanal Fishery

'Artisanal fishery' is not a term commonly used by the Egyptian authorities. What we categorized as 'artisanal fishery', based on its mode of operation, is called 'semi-industrial fishery' in Egypt. The artisanal fishery uses motorized boats locally called launches, whose number varied between 93 and 178 in the Gulf of Suez and about 415 outside, with lengths ranging between 10 and 15 m, and inboard engines of 50–200 hp. The fishing trips take about 10 days, and the crew ranges from 2 to 10. The fishers on these boats use sev-

Fig. 2.3 Egyptian industrial fishing vessels (Photo: Sahar F. Mehanna)

eral fishing gears such as long-line, hand-line, gillnet, trammel net and beach-seine (Mehanna 1999).

The crew of the artisanal boats frequent fishing grounds within and adjacent to Gulf of Suez, and land their catch in Ataka and Salakhana (in Suez City), Sakkala in Hurgada and El-Tor in South Sinai. Their main targets are reef-associated fishes and they use seines and cast nets to catch bait. Part of their catch is stored in ice boxes with ice and transported to Cairo. The reef-associated catch from Foul Bay used to be done by purse seines landed in Berenis, Shalateen and Abo Ramad and transferred directly to Hurgada and Cairo (Chakraborty 1984a, b). Currently, the purse seiners are prohibited from catching reef-associated species.

Subsistence Fishery

As in the case of the artisanal fishery, Egyptian authorities generally do not use the term 'subsistence fishery', but use instead 'traditional fishery', who fish for their own consumption, but who may sell a small portion of their catch at the local markets. These fisheries include foot-fishers, sailing boats (non-existent at the present) and a few motor boats, equipped mainly with outboard or small inboard engines. These fisheries operate in the near-shore waters along the Egyptian coast, including in South Sinai, which is known for its touristic attractions, but not for fishing. All the fisheries in Sinai are categorized as 'traditional' except in El-Tor (Chakraborty et al. 1983). The subsistence fishery targets mainly fishes in shallow waters and coral reef areas using handline, gillnet, trammel, seine nets and cast nets. In some areas, e.g., Quseir, large sail boats locally called '*Katira*' are used to catch sardines, mullets and goatfish. The traditional fishers may occasionally use trucks and camels to transport food and fishing equipment from place to place (Chakraborty 1984a, b).

Recreational Fishery

Some form of recreational fishing has been practiced in Egypt for a very long time, starting from the time of the Pharaohs (Pitcher and Hollingworth 2002). The more contemporary recreational fishery started with the growth of tourism in the Egyptian Red Sea, i.e., after the 1967 Arab-Israeli war (Hawkins and Roberts 1994). Both tourists and the local population are involved in recreational fishing. They usually use line fishing (simple hook and line, longline and trolling) and more rarely nets (mainly to catch bait). The catch is usually consumed by the fishers, some given to friends and very rarely, in the case of a big catch, a small portion may be sold in the market. The number of boats involved in the recreational fishery has grown very fast in the last few years (FAO 2004, 2010). Some of the artisanal boats in Suez Bay, El Tor and Hurgada operate as recreational fishing operators as well during the fishing closed season.

Fisheries Catches

The reconstructed total Egyptian Red Sea catch deemed taken within the Egyptian EEZ or EEZ-equivalent waters (i.e., prior to the formal declaration of the EEZ) is shown in Fig. 2.4, together with the catch Egypt reported to FAO from statistical area 51, to which the Red Sea belong. The reconstructed and FAO catches were more or less the same in the 1950s. Then two distinct patterns appeared, before and after 1993. The earlier pattern, from the early 1960s–1993, was where the annual reconstructed catches for Egyptian waters were clearly higher than the FAO reported. From 1993 on, the opposite occurred, i.e., the FAO reported catches were much higher than the reconstructed catch. Our explanation, given all the information we have, is that the catch was underreported from the early 1960s to the early 1990s, but the latter part included catches made outside Egypt's own waters, which are not included here in the reconstructed catch, pertaining only to the Egyptian EEZ. Egyptian boats, mainly industrial, have been expanding their operations to the other Red Sea countries mainly since the early 1990s and Egypt's catch from those waters were presented in their respective country (see Chaps. 3, 4 and 5).

Here, we are primarily interested in getting a better estimate of what is caught in the ecosystem, i.e., where the catch comes from, and secondly who (or which country) is catching it. Thus, what Egypt caught in Sudanese, Eritrean and Yemen is reported in the respective countries chapters with clear indication that it was caught by Egyptian vessels. Nevertheless, adding the values of Egyptian catch outside its EEZ to the reconstructed catch will not fill the gap between the reconstructed catch and FAO data of Fig. 2.4. In order to understand the difference, we dug deeper into Egyptian catch by area, which we were able to obtain for few years from the Egyptian authority (GAFRD 2012). The overall total Egyptian catches in the Red Sea match what is reported in the FAO data. However, when the data were dissected by region, a different picture appears. The areas reported include the 'Gulf of Suez' (or catch landed in the ports in Suez, the main landing sites), 'Aqaba', and what is vaguely labeled as 'south' or 'southern Red Sea', and 'outside'. These terms sometimes refer to the catch caught 'outside Gulf of Suez' or fish caught in the southern Red Sea including areas outside Egypt's EEZ. For comparison, those catches were taken out from what Egypt reports to FAO for area 51; in this case, we obtained estimates closer to our reconstructed catch, corresponding to the open circles in Fig. 2.4.

The sudden sharp dip in the reconstructed total catch in 1973 was the consequence of the 1973 Arab-Israel war, which destabilized the region and destroyed infrastructure. The effect of the previous 1967 Arab-Israel 6 day war is also shown by the smaller dip in the late 1960s. The first rapid

Fig. 2.4 Reconstructed catch of the Egyptian fishery in the Egyptian Red Sea EEZ by sector and the data Egypt submitted to FAO from 1950 to 2010 for FAO area 51. Open circles indicate Egyptian catch reported to FAO without what is possibly caught outside its EEZ put for comparison purpose

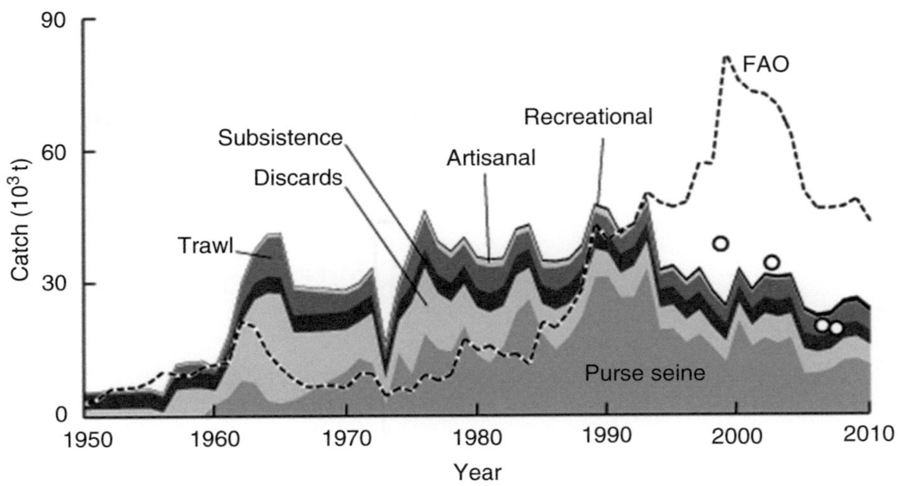

increase in the early 1960s was mainly due to the trawl fishery, which was expanding at the time. The reconstructed total catch remained stable, albeit with minor fluctuations until it declined in 1993, which may be due to Egyptian vessels sailing outside their EEZ for better returns on their operation than staying in the heavily-fished Egyptian EEZ. Logically, it makes sense that they would put more effort (fuel and time) to go further south and thus decrease the total catch from the Egyptian EEZ. It is also important to note that Egypt has a very high concentration of fishing vessels, and the fishery authority banned new entry, thus even existing vessels were forced to explore new grounds as the catches from the traditional fishing grounds began to decline.

Overall, from 1950 to 2010, purse seining contributed the largest share to the total catch (42 %), followed by trawling (27 % discarded and 13 % retained catches). The trawl fishery was dominant in the early years, i.e., the 1960s until the mid-1970s, after which it was more or less at par with purse seining, which took over in recent years. The subsistence fishery was the third most important fishery by total catch (14 %). The artisanal fishery and recreational fisheries had low contributions (3 % and 1 %, respectively). The subsistence fishery kept more or less the same level throughout, which is the case in the other Red Sea countries as well (see Chaps. 3, 4 and 5). The subsistence fishery is less affected by external factors, such as international market demand. They are conducted mainly for subsistence and they operate as long as there are people, mainly their families and communities, to consume their catches.

The reconstructed total catch in Egypt's EEZ can be divided into a component that can be found in the official report to FAO (reported catch), unreported landed catch and the discarded catch, which is also not reported to FAO. The unreported landed catch contributes 43 % of total catch, while reported catch is 30 % and discarded catch 27 %.

For all fisheries combined, the composition of the reconstructed total catch is dominated by jacks (Carangidae; 13 %), red-eye round herring (*Etrumeus sadina*; 11 %)[1], scads (*Decapterus* spp.; 10 %) and sardinella (*Sardinella* spp.; 6 %) from the dominant purse seine fishery. The other dominant taxa are Berber ponyfish (*Leiognathus berbis*; 7 %) and lizardfishes (Synodontidae; 5 %) (Fig. 2.5), both from the trawl fishery; the former is a discarded species, but the latter is retained. The total number of taxa identified in the Egyptian fishery was quite large (42), which accounted for 89 % of the total catch. The remaining 11 % was a mix of many taxa not identified separately, as their contribution was small. They were lumped in the 'others' group. In Fig. 2.5, only the taxa that have major contributions to the total catch are presented.

Looking at the fishery sectors separately (Fig. 2.6), the industrial purse seine fishery had a continuous, although fluctuating, upward trend from its beginning until it peaked in 1993. After 1993, it declined, but again with fluctuations. The trend of the purse seine fishery shaped the trend of the overall reconstructed catch for Egypt, especially in the later years because it was the fishery with the highest contribution (Fig. 2.6a, also compare it with Fig. 2.4). As far as the composition of the purse seine fishery is concerned, three taxa contributed more than 70 % of the total catch. They were red-eye round herring (*Etrumeus sadina*, 26 %), jacks (Carangidae, 23 %) and scads (*Decapterus* spp., 23 %).

The industrial trawl fishery expanded very quickly at the beginning of the 1960s after its exploratory phase throughout the 1950s; then there was a sudden decline during both the

[1]According to FishBase (www.fishbase.org) the distribution of *Etrumeus sadina* does not include the Red Sea; however it is reported both by FAO and GAFRD by its common name as 'red-eye round herring' and 'round herring', respectively, both common names for *Etrumeus sadina*.

Fig. 2.5 Composition of the total Egyptian fishery catch in the Red Sea from 1950 to 2010

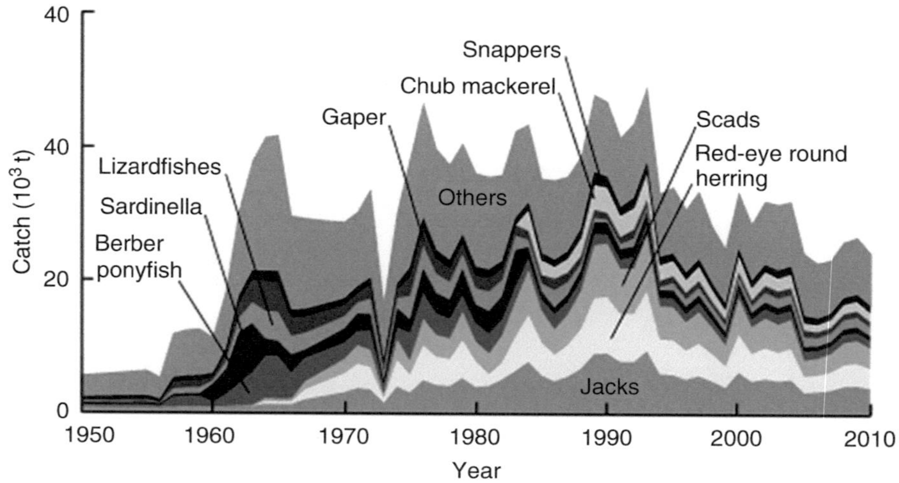

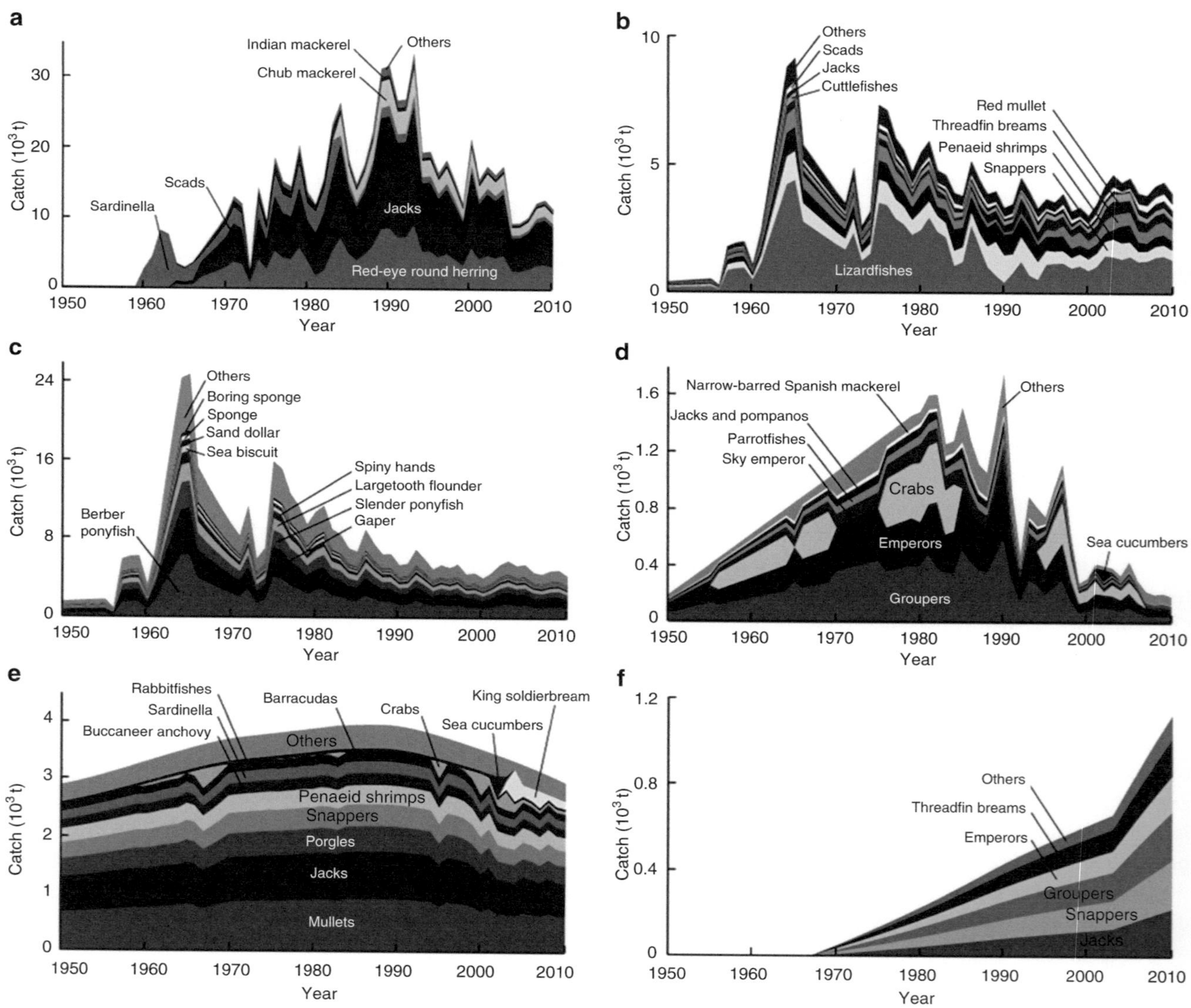

Fig. 2.6 Catch composition of (**a**) purse seine (**b**) trawl-retained (**c**) trawl-discarded (**d**) artisanal (**e**) subsistence (**f**) recreational fisheries in the Red Sea EEZ of Egypt from 1950 to 2010

1967 and 1973 Arab-Israel wars. Since its recovery in 1974, the trawl fishery exhibited fluctuations with a slight declining trend. In the earlier period, the trawl fishery had the highest contribution to the total catch until it was taken over by purse seine. Although the main prized target of trawl fishery is shrimp, lizardfishes (Synodontidae) had the highest contribution by far (40 %), followed by snappers (Lutjanidae; 14 %), while shrimp was the third with 12 % of the retained catch. The three taxa contributed more than 66 % of the total landed catch from 1950 to 2010 (Fig. 2.6b). The discarded trawl catch followed a similar pattern, except that the decline was stronger (Fig. 2.6c). This is because the percentage of discarded catch to the total catch was lower in the later years as more and more of the lower-grade fishes were retained in the catch when the most sought-after fishes started to decrease. Berber ponyfish (*Leiognathus berbis*) had the highest contribution to the discarded catch (26 %).

The artisanal fishery (sometimes described as 'semi-industrial' in Egyptian national reports), reached its peak in 1990 (Fig. 2.6d). There was a continuous increase from 1950 to 1979, largely driven by our estimation process, and thus ignoring potential inter-annual variability. We did not have data for that period for the fishery and we assumed the lowest catch that was realized, which was the catch of 2010, for 1950 and interpolated the rest of the period. For the period we had data, the fishery had a declining trend with fluctuations. In terms of the composition of the catch, groupers (*Epinephelus* spp.) had the highest share (31 %) and second was emperors (Lethrinidae; 25 %).

The subsistence (traditional) fishery had the most stable pattern (Fig. 2.6e). Because of its diffuse and unregulated nature, this fishery sector was the most difficult in terms of getting estimates as to its size or catch. This is the catch taken by the smallest boats, some even by fishers without boats, who mostly fish for their own family consumption and is the least affected by markets and other external factors. Mullets (Mugilidae) had the biggest share of the total catch (23 %), followed by jacks (Carangidae; 20 %) and porgies and seabreams (Sparidae; 11 %).

The recreational fishery was the sector with the least contribution to the total catch and also that which started last. Since then, however, it is the only fishery with a continuously increasing trend (Fig. 2.6f). The number of boats involved in recreational fishing has increased considerably as sea-based tourism (both international and local) has increased in Egypt and became economically quite important. As compared to the other sectors, there was no dominant taxon in this fishery. The recreational fishery does not seem to get much attention from the Egyptian fishery authorities, possibly because of its limited size and/or that it does not fall into the commonly accepted fisheries in the country: fishing solely done for commercial purposes or to feed one's family.

This reconstructed catch of Egypt in the Egyptian Red Sea EEZ is quite detailed in terms of its comprehensive coverage of the sectors and the composition of their catches, incorporating all the information available to us. To our knowledge, this is the only attempt made to clarify and standardize the Egyptian fisheries catch data for the Red Sea. When assumptions were made, they are clearly stated. The methods and results are standardized and can be used for any further analysis of the fishery. We believe this work will help to better understand the Egyptian fishery in the Red Sea in assessing, managing and conserving the livelihoods and the ecosystem and improve the practice of data collection and presentations of fishery data in the future. Most of the assumptions we had to make can be, in the future, replaced by some simple procedures in the data collections systems.

Sources and Methods

As this research deals with historic statistics on the fisheries of Egypt, the methodology required the compilation of data from different sources, cleaning, and standardizing them for presentation in usable form (Tesfamichael and Pauly 2011). Whenever there were data gaps, they were filled using assumptions, which are stated openly for criticism and further refinement. An extensive search was made for fishery catch data of Egypt in the Red Sea in journal articles, gray literature, reports and databases. In addition, the knowledge of people (fishers and administrators) familiar with the statistical system was sought to fill in gaps and interpret data and the results of our analysis. We found that there are many published reports, mainly in national journals, on the Egyptian fisheries. After a close scrutiny of the literature, the main sources of Egyptian fisheries catch data can be categorized into three sources. The earlier ones were published (mainly in the 'Bulletin of the Institute of Oceanography and Fisheries') from the 1950s until the 1970s by local researchers in order to identify the potential of the resource using traditional stock assessment tools such as surplus-production models. The next category consists of reports from an FAO project, the 'Development of Fisheries in Areas of the Red Sea and Gulf of Aden', which ran from the late 1970s to the mid-1980s. The reports were very detailed and gave very good insight into the fisheries of Egypt by gear type, taxon and place. They were written mainly by foreign experts, sometimes with local co-authors. This phase is data rich and several similar reports, sometimes with overlapping contents, were written by different authors, mostly technical consultants to the project. The third category of data source is the database of GAFRD (2012), which contains data from 1979 to the present. The GAFRD database is informative, but it has gaps. The catch reconstruction was done separately for each major sector of the Egyptian fisheries in the Red

Sea. A spatial distribution of the catch is given at www.seaaroundus.org.

Industrial Fisheries

Purse Seine Fishery

The earliest report for the purse seine fishery in the Egyptian Red Sea available to us was from Rafail (1970). Landings were given for Gulf of Suez from 1960 to 1965 for sardinella, *Sardinella gibbosa*, reported as *S. jussieu*, but also as goldstripe sardine in some reports (Chakraborty 1984a). The geographic distribution for *S. gibbosa* in Fishbase (Froese and Pauly 2012) included the Red Sea, but this was not the case for *S. jussieu*. The report gave the proportion of the taxon from the total purse seine fishery landings of the Gulf of Suez, which were used to calculate the total catch. For Hurgada, landings of sardinella, which accounted for 25 % of the total purse seine catch, were given for two seasons (September–May) 1964/1965 and 1965/1966. First, the catches of Hurgada were converted to calendar year assuming that 63 % was caught from September to December and the rest from January to May; this ratio is an average calculated based on reports from Sanders et al. (1984a) and data from GAFRD (2012). Catches were given for the Gulf of Suez and Hurgada separately, but not for Foul Bay, because the boats that fished in Foul Bay were from the Gulf of Suez and Hurgada, and it is probable that their catch was included in the reports from those two areas (Rafail 1970). For 1966, it was possible to calculate total catch only for Hurgada, as there was not any data given for the Gulf of Suez. Thus the total was calculated using the ratio of Hurgada to the Red Sea from 1965 and the total of Hurgada for 1966.

Prior to the industrial purse seiners, non-motorized sailing boats had already been purse seining, although at a smaller scale. Their catch was not recorded (based on interviews with fishers and administrators), thus it was estimated to be about 5 % of the total purse seine catch. Thus, 5 % of the purse seine catch for 1960, which was 133 $t \cdot year^{-1}$ annually, was assumed from 1950 to 1959. In addition, trawlers were also involved in purse seining starting 1960 (Sanders et al. 1984b) and their catch is assumed to be 1 % of the total purse seine catch. Accordingly, 6 % (5 % from non-motorized boats and 1 % from trawlers) was added to the catch from 1960 to 2010. There were no data from 1967 to 1969, and the catch was interpolated between 1966 and 1970.

From 1970 to 1978, the annual purse seine catches were available (Chakraborty 1984a) and from 1979 to 1982 (Sanders et al. 1984a). Unlike other reports, these catches were for the whole Red Sea, not only for the Gulf of Suez. Thus, these values were used as they were, but 6 % was added to account for the unreported catch by sailing boats and others.

From 1980 to 2010, data were available from GAFRD (2012), but it was only for the Gulf of Suez. The data from 1979 to 1989 were presented by fishing season only, i.e., September to May. These data were first converted to calendar year using the same procedure as described above. From 1990 to 2004, there were two data sets: one presented by season, which was converted to calendar year, and another by calendar year. Comparison of the two data sets resulted in correlation coefficient of $r = 0.91$, which showed our conversion process to be reasonable. From 2005 to 2010, the only data available were presented by calendar year. The converted data were used from 1983 to 1989 and the other set from 1990 to 2010.

Once the annual total catches were calculated for the Gulf of Suez, they were scaled up to the whole Red Sea based on Sanders et al. (1984a), who reported that for 1980 and 1981 the Gulf of Suez accounted for 75 % of the total Egyptian Red Sea purse seine catch. The total annual catch calculated from 1979 to 2010 were used except from 1980 to 1982, where the more detailed data from Sanders et al. (1984a) were used instead of those from GAFRD (2012). To all the totals, 6 % unreported catch was added.

The earliest catch composition data available for Egyptian purse seine fishery was the percentages of only *S. gibbosa*, which accounted for up to 95 % of the total of the Gulf of Suez purse seine catch starting in 1960 (Rafail 1970). The difference between 100 and the percentages for *S. gibbosa* were allocated to 'others'. Since, from 1960 to 1964, purse seining occurred only in the Gulf of Suez, those percentages were used for the whole Red Sea. The ratios of 1960 were used from 1950 to 1959. From 1964 to 1966, catch composition ratios were given for both the Gulf of Suez and Hurgada landings (Rafail 1970). For 1965, the unidentified catch of 'others' was too high (72 %) and for 1966 the data were incomplete; thus, for these 2 years, the ratios of 1964 were used. For 1964, the group 'others' had a value of 33 % and it was reduced by dividing it to the taxa not represented for 1964, but were represented in a more detailed catch composition for 1980 (Sanders et al. 1984a). Because of the uncertainty in identifying the sardinellas into species level, we presented them here as *Sardinella* spp.

Better catch composition data were available from 1980 to 1982 (Sanders et al. 1984a) and 1983 (Chakraborty 1984a). In both sources, horse mackerels and scads were reported as one group. However, the two belong to two different genera and are reported separately in the FAO database. Hence the ratio given for the two together was divided equally between the two. In addition, there was a group called 'miscellaneous' and another one called 'others', which were combined. The ratios of 1980 were used from 1967 to 1979. From 1984 to 2010, GAFRD (2012) data presented the catch by taxonomic components. The data were first converted to calendar year. For most of the years, a large proportion was categorized in the group 'others'; this was disaggregated using the average catch composition ratios from the years where the group 'others' was less than 10 % (1992–1994, 2000 and 2001). Similar to 1980–1983, the group 'horse mackerel and scads' was divided into two equal separate groups.

Trawl Fishery

The earliest Egyptian trawl fishery record available was the catch of shrimp from 1921 (Al-Kholy and El-Hawary 1970) from the Gulf of Suez. Although the catch was given only for shrimp, we believe that at least some of the fishes were also retained, as implied in other reports on trawling (Latif and Shenouda 1972), which, however, presented data only for 1963. In addition, there are reports of demersal fish caught by trawling in Egyptian Red Sea data submitted to FAO. Thus, the total trawl catch from 1950 to 1961 was calculated based on reported shrimp catch. First, a continuous series of shrimp catches was established because the data from Al-Kholy and El-Hawary (1970) were intermittent. Then the gaps in the shrimp catch from 1945 to 1955 were filled by interpolation, and scaled up to total retained catch for the Gulf of Suez, based on the ratio of shrimp from total catch, which was reported to be 10 % (Chakraborty 1984a). The Gulf of Suez was later scaled up to the whole Red Sea based on Sanders et al. (1984b), who reported that the Gulf of Suez accounted for 90 % of the total Egyptian Red Sea trawl catch. We used this procedure to reconstruct the catch from 1950 to 1961.

The next data set available was from 1963 to 1966 (Latif and Shenouda 1972). The value for 1963, however, was not used as it did not correspond to the other figures given in the same report. First, the Gulf of Suez catch was scaled up to the whole Red Sea, based on Sanders et al. (1984b), who reported that it accounted for 90 % of the Red Sea total. The totals for 1962 and 1963 were interpolated using data from 1961 and 1964.

Chakraborty (1984a) presented the total catch of trawl fishery for the whole Red Sea from 1970 to 1983 with some years missing. However, only the data from 1970 to 1978 were used, as the years after that were not complete. Besides, there was a continuous data set from 1979 to 2010 from GAFRD (2012). In addition to Chakraborty (1984a), there were more sources for the years from 1980 to 1983, thanks to the FAO funded project, which employed many experts, e.g., Sanders et al. (1984b), who even presented monthly catches. However, all those data sets were very similar and because of its completeness and continuity, the data from GAFRD was used. It consisted of two sets: one only for the Gulf of Suez from 1979/1980 to 2007/2008 (except 2005/2006), by fishing season (September–May). The catch data therein were converted to calendar year using similar procedures we employed for the purse seine fishery (the value for 2005/2006 was calculated as an average of 2004/2005 and 2006/2007). The annual total of the Gulf of Suez was then scaled up to the whole Red Sea, based on data from Sanders et al. (1984b), who reported that the Gulf of Suez accounted for 90 % of the total Egyptian trawl fishery catch in the Red Sea. The second data set were from 1990 to 2010, by calendar year. Comparison was carried out between the two datasets over the overlapping years, 1990–2008, and resulted in a correlation coefficient of $r=0.83$, which indicated that our conversion process of the seasonal data to calendar year is reasonable. From 1979 to 1989, the converted data were used and from 1990 to 2010, data given by calendar year were used. The main unreported portion of the trawl fishery is discarded catch. Since this was a substantial amount, it is treated separately.

A good proportion of trawlers' catch is thrown back as discard. The only Egyptian study we found on this topic was El-Ganainy et al. (2005), who found, based on trawl surveys in 2003, that discards amounted to 56.1 % of the total catch. Given that the trawl fishery in 1950 was mainly for shrimp, which has a discard amount of up to 90 % in the Red Sea (Sanders and Morgan 1989; Tesfamichael and Pitcher 2007), a conservative estimate of 80 % of total catch was assumed for 1950. Using these data points, the discard proportion was interpolated from 1950 to 2010, to mimic the behaviour of the fishers who tend to retain more and more of the less valued fishes as the premium species, usually high-trophic level ones, start to decrease or disappear altogether in their catch (Pauly et al. 1998). Once the discard proportions were established, the discard amount was calculated based on the reconstructed retained catch.

The composition of retained catch was calculated based on data from Sanders et al. (1984b) for 1980–1982, Chakraborty (1984a) for 1983, and GAFRD (2012) for 1984–2010. The categories 'miscellaneous' and 'pony fish' (i.e., Leiognathidae), which were less than 1 % each, were added to the category 'others' from 1980 to 1983. From 1950 to 1979, the average proportions of 1980–1983 were used. All the sources put 'horse mackerel and scads' together. However, given that the two belong to different genera and are reported separately in the FAO database, the group was divided into two equal portions.

The catch composition of the discarded catch was calculated using data from El-Ganainy et al. (2005), where the ratio of fish species and crustaceans were given separately. However, the proportion of each category to total discards was not given. It is assumed, by comparison with other Red Sea trawl fisheries, that fish discards contributed 75 % and crustaceans 25 %. For a few taxa, the scientific names presented were not the valid names. For those taxa, valid scientific names were obtained from FishBase (www.fishbase.org) and SeaLifeBase (www.sealifebase.org).

Artisanal Fishery

The earliest record available for the artisanal (semi-industrial reef) fishery was for the 1979/1980 season, lasting from October 1979 to September 1980, and for the next three seasons, i.e., until September 1982, for the whole Red Sea (Chakraborty 1984a). GAFRD (2012) also had data from 1979/1980 to 2004/2005 pertaining, however, only to the Gulf of Suez. The data from Chakraborty (1984a) were used because they were based on extensive frame surveys, detailed and included catches from areas that are not included in the GAFRD data set. For the rest of the years, the GAFRD database was used. First, the GAFRD data, 1982/1983–2004/2005, were scaled up to the whole Red Sea by comparing the data for the overlapping years (1979/1980–1981/1982) between Chakraborty (1984a) and GAFRD (2012), which resulted in the Red Sea total being 2.29 times that of the Gulf of Suez catch. Then the data were adjusted to calendar year using data from Sanders et al. (1984c): October to December (22 %) and January to September (78 %). From 1990 to 2010, the GAFRD database, which presented the Gulf of Suez catch by calendar year, was used. Comparison of the overlapping years, 1990–2005, between the seasonal data converted to calendar year and the data already presented by calendar resulted in correlation coefficient of $r=0.98$, which indicates that the conversion process performed well.

Based on our observation of the operation of this fishery, we believe that part of the catch of this fishery is not fully reported. Some of the catch is sold in informal markets where data recording does not occur. A conservative estimate of 10 % was added to the totals to account for the unreported catch.

No records were available for this fishery from 1950 to 1979. Approximate catches were derived for this period by assuming the level of catch in 1950 to be the same as 2010, where the catch declined to its lowest level, and interpolating between 1950 and 1979.

The catch composition from 1980 to 1983 was calculated using data from Sanders et al. (1984c) and for 1984 from (Chakraborty 1984a). The only change made to the ratios given in those reports was that 10 % of the category 'others' from the Gulf of Suez catch from 1980 to 1983 was deducted and allocated to Spanish mackerel, which appeared in other reports. Data from GAFRD (2012) gives the catch composition from 1979 to 2005, in which Spanish mackerel was presented separately and the average for 1980 to 1983 was 10 % of the 'others'. For 1984, 10 % of 'others' from the overall Red Sea catch composition was assigned to Spanish mackerel. The GAFRD database catch composition from 1984 to 2005 is highly aggregated under the category 'others'. Thus the catch composition of GAFRD is ignored and the average of 1980–1983 is used for 1984–2010 and 1950–1979.

Subsistence Fishery

Although the presence of a subsistence (traditional) fishery is acknowledged by many authors, we found only one estimate of its catch, for 1983, from a frame survey (Chakraborty 1984a). In the absence of other data, population size, based on the United Nations, Population Division (http://esa.un.org/unpd/wpp/Excel-Data/population.htm) was used as a proxy to estimate the subsistence fishery catch. This is a reasonable assumption, because catch of traditional fisheries is usually consumed locally, hence related to population size. Because the catches of the traditional fishery are almost exclusively consumed by local communities, this fishery is categorized as subsistence fishery in our reconstruction. First, the catch for 1983 was divided by total population. This per capita ratio was multiplied by 1.5 for 1950, assuming per capita rate of subsistence catch was 50 % higher in 1950 when the resource was more abundant and the population size small. For 2010, the 1983 ratio was halved (50 % less), reflecting the overall decline in fish abundance and increase in population size. Once these three points were established, the ratios for the rest of the years were interpolated. Then the total catch was calculated by multiplying these ratios by the population size. This is a very conservative estimate; as most of the estimates are less than the only report available, for 1983. Traditional fishers are known to use a large proportion of their catch to feed their families and give to relatives and friends. Based on interviews in the Red Sea, this can be up to 50 % of their total catch. It was also observed during the interviews that fishers used to give a larger proportion of their catch in the past, when there was less marketing of their catch, but later the ratio they give freely decreased. In order to estimate the proportion of the unreported catch of the traditional fishery, we used a conservative 30 % for 1950 and 10 % for 2010. The ratios were interpolated between the 2 years.

The catch composition of the subsistence catch was calculated using the ratios given in Chakraborty (1984a) with minor modification. In the report, sardines and anchovies were reported as one group. However, they are reported separately in the FAO database. Thus, we presented them separately, with each having a composition of 4.4 %.

Recreational Fishery

The presence of a recreational fishery in Egypt is mentioned in many reports, but catch data are very scarce. Indeed, the only quantitative information available for the recreational fishery of Egypt in the Red Sea was that there were 3013 recreational fishers in 2003 (FAO 2004) and 5079 in 2008 (FAO 2010). The term 'recreational fishers' in Egypt refers to full-time operators of boats taking tourists (local and foreign) on day trips for recreational fishing. The 2008 report also gives the number of boats involved in recreational fishing activity. In order to estimate the total catch of the recreational fishery, the numbers of operators from 2003 and 2008 were used. First, the proportions of recreational fishers in the total population (participation rate) was calculated for 2003 and 2008, which were 0.0043 % and 0.0067 %, respectively. The recreational fishery was assumed to start in 1968, after Egypt's war with Israel in 1967. The tourism industry in the Red Sea started after the war (Hawkins and Roberts 1994) and tourism has major impacts on the recreational fishery. The participation rate for 1967, therefore, was assumed to be zero. The rates were interpolated between 1967 and 2003 and again from 2003 to 2008. The slope of change from 2007 to 2008 was used to calculate the participation rate for 2009 and 2010.

Once the participation rates were estimated from 1967 to 2010, the total number of participants was calculated by multiplying the participation rates by the population of Egypt, obtained from the United Nations, Population Division (http://esa.un.org/unpd/wpp/Excel-Data/population.htm). In addition to the number of participants, data on number of days per year and the catch rate per day are needed for the estimation of total catch. The number of days per year we assumed a conservative 250 days $\cdot$ year^{-1} based on the report from FAO (2010), stating that the recreational boats sail on more than 280 days per year. For the catch rate per operator, we assumed 2 kg $\cdot$ day^{-1} for 1968 and 1 kg $\cdot$ day^{-1} for 2010. This is again a conservative catch rate assumption. The catch rate was interpolated between 1968 and 2010, to mimic the change in catch rate as the intensity of fishing increases and abundance decreases. The total catch was then calculated by multiplying the number of operators by the number of days per year they fish multiplied by the catch rate. Since the estimated number of recreational operators was for the whole of Egypt's EEZ, i.e., both the Mediterranean and the Red Sea, the Red Sea part was calculated by assuming that 75 % of the recreational fishery occurs in the Red Sea, based on the report that most of the recreational fishing takes place in the Red Sea (FAO 2010).

The catch composition of the Egyptian recreational fishery in the Red Sea is estimated based on the field observation by the second author. The dominant taxa in catch are groupers (Serranidae), pelagic jacks and mackerels (Carangidae) and snappers (Lutjanidae). Emperors (Lethrinidae) and large-sized threadfin breams (Nemipteridae) are also common in the catch, although not as dominant as the three previously listed ones. We assumed a contribution of 20 % each for the three dominant taxa, 15 % each for emperors and threadfin breams and 10 % was allocated to the 'others'.

Comparing Reconstructed Catch with FAO Statistics for Egypt in the Red Sea

The catch data we reconstructed from different sources was compared with the Egyptian Red Sea catch as reported in the FAO database. Although Egypt has access both to the Red Sea and Mediterranean Sea, they belong to two different FAO statistical areas; hence, the FAO data for Egypt is separate for the two seas. To compare our data with Egypt's FAO Red Sea data, we first checked for taxa not included in the different sectors of the reconstructed catch, but reported to FAO, and vice versa. Some groups: 'silversides (sand smelts) nei', 'largehead hairtail' and 'flatfishes nei' were reported by Egypt to FAO for a few years, but not represented in the reconstructed catch. We assigned their low catch to the group 'others'. Other taxa, found in the FAO data but not the reconstruction, were allocated to the appropriate sector in the reconstruction. Their amount in the reconstructed catch was taken to be the same proportion they had in the total FAO catch. The amounts were later deducted from the 'others' of the sector to which they were allocated. For each sector, for the years in which the group 'others' was higher than 10 %, it was reduced to 10 % and the rest distributed to the taxa already identified according to their proportion in each sector. These procedures changed the original taxonomic composition of the catches described in each sector in the above. Each taxon in the reconstructed catch was compared with its corresponding value in the FAO data. The part of the reconstructed catch that is accounted in the FAO data is referred as 'reported catch' in our result. The difference was presented either as unreported or over-reported catch depending whether the reconstructed value was higher or lower than the FAO value.

Egyptian Fishery outside Its Waters in the Red Sea

Egyptian fishery is the most developed in the Red Sea and the earliest to expand and use advanced technology. The concentration of fishing fleets is so high that the government banned new entry to the fishery (Mehanna and El-Gammal 2007). The Egyptian fishing fleet has been known to roam the whole Red Sea, i.e., outside Egypt's Exclusive Economic Zone (EEZ) with or without formal access agreements with the other Red Sea countries. There are even reports that Egyptian vessels fishing outside the Red Sea, besides Egyptian waters in the Mediterranean, as far as the eastern Atlantic (Feidi 1976). The Egyptian fleets that venture out of

Egyptian waters are mainly large trawlers and purse seiners. Based on the catch reconstruction of Sudan (see Chap. 3), Eritrea (see Chap. 4) and Yemen (see Chap. 3), we were able to estimate the total catch of Egyptian vessels in those countries. It does not include the illegal fishing activities carried out by Egyptian vessels, which are not uncommon occurrences. Several Egyptian vessels and their crew have been arrested in Eritrea and interviews with fishers in Sudan revealed their grievance of their resources being exploited illegally by Egyptian vessels and how that had affected their catch.

Acknowledgements We would like to thank the fishery authorities and research institutes in Egypt for their collaboration during this research, and Hatem Hanafy Mahmoud for reviewing the chapter. Thanks also go to Ms. Eden Mesfin and Ms. Emnet Mesfin for hosting the first author during his field trip to Egypt. This research was supported by the *Sea Around Us*, a scientific collaboration between the University of British Columbia and The Pew Charitable Trusts.

References

Al-Kholy AA, El-Hawary MM (1970) Some penaeids of the Red Sea. Bull Inst Oceanogr Fish Cairo 1:339–377

Amin AM (2011) Population dynamics of Por's Goatfish (*Upeneus pori*, Ben-Tuvia and Golani, 1989) inhabiting Red Sea, Egypt. Egypt J Aquat Res 37(3):283–292

Amin AM (2012a) Biology and assessment of thread fin bream *Nemipterus japonicus* in Gulf of Suez, Egypt. Egypt J Aquat Res 16(2):47–57

Amin AM (2012b) Population dynamics of *Lutjanus lineolatus* (Family:Lutjanidae) Egypt. Egypt J Aquat Res 10(2):79–92

Barrania AA, El Shennawi MA (1979) Report on the exploratory socio-economic survey of the Egyptian Red Sea fisheries. FAO/UNDP Project for Development of Fisheries in Areas of the Red Sea and Gulf of Aden, Suez

Bebars MI, El-Etreby SG, Bayoumi AR, Zakaria MS, Abdulla M (1993) Fisheries and growth studies on the spotted sardine (*Sardinella sirm*, Walbum, 1792) in the north-western part of the Red Sea. Bull Fac Sci Zagazig Univ 15(2):655–668

Chakraborty D (1984a) Fish landings on the Red Sea and Gulf of Aden coast of the member countries: a preliminary estimate. Development of Fisheries in Areas of the Red Sea and Gulf of Aden, United Nations Development Programme (UNDP)/Food and Agriculture Organization (FAO) of the United Nations, Suez, 29 p

Chakraborty D (1984b) Fishery statistics in the Egyptian Red Sea coast: an expanded plan. FAO/UNDP, Suez, 55 p

Chakraborty D, Moharram M, Waheed A (1983) An account of basic fisheries information of the South Sinai Governorate in Egypt. Development of Fisheries in Areas of the Red Sea and Gulf of Aden, United Nations Development Programme (UNDP) and Food and Agriculture Organization (FAO) of the United Nations, Cairo, 26 p

El-Gammal FI, Mehanna SF (1999) Maximum sustainable yield of the demersal fish resources exploited by trawling in the Gulf of Suez with special reference to shrimp fishery. The role of Science in the Development of Egyptian Society and Environment, 23–24 Oct 1999, Benha, Egypt, pp 198–210

El-Gammal FI, Mehanna SF (2002) Purse-seine fishery in the Gulf of Suez with special reference to sardine fishery. Asian J Fish 15(1):81–88

El-Ganainy AA (2005) Biological characteristics and fishery assessment of the lizard fish *Saurida undosquamis* from the Red Sea, Egypt. Egypt J Aquat Res 8(2):93–113

EL-Ganainy AA, Sabra MM (2008) Age, growth, mortality and yield per recruit of the filefish *Stephanolepis diaspros* (Fraser-Brunner, 1940) (Pisces: Monacanthidae), in the Gulf of Suez, Egypt. J Fish Aquat Sci 3(4):252–260

El-Ganainy AAH, Yassien MH, Ibrahim EA (2005) Bottom trawl discards in the Gulf of Suez, Egypt. Egypt J Aquat Res 31(Special issue):240–255

FAO (2004) The Arab Republic of Egypt: fishery country profile. Food and Agriculture Organization of the United Nations (FAO), Rome. Available at. http://www.fao.org/fi/oldsite/FCP/en/EGY/profile.htm

FAO (2010) Fishery and aquaculture country profile: the Arabic Republic of Egypt. Food and Agriculture Organization of the United Nations (FAO), Rome. Available at http://www.fao.org/fishery/facp/EGY/en

Feidi IH (1976) Egyptian high seas fisheries development project. Food and Agriculture Organization (FAO) of the United Nations, Cairo, 58 p

Froese R, Pauly D (2012) FishBase. Available at www.fishbase.org

GAFRD (2012) General Authority for Fisheries Resources Development database. Suez

Gladstone W, Krupp F, Younis M (2003) Development and management of a network of marine protected areas in the Red Sea and Gulf of Aden region. Ocean Coast Manag 46(8):741–761

Hawkins JP, Roberts CM (1994) The growth of coastal tourism in the red sea: present and future effects on coral reefs. Ambio 23(8):503–508

Ibrahim MA, El-Ganainy AA, El-Etreby SG (1998) Status of the trawl fishery in the Foul Bay, Red Sea. Bull Inst Ocean Fish UAR 24:339–348

Jameson SC, Ammar MSA, Saadalla E, Mostafa HM, Riegl B (1999) A coral damage index and its application to diving sites in the Egyptian Red Sea. Coral Reefs 18(4):333–339

Kotb MMA, Hanafy MH, Rirache H, Matsumura S, Al-Sofyani AA, Ahmed AG, Bawazir G, Al-Horani FA (2008) Status of coral reefs in the Red Sea and Gulf of Aden Region. In: Wilkinson C (ed) Status of coral reefs of the world: 2008. Global Coral Reef Monitoring Network/Reef and Rainforest Research Centre, Townsville, pp 67–78

Latif AFA, Shenouda T (1972) Biological studies on *Rhonciscus striatus* (Fam. Pomadasyidae) from the Gulf of Zuez. Bull Inst Oceanogr Fish UAR 2:105–134

Mahmoud HH, Altaf EA, Ibrahim MA, Abaza KM (2009) Fisheries status of Halaieb/Shalatien area "RED SEA", Egypt. Egypt J Aquat Res 35(3):335–353

Marshall NA, Marshall PA, Abdulla A, Rouphael T (2010) The links between resource dependency and attitude of commercial fishers to coral reef conservation in the Red Sea. AMBIO 39(4):305–313

Mehanna SF (1999) An assessment and management of the coral reef fish stocks in the Gulf of Suez. Egypt J Aquat Biol Fish 3(2):103–114

Mehanna SF (2007) Fisheries assessment of Lake Nasser, Egypt with special reference to tilapia fishery. J Egypt Acad Soc Environ Develop 8(4):27–35

Mehanna SF, El-Gammal FI (2007) Gulf of Suez fisheries: current status, assessment and management. J KAU Mar Sci 18:3–18

Pauly D, Christensen V, Dalsgaard J, Froese R, Torres F Jr (1998) Fishing down marine food webs. Science (Washington) 279(5352):860–863

Pitcher TJ, Hollingworth CE (2002) Fishing for fun: where's the catch? In: Pitcher TJ, Hollingworth CE (eds) Recreational fisheries: ecological, economic and social evaluation. Blackwell Science Ltd, Oxford, pp 1–16

Rafail SZ (1970) Studies of populations and exploitation status of Egyption Red Sea abundant sardines. Bull Inst Oceanogr Fish UAR 1:129–148

Rafail SZ (1972) Studies of Red Sea fisheries by light and purse-seine near Al-Ghardaqa. Bull Inst Oceanogr Fish UAR 2:25–49

Riegl B, Velimirov B (1991) How many damaged corals in Red Sea reef systems? A quantitative survey. In: Williams RB, Cornelius PFS, Hughes RG, Robson EA (eds) Coelenterate biology: recent research on Cnidaria and Ctenophora, vol 66, Developments in Hydrobiology. Springer, Dordrecht, pp 249–256

Riegl BM, Bruckner AW, Rowlands GP, Purkis SJ, Renaud P (2012) Red Sea coral reef trajectories over 2 decades suggest increasing community homogenization and decline in coral size. PLoS One 7(5):e38396

Sabrah MM, El-Ganainy AA (2009) Observation on biological traits of striped goatfish (*Upeneus vittatus*) and freckled goatfish (*Upeneus tragula*) from the Gulf of Suez, Egypt. World J Fish Mar Sci 1:121–128

Sanders MJ, Morgan GR (1989) Review of the fisheries resources of the Red Sea and Gulf of Aden. FAO, Rome, 138 p

Sanders MJ, Kedidi SM, Hegazy MR (1984a) Catches, fishing efforts, catches per fishing effort, and fishing locations for the Gulf of Suez and Egyptian Red Sea coast purse seine fishery during 1979 to 1982. FAO/UNDP, Suez

Sanders MJ, Kedidi SM, Hegazy MR (1984b) Catches, fishing efforts, catches per fishing effort, and fishing locations for the Gulf of Suez and Egyptian Red Sea coast trawl fishery during 1979 to 1982. FAO/UNDP, Suez

Sanders MJ, Kedidi SM, Hegazy MR (1984c) Catches, fishing efforts, catches per fishing effort, and fishing locations for the Gulf of Suez and Egyptian Red Sea fishery for reef associated fish during 1979 to 1982. FAO/UNDP, Suez

Tesfamichael D, Pauly D (2011) Learning from the past for future policy: approaches to time-series catch data reconstruction. West Indian Ocean J Mar Sci 10(2):99–106

Tesfamichael D, Pitcher TJ (2006) Multidisciplinary evaluation of the sustainability of Red Sea fisheries using Rapfish. Fish Res 78(2–3):227–235

Tesfamichael D, Pitcher TJ (2007) Estimating the unreported catch of Eritrean Red Sea fisheries. Afr J Mar Sci 29(1):55–63

Yassien MH (2003) Some biological aspects and dynamics of the fiddler shrimp *Metapenaeopsis stridulans* (Alcock, 1905) from the Gulf of Suez. Bull Inst Oceanogr Fish UAR 29:71–86

Sudan

3

Dawit Tesfamichael and Abdalla Nassir Elawad

Abstract

The fisheries catch in the Red Sea Exclusive Economic Zone (EEZ) of Sudan are presented for the years 1950–2010, by major fisheries and taxa. Sudanese fisheries went through major shifts. The catches were relatively low about 2,000 t·year^{-1} in the 1950s, remained low, then took off at the end of 1970s, mainly due to development projects funded by foreign organizations, which led to a massive increase in artisanal fishing effort, and ultimately, to catches of more than 5,000 t·year^{-1} in the 2000s. The fisheries also shifted from being dominated by a shellfish fishery in the early years to one dominated by finfish fishery in the later years. The contribution of the industrial fishery is generally low. The reconstructed catch was at first higher than the catch reported by the Food and Agriculture Organization of the United Nations (FAO) on behalf of Sudan in the early years. However, in later years, the opposite occurred, i.e., the catch by FAO was higher than the reconstructed catch. This was deemed to be due to over-reporting by Sudan.

Keywords

Catch time series • Fisheries • Catch composition • Mis-reported catch • Foreign aid

Introduction

Sudan borders the Red Sea and its shore is characterized by a relatively narrow shelf (as compared to other Red Sea countries) of about 4,000 km^2 (Fig. 3.1). Overall, Sudanese waters are deep; indeed, the deepest part of the Red Sea, about 3,040 m, is off Port Sudan (Morcos 1970). Inlets, or '*marsas*' in Arabic, are common on the coast, and have deep narrow entrances with shallow fringing coral reefs, which drop rapidly to greater depth. These inlets are used as shelters by the artisanal fishers. They sail from their bases, usually large ports and human settlements, to the inlets which they use as fishing camps, from which they venture out to the open water to fish. Sometimes, they stay in the inlets for months and their catches are collected by trucks which come through dirt roads. About half a mile from the shore, there are what are commonly referred to as 'boat channels'. They are relatively shallow, up to 6 m deep, and are the navigation routes of local fishing boats. The boat channels are bordered by fringing reefs, which are an important area for fishing mainly by the small boats. Further from fringing reefs are deeper channels bordering barrier reefs. Fishing occurs mainly in the reef structures. Most of the fishing in Sudanese Red Sea is performed by small-scale artisanal fisheries and handlining is the main fishing technique for finfish in Sudan. Sudan had a well-established shell fishery in the past, which has become less important in recent years. Industrial fishing in Sudanese Red Sea is almost exclusively performed by foreign vessels.

D. Tesfamichael (✉)
Fisheries Centre, University of British Columbia, Vancouver, BC, Canada

Department of Marine Sciences, University of Asmara, Asmara, Eritrea
e-mail: dawittes@gmail.com

A.N. Elawad
Animal Resources Research Corporation, Ministry of Animal Resources and Fisheries, Khartoum, Sudan

D. Tesfamichael, D. Pauly (eds.), *The Red Sea Ecosystem and Fisheries*, Coral Reefs of the World 7,
DOI 10.1007/978-94-017-7435-2_3

Fig. 3.1 The Red Sea coast of Sudan, its Exclusive Economic Zone (EEZ) and shelf waters to 200 m depth

Coral Reef Ecosystems

A prominent feature of the Sudanese coast is the extensive reef development, one of the most developed reef structures in the Red Sea. Fringing reefs run parallel to the coast, about half mile from the shore, and create safe navigational channels used by the small boats. These channels are cut off by inlet channels, which are perpendicular to the shore and allow the boats to approach the shore from the channels. The fringing reefs separate the shallow boat channels from the deep channels, which are from 80 to 400 m deep. They are the migratory routes of some fishes and some fishing takes place in those areas as well. Offshore from the deep channel is the barrier reef, about 3–6 nautical miles from shore, with most of the artisanal fishery catch originating from this area. Most of the commercially important coral reef species, which are the target of the hand line fishery, are found in the barrier reef area (Barrania 1979; Vine and Vine 1980). The barrier reef is virtually continuous along the coast. The Sudanese coast also boasts the famous Sengenab Atoll, the only atoll in the Red Sea (Fig. 3.2). It has very high biodiversity and has been nominated as a UNESCO world heritage sites. In addition, Rumi Reef (‘*She*’*ab Rumi*’ in Arabic), was the site Jacques-Yves Cousteau chose for his Oscar award-winning documentary ‘World without Sun’, where an underwater house was built and the life in and around the underwater habitat of several ‘oceanauts’ was documented for 1 month.

The coral cover of Sudanese coast is in reasonably good shape, with a life cover of 40 % (25 % stony corals and 15 % soft corals). Butterflyfish are the most abundant of the fishes in the reefs, while groupers and parrot fishes are common as well (Kotb et al. 2008). The main anthropogenic threats to the health of coral reefs in Sudanese waters are mainly associated with the port activities (e.g., dredging) in major coastal cities such as Port Sudan and Suakin, and unregulated tour-

Fig. 3.2 Sangenab Atoll, Sudan (Photo: Mauro Serafini)

ism related activities such as waste disposal and desalination plant (Kotb et al. 2008; Elamin et al. 2014). Life coral coverage is changing in Sudan and is having impact on the fish species associated with the habitat. For example, the abundance of butterflyfish declined from 2003 to 2013 and in some areas two species disappeared during that time (Elamin et al. 2014). Dungunab Bay and Senganab Atoll have been declared marine protected areas (MPAs), but management measures have not been implemented.

Fisheries

Despite the high biodiversity in Sudanese marine waters, the fish catch is not high and more than 95 % of the fish supply of Sudan originates from inland waters, i.e., the Nile river, and lakes and dams (Chakraborty 1983; Tesfamichael and Pitcher 2006). Most marine fishing in Sudan is handlining using different kinds of boats. Common are dugout canoes of 2–3 m that take 1–2 fishers using paddles (or sometimes sail) to fish in the inlets and behind the fringing reef. They are also very important for shell collection (see below). Other common boat types are *huries* of 3–5 m, and used mainly for handlining along the fringing reefs and in deeper waters just off the reefs. They can take 2–3 fishers and usually have sail and paddles. Bigger than *huries* are the rare *felukas* of 5–7 m length, which unlike *huries*, are fitted with transom stern and usually have sail. Both *huries* and *felukas* are sometimes equipped with outboard engine of 3–8 hp (*huries*) and 10–12 hp (*felukas*). The biggest boat types used by Sudanese fishers are *launches* ranging from 7 to 11 m and are usually fitted with an inboard engine of 30–100 hp. They are used for handlining further offshore, and also, albeit to a lesser extent, in gillnet fishing. The length of a single trip is proportional to the size of the boats. Canoes and *huries* usually spend a single day per trip, while launches can spend up to 6 days (MEPI 1993).

Fig. 3.3 Weighing of catches at the fish market in Port Sudan (Photo: Dawit Tesfamichael)

There are many fish landing sites along the Sudanese coast; the most important are Abu Hashish, Salobona and Khor Kilab (the last not used much anymore) around Port Sudan; Mohammed Qol in the north and Suakin in the south. The main fish market is in Port Sudan (Fig. 3.3). Suakin has better facilities to deliver fish to Port Sudan, and is also closer (Fig. 3.1). Fish from the Red Sea is usually consumed on the coast, i.e., do not supply the inland markets, where the population concentration is higher. This is mainly due to poor transportation facilities and the fact that fresh water catches tend to cover the demand of the inland population. Sometimes, a small portion of fish caught by artisanal fishers is exported to neighboring countries such as Saudi Arabia and Egypt. In addition to the artisanal fishers, trawlers (and to a lesser extent purse seiners) from foreign countries operate in Sudanese waters.

Prior to the prominence of finfish in catches, shells were, for centuries, the main target in Sudan (MEPI 1993), which was a major exporter of shell (Eltayeb 2004). The shells collected were mainly trochus and mother-of-pearl. Though the catch of the former was much larger, the latter had greater economic importance as its unit price was quite high. Sudan was an important market for shell fisheries of the whole Red Sea in the 1950s and 1960s, and fishers would come from all countries in the region, including from as far as Somalia and Yemen, to land their catch. More than half, and sometimes up to 90 %, of the shells landed in Sudan used to come from other neighboring countries (Kristjonsson 1956; Reed 1962). The fishers were sea nomads, who used to land their catches in either Port Sudan or Suakin, though they also used Massawa, in Eritrea or Jeddah, in Saudi Arabia, when they received better prices and marketing conditions. They would sail for days for a small difference in price, because as they were using sails, it did not cost them much to go from one port to another (Reed 1962). However, in the 1970s, marketing in Sudan became difficult for foreigners, and the importance of that country as a destination for shell collectors declined (Eltayeb 2004).

The fishery for finfish started to gain momentum after mass mortality of shells in 1969; and many of the shell fishers converted to finfish through various development projects (Barrania 1979). Two major projects had major effect on Sudanese fisheries. One was run by the British Overseas Development Agency (ODA) from 1975 to 1990, targeting the southern part of Sudan (around Suakin). The main aim was to motorize the sailing boats of the artisanal fishers and to enhance the infrastructure at the landing site by providing storage and workshops (ODA 1983). The second important project was an FAO/UNDP project (1979–1985), which emphasized the northern part of Sudan, around Mohammed Qol and Dungunab, where the shell industry faced a major crisis. The objective was to organize the fishers into cooperatives so that they could access technical and financial facilities (Barrania 1985). The Canadian International Development Agency (CIDA) was also involved in shell fish culture research. These projects changed the landscape of Sudanese fisheries, especially the artisanal sector.

In this chapter, the catch of the major Sudanese fisheries are presented from 1950 to 2010. The major fishery sectors were treated separately and a few minor fisheries were also examined. The catch compositions of the different fisheries were then estimated. First, we introduce the fisheries with respect to their operation, the fish they target and their marketing. Then the catches are presented and discussed. The sources and methods to estimate the catches and composition are presented at the end.

Artisanal Finfish Fishery

The two major sectors of the artisanal fisheries are shell collection, which is treated separately in this chapter, and the finfish fishery, which is the main artisanal fishery and sometimes referred as 'the artisanal fishery'. Until recently, the artisanal fishery mainly operated dugout canoes in the shallow waters near the shore, using handlining, which accounts for up to 80 % of the fish caught (Kristjonsson 1956; MEPI 1993). The fishers target predatory fishes on rocky grounds or near coral reefs up to 200 m deep using stones as sinkers, which are tied in such a way that they are released by jerking the line when it gets to a suitable depth. Knowing the right depth for releasing the sinker is gained through experience. In the past, fishing trips would be single day-trips using sail-power and no ice was used, because it was expensive and the fish did not need to be held more than a few hours as the fishers could get a reasonably good catch in a single day. The fish commonly caught were groupers (Serranidae), snappers (Lutjanidae), emperors (Lethrinidae), jacks (Carangidae), and sharks (Elasmobranchii). Most of the catch was consumed within Sudan and a large fraction of the catch was delivered to Port Sudan by trucks owned by merchants, with only a small portion consumed locally. Mullet (Mugilidae) catch used to be salted in barrels, locally called '*fissik*' and exported to Egypt; lately some is consumed locally. Until the end of the 1960s, the traditional artisanal fishery was shell collection. However, in 1969 there was a mass mortality of shells in Sudan and most of the fishers that were active in shell fishery switched to finfish fishing or other activities such as farming and trading.

Artisanal Invertebrate Fishery

The artisanal invertebrate fishery in Sudan includes collection of shells in shallow waters and diving in deeper waters, for example for sea cucumber. Shell collection has a very long tradition in Sudan, where shell middens can be found all over the coast and on the islands. One oyster midden, estimated to consist of 3,000 t of shells, was dated as 1,500 years old (MEPI 1993). In the contemporary history of Sudanese artisanal fishery, shell collection was dominant and the tradition continued until the end of the 1960s. However, most fishers previously active in the shell fishery switched to finfish fishing or other activities after the mass mortality of shells in the late 1960s. The major shells collected in Sudan are mother-of-pearl (*Pinctada* spp.) and trochus (*Trochus* spp.). Dungunab Bay, in the northern part of Sudanese Red Sea coast, is famous for its mother-of-pearl oysters.

Shells are collected by skin diving in shallow waters (down to 10 m). In the past, trips were usually done by bigger *sambuks* with a crew of 6–8. When they reached a protected bay (*marsa*), they would anchor the *sambuk*; the divers then would take the *huries* they carried onboard to the diving sites and brought back their catch. Shell collection consisted of drifting with the *huries* over the shell beds, while monitoring the sea floor with a glass-bottomed bucket. The fishers then dove when they spotted oysters (Kristjonsson 1956). *Huries* with sail and motorized launches were also used to sail to the fishing ground. Some shells were collected by walking in the shallow waters near settlements. In the past, the fishers hardly used any snorkeling equipment for shell collection, but eventually, they started using masks, snorkels and flippers. The fishing season is the warm summer months. The season usually started around April but the main activity occurs between July and November, when the water is warm, calm and thus provides better visibility. There was no shell collection from January to March and the fishers switched to finfish fishing or other activities such as farming or trading (MEPI 1993).

The most important species in the shell fishery are black lip mother-of-pearl oyster *Pinctada margaritifera* var. *erythreaensis* and *Trochus dentatus*, in Arabic '*sedaf*' and '*kokian*', respectively (Reed 1962). Previously oysters were collected mainly for their pearl. Later, oyster shells started to be exported to Europe for the manufacturing of fancy buttons, inlay work, jewelry and artifacts. A small proportion is used locally for poultry feed as a source of calcium. Trochus is also exported for production of small buttons. Oysters and trochus were also used for making plaster, used for whitening walls. However, because it was very expensive, it is now abandoned; instead fossilized corals and other shells are used. Other mollusks are collected either for their meat and/or their opercula. The sun dried meat of mainly conchs (Family Strombidae, specifically *Stombus* spp. and *Lambis* spp.), locally called '*surumbak*', is consumed locally and some of it used to be exported to Saudi Arabia. The nail (opercula) of conchs and other gastropods such as *Fasciolaria* spp. locally called '*tsifri*' were also collected and used as fixatives in the local production of perfumes (MEPI 1993).

Sea cucumber has been collected for export to Asian markets. Fishers started collecting sea cucumbers in shallow waters, and later moved to deeper water using skin diving. As those resources started to get scarce, SCUBA diving was introduced, and finally, fishers used boat *hookah* diving with compressors onboard so that they can stay longer in the water to collect more sea cucumbers. Some of these practices lacked enough safety measures and training and many serious accidents occurred. Once the sea cucumbers are collected, they are usually boiled and dried for export to Asian markets. The local fishers sometimes fish for shrimp in the shallow lagoons using cast net (Elnaiem 2002).

Subsistence Fishery

The definition of subsistence fisheries may vary. FAO defines it as "a fishery where the fish caught are consumed directly by the families of the fishers rather than being bought by middle-(wo)men and sold at the next larger market".[1] Here, in line with the FAO definition, we considered catch that is consumed by the crew, given to family, friends and part of the community who need the support (e.g., widows) as part of the subsistence fishery. In Sudan, like other Red Sea countries, it is a common practice for fishers to share part of their catch with the community before they land their catch. This portion of the catch is significant, sometimes half of the total catch, and never gets reported. It is a social obligation for everyone to do their duty and not contributing has social consequences. The part of the catch that is given freely to the community is the fish that is not directed for export, and thus excludes items such as shark fins.

Industrial Fishery

Only a small fraction, about 700 km^2, of the narrow continental shelf along the Sudanese coast, which is dominated by coral reefs, is suitable for bottom trawling (Sanders and Kedidi 1981). Most of this is in the southern part of the coast, including the Toker Delta. The Gulf of Agig, on the border with Eritrea, is also frequently visited by trawlers. The earliest trawling survey to explore trawl suitable grounds started intermittently in the late 1950s and the early 1960s (Reed 1962). On the other hand, the more concerted surveys performed from 1976 to 1981 by ODA and private companies ushered the beginning of commercial trawling operation in the early 1980s. Since the beginning, trawling has been undertaken by foreign companies from Egypt, China and others, targeting shrimp and some fish. Although trawling is done during the day and at night, the catch is found to be higher at night between 8 pm and 4 am (Elnaiem 2002). The shrimp catch usually accounts for only 3–10 % of the total catch, and most of the bycatch ends up discarded at sea (Elawad 2002; Elnaiem 2002).

Purse seining is another industrial fishery that operates in Sudan, but less common than trawling. The main fishing areas are in the northern part of the Sudanese coast in and around the area also claimed by Egypt (Fig. 3.1), where Egyptian purse seiners have been fishing for decades without any permit from Sudan, which hardly has any fishing activities of its own in the area (MEPI 1993). The report estimated that half of the Egyptian purse seine catch comes from the Sudanese part. Starting in 2002, purse seiners from Egypt started fishing in the southern part of the Sudanese coast with permits. All their catch was landed in Egypt.

[1] http://www.fao.org/docrep/003/X2465E/x2465e0h.htm

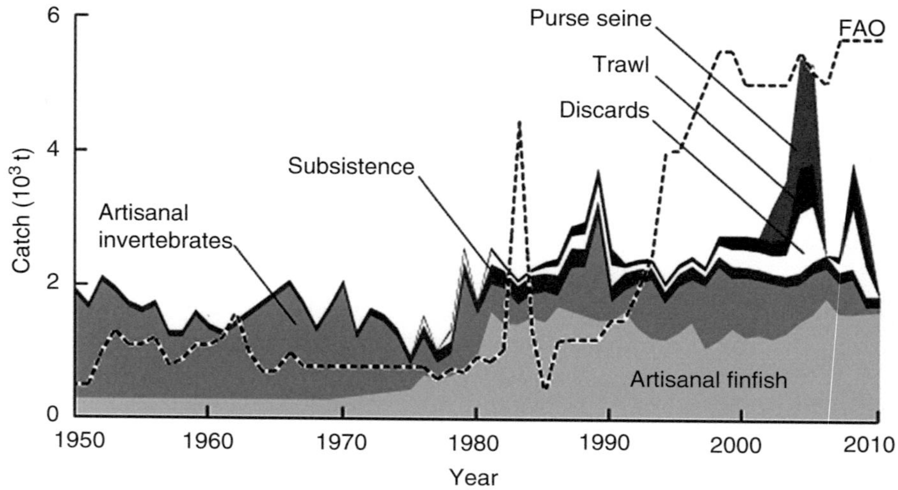

Fig. 3.4 Catch in the Red Sea EEZ of Sudan by sector type and what is reported to FAO, 1950–2010

Fisheries Catches

The total catch of all the fisheries in the Sudanese EEZ in the Red Sea was generally low, below 2,000 t, until the end of the 1970s (Fig. 3.4). The total catch includes the amount caught by foreign vessels, trawling and purse seining as explained above, in the Sudanese EEZ. Between the 1980s and end of 2000, it was somewhat higher until it showed a rapid increase following the turn of the century. The FAO database provides the total catch reported by Sudan (http://www.fao.org/fishery/statistics/software/fishstat/en). Until 1992, the reconstructed catch was higher than the FAO data except in 1962 and 1983. After 1992, the FAO data were markedly higher except in 2005. This is not due to either Sudan fishing in other countries' EEZs, for example Egypt fishing in Sudanese, Eritrea and Yemeni waters (see Chap. 2), or vessels from other countries fishing with Sudanese flag of convenience. Sudan is not listed as a country of flag of convenience (see http://www.itfglobal.org/flags-convenience/flags-convenien-183.cfm). The most reasonable explanation is that Sudan has misreported its catch to FAO as is the case with many countries (Watson and Pauly 2001; Pauly and Froese 2012).

The shell fishery was the main contributor to the total catch until the mid-1970s, where it started to decline and the catch of artisanal finfish fishery started to increase. The artisanal finfish fishery then became dominant, and remained more or less stable until the present. The rapid increase in the total catch after 2001 was mainly due to the large catch from industrial fisheries, trawling and purse seining (Fig. 3.4).

The reported catch (part of the reconstructed catch accounted in the FAO data) has the highest contribution to the total catch, with 64 %. The unreported landed catch was relatively higher until the end of the 1980s. Overall the unreported landed catch accounted for 30 % of the total catch from 1950 to 2010. The decline in the proportion of the unreported landed catch, starting in the 1990s, corresponds to the decline of the shell fishery, which was not reported in the FAO database and accounted for the highest ratio of the unreported landed catch when it was active. The discarded catch, almost exclusively by trawling, was generally low, 5 % from 1950 to 2010, and appears only later, starting the mid-1970s.

Trochus was the main species in the total catch until the beginning of the 1970s. As soon as the artisanal finfish fishery started growing, in the mid-1970s, the number of species in the catch increased. The catch of uncategorized species (i.e., 'others') also became very large. This pattern continued until the catch of industrial fisheries also became very important, i.e., after 2001. However, what became most prominent from the industrial fisheries was the high proportion of discards such as ponyfishes (Fig. 3.5). The catch composition of the different sectors are given in Fig. 3.6.

The total catch in the EEZ of Sudan is the lowest in the Red Sea (Tesfamichael and Pitcher 2006) after Israel and Jordan (see Chaps. 7 and 8). To a large extent, this can be attributed to the very narrow continental shelf along the Sudanese coast. Nevertheless, the Sudanese fishery has not developed well to fish in its shelf; for example, there has not been a continuous trawl fishery even on suitably grounds, with all the trawling has been done by foreign vessels (MEPI 1993). Most of the catch of the other countries bordering the Red Sea comes from bottom trawling and pelagic species, which are not well developed in Sudan due to characteristics of its shelf. Also, the nutrient rich water that flow into the Red Sea from the Gulf of Aden through the narrow straight 'Bab al Mendab' does not reach the Sudanese coast. Thus, the marine fishery sector in Sudan is very small, although still important to the livelihood of the coastal communities. The contribution of Red Sea fisheries to the total fish supply of the country is only around 5 %. The major supply comes from inland waters: the Nile River, lakes and reservoirs

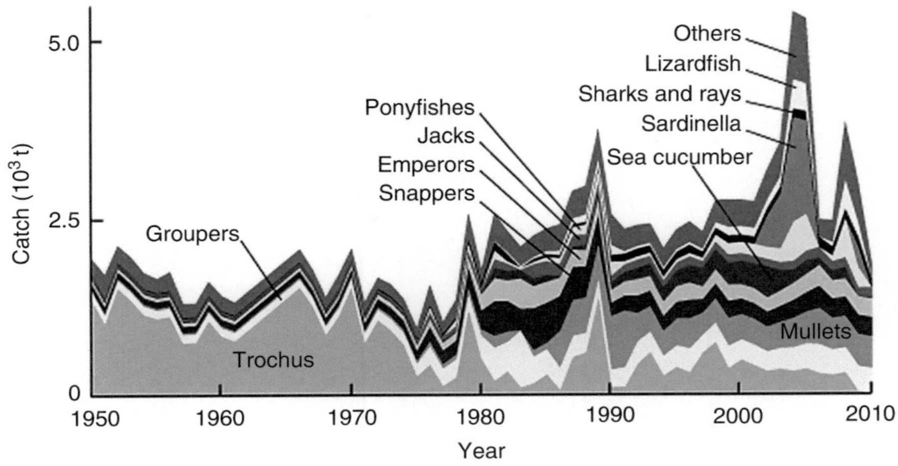

Fig. 3.5 Composition of the total catch in the EEZ of Sudan, 1950–2010

(Chakraborty 1983). The coastal communities of Sudan are traditionally involved in livestock, fishery is not the main socio-economic activity.

The catches reported by the FAO on behalf of Sudan are thought to be overestimated and Sudan appears to submit not accurate estimates of its annual catch data on a regular basis. It has been a common problem for FAO to compile an accurate data of global fisheries catch (Garibaldi 2012). The big spike of total catch according to data submitted to FAO in 1983 is probably an error due to Kedidi (1984), who wrote the potential annual catch (not reported) to be 4,550 t, which is the value reported in the FAO database. However, the catch for 1983, also in Kedidi (1984), was 1,443 t and this value is used as a basis for the reconstructed catch.

The exploitation levels of Sudanese fisheries may not be very high (Tesfamichael and Pitcher 2006), as the artisanal sector has not expanded in recent years (Tesfamichael and Pauly 2011). Thus, it is not uncommon to still see large sized fishes in the catch and in markets. However, there is no reliable catch survey and data recording system in place to provide reliable estimations as to the status of the fisheries. We believe the present reconstruction to be a good starting point to assess the fishery, as it provides a more comprehensive and accurate estimate of catches, which is key for assessment and management (Pauly and Zeller 2003; Tesfamichael 2012). The general trends are clear, and the major shifts in the fisheries are the decline in the shell fishery, replaced by an artisanal finfish fishery, with a foreign industrial fishery added in recent years. The artisanal fishery, whose main market is the local population, has remained stable. However, the fast growth of the industrial fishery calls for serious assessment of the states of the stocks.

This is the first comprehensive catch reconstruction and fishery review for Sudan. Because there were some clear gaps in data, we had to make many assumptions, based on our knowledge of the fishery. These assumptions (see below) are open to criticism and can be replaced whenever better information is available.

Sources and Methods

Published papers, technical reports, government reports, archives, theses and research reports were searched for data and information on Sudanese marine fisheries. Early reports, from the mid-1950s to the late 1980s were authored mainly by FAO personnel who visited the country as expert advisors. The ODA project reports also provided some valuable information for the 1970s and 1980s. Information referring to later periods (after 1990) was available mainly from local reports and files of the Fisheries Administration in Port Sudan (FA 2007, 2012). On-site interviews were also carried out by the first author in 2007 to fill in information gaps (Tesfamichael et al. 2014). Where information was not available at all, interpolation was used to estimate the catch given the best knowledge available at the time of the research. Since different procedures were used for the different fishery sectors, the method for each fishery is given separately. A spatial distribution of the catch is given at www.seaaroundus.org.

Artisanal Finfish Fishery

According to reports and the Fisheries Administration of Sudan, the small-scale finfish fisheries are divided into artisanal and semi-industrial. The main difference is that the former uses smaller boats and usually no motors, while the latter uses engines and larger boats. For the purpose of this research, they are all considered part of the artisanal fishery (the shell fishery is treated separately).

Kristjonsson (1956) was the first to publish an estimate of Sudanese finfish catch, i.e., 300 t·year^{-1}. At the time, this

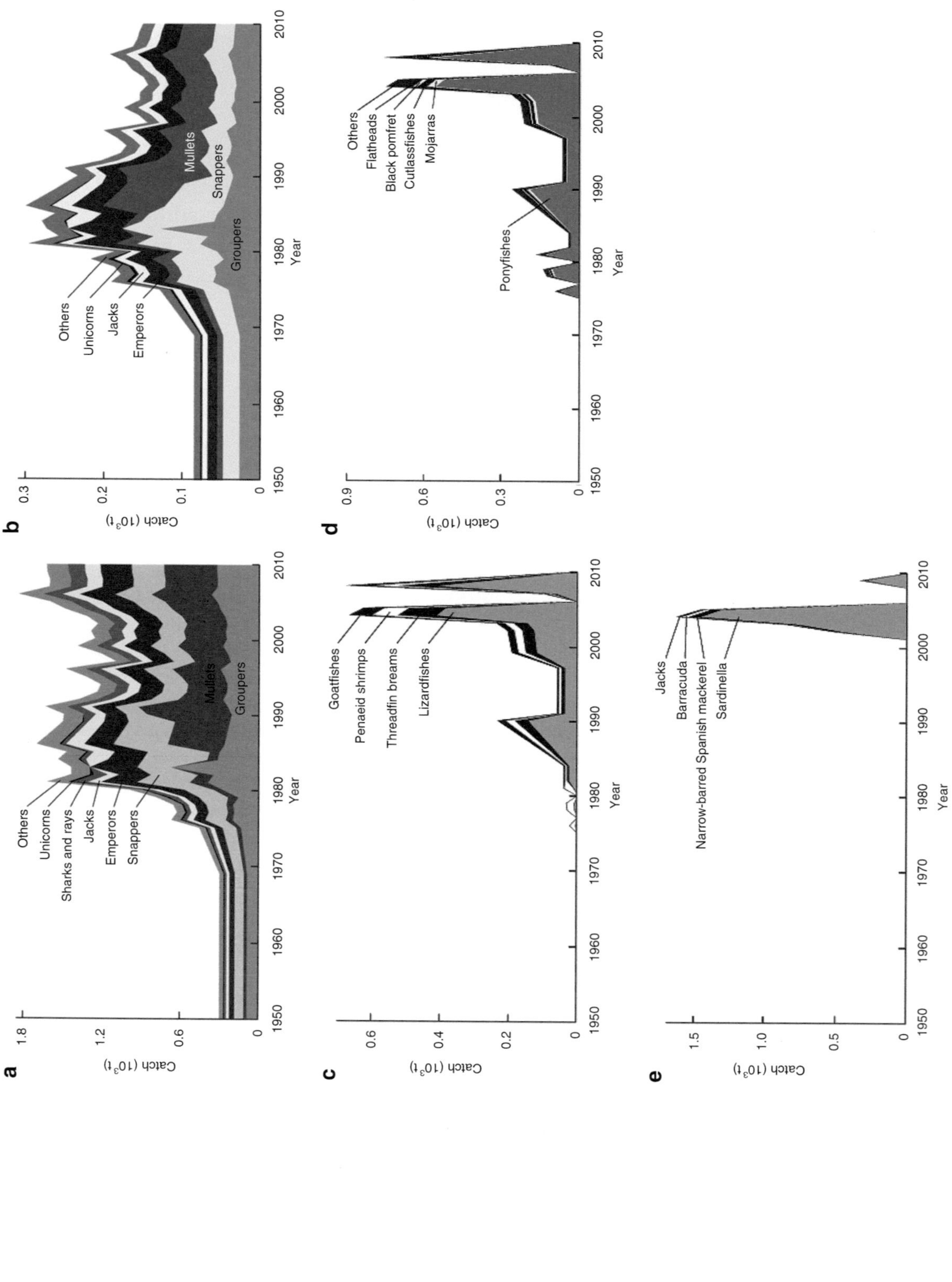

Fig. 3.6 Catch composition of **(a)** artisanal **(b)** subsistence **(c)** trawl-retained **(d)** trawl-discarded and **(e)** purse seine fisheries in the Red Sea EEZ of Sudan from 1950 to 2010

fishery was mainly for local consumption and to supply a small market in Port Sudan. This estimate was later adopted by Oswald (1958) and Reed (1962). In the 1950s and 1960s, Sudanese fishers were more interested in shell collection than finfish fishing, as there was a lucrative market for shells. However, this changed in 1969 when a massive mortality of shell fish occurred. Thus, the estimate of 300 t·year^{-1} is used here as the total catch of artisanal fishery from 1950 to 1969. The next catch estimates were from Barrania (1979), who quantified the annual amount of fish sold in the fish market of Port Sudan from 1975 to 1978. A later survey showed that about half of the fish consumed in Port Sudan did not go through the formal market channels; in addition, 300 t of fish were consumed every year outside Port Sudan (Chakraborty 1983). So, to estimate the total annual catch of the artisanal fishery from 1975 to 1978, the amount of fish sold in Port Sudan market is multiplied by two and 300 t added to it. An independent detailed fishing effort and catch survey for 1976 along the Sudanese coast was published by ODA (1983). The ODA estimate, 671 t for 1976, was higher than the one calculated using the data of Barrania (1979) and the method described above, i.e., 609 t. Thus the total catches for all years calculated using Barrania (1979) data from 1975 to 1978 were corrected using the ODA survey. Catch estimates from 1970 to 1974 were interpolated so as to reflect the slight increase in total catch as fishers moved from shell collection to finfish fishery.

The Fisheries Hydrobiological Administration of Sudan (FHAS 1984) reported Sudanese catch for 1979 and 1980 using Port Sudan reported catch and the procedure described above, while Chakraborty (1983) also estimated catch for 1979, which was similar to FHAS estimate. The FHAS (1984) data were used here because they provided more details on the data collection procedure, including the taxonomic composition of the catch.

For the years 1981–1983, a number of catch estimates were published by experts and consultants hired by the ODA and FAO/UNDP projects. The estimates of Chakraborty (1983) were obtained using the reported catch of Port Sudan. Kedidi (1984) based his estimate on effort data (number of fishers for 1981 and number of boats for 1982 and 1983), multiplied by the catch per unit of effort taken from a sample of boats, and ODA (1983) based their estimate on a survey of fishing villages for 1981 and the number of full time and part time fishers for 1982. The ODA estimates were used for 1981 and 1982, while Kedidi's estimate was used for 1983. These data points were chosen because they provided details that were missing from the other reports.

Data for 1984 and 1989 were available from MEPI (1993). Again, the total catch was calculated using Port Sudan reported catch multiplied by two and adding 300 t to it. However, the estimate of 300 t·year^{-1} for fish consumed outside Port Sudan used in the 1970s calculation, which was still being used in the 1980s and 1990s, was here assumed to change with increase in population size. Thus, we assumed it to be 300 t·year^{-1} in 1980; for the other years, this figure was increased based on the ratio of population of the respective year to the population of 1980. Data from 1985 to 1988 were interpolated. From 1990 to 2010, catch data of fish sold in the Port Sudan market were compiled by the authors from the files of the Fisheries Administration (FA 2007, 2012) in Port Sudan. Those values were multiplied by two and the catch consumed outside Port Sudan was added based on the population size adjustment outlined above.

There was a mullet fishery using veranda nets by Egyptian fishers along the Sudanese coast all the way south to the Eritrean waters, which caught an estimated 1,000 t·year^{-1} (Sanders and Morgan 1989). The fish were salted in barrels, producing what is locally called '*fissik*'. According to records and experts from the Fisheries Administration in Port Sudan, this fishery existed only from 1986 to 1991 and was experimental. There were about 4–6 boats operating only 4 months per year. It is believed the 1,000 t·year^{-1} is an overestimate for the Sudanese coast, partly because the fishers were fishing all over the Red Sea. Thus the annual 1,000 t was not used for Sudan, instead the annual average catch was calculated taking the average of 5 boats operating 4 months and a catch per day of half a tonne, resulting in 300 t·year^{-1}, which was added to the totals of 1986–1991.

The catch of the artisanal fishery consists of many species; however, only a few are dominant. Barrania (1979) estimated the catch composition for 1978, which was used for the period 1950–1978. Data files from the Fisheries Administration, Port Sudan office (FA 2007) provided the composition of the fish landed in Port Sudan for 2006 and were used for 1990–2010. For 1981–1982, the weighted average of 1979–1980 was used. For the calculated total catch composition 300 t·year^{-1} of mullet was added from 1986 to 1991. The blue-spotted sea bass reported in Chakraborty (1984) for 1983 was lumped together with the grouper category.

Artisanal Invertebrate Fishery

The trochus shells landed in Sudan were almost exclusively destined for export; as a result, reliable records were available starting in the early twentieth century. Export data were available for 1950–1961 (Reed 1962), 1966–1989 (MEPI 1993), 1992–2002 (Eltayeb 2004), 2003–2006 (FA 2010) and 2007–2010 (FA 2012). Not all shells that were landed were exported; some of them were discarded simply because the shells had lost their nacreous layer making them unsuited for export (Eltayeb 1999). Data of landed and export amounts were available from 1997 to 1999 (Eltayeb 1999) and the average discarded amount was 42.5 % of the exported shell weight. This ratio was used to scale up the export amount to total landing. There was no export of trochus in 1990 and 1991 (Eltayeb 1999). However, MEIP (1993) reported a total catch of 114.8 t annually for the early 1990s, and this value was used for 1990 and 1991. Data were not available from 1962 to 1965 and an interpolation was used to fill in the gap.

The oyster fishery in Sudan was also almost exclusively for export, thus records existed starting in the early twentieth century. Data were available from 1950 to 1961 (Reed 1962), while MEPI (1993) had export data for 1966–1979, 1984, 1987, 1989 and 1992. Records from the Fisheries Administration indicated a catch of 1 t for 2003, and none thereafter (FA 2010, 2012). Reed (1962) reported that the fishers could not tell if the oysters they picked were large enough for the export market. Thus, about 20 % of the oysters collected were undersize and they had to be discarded at sea, with many eventually dying. Later, a method was introduced for planting the small oysters in wired trays to grow them until they were large enough for the market. We assumed that about half of the 20 % undersized oysters died. So, 10 % was added to the export value to calculate the total catch of oysters. For the years data were not available, interpolation was applied to estimate the catch.

There were few data points for the export of dried sea cucumber: MEPI (1993) reported 15 t in 1981, and FA (2007)) 10 t and 3 t for 1985 and 1986, respectively. A continuous dataset was also available from 2001 to 2010 from the files of the Fisheries Administration of Sudan in Port Sudan (FA 2012). These values were converted to wet weight based on dry weight corresponding to 10 % of wet weight (see Chap. 4). The estimates from 1982 to 1984, and from 1987 to 2000 were interpolated. From 1950 to 1980, we assumed a catch of 27 t·year^{-1}, the lowest of available data points. There was only one data point for the sun dried meat of conchs (*Strombus* spp. and *Lambis* spp. of family Strombidae), locally called '*surumbak*' for 1992 from MEPI (1993), which estimated the annual total catch to be 44 t. In the absence of other data points and the knowledge that this activity has been going on for a long time, this estimate is assumed to be the annual value from 1950 to 2010. Similarly, only one data point was available for the opercula of gastropods for 1992, i.e., 0.55 t (MEPI 1993). This was rounded to 1 t and assumed as the annual value from 1950 to 2010.

Subsistence Fishery

The total catch and composition of the subsistence fishery was calculated based on the information and knowledge obtained through the reconstruction of the finfish artisanal fishery. Data on the artisanal shell and shark fisheries were not included, because they are exclusively for export, hence it does not satisfy our definition of a subsistence fishery, which is catch given freely to family and friends and which not recorded. To estimate the extent of the subsistence fishery, the time line of the artisanal fishery was divided into two periods, 1950–1979 and 1980–2010, based on the motorization of boats, which even if started earlier (Barrania 1979), accelerated in 1980 (Chakraborty 1983). We assumed the subsistence fishery to be 30 % of the artisanal fishery from 1950 to 1979. This is a reasonably conservative estimate, because interviews with fishers indicated that before motorization and strong commercialization of fishery, they used to give up to half of their catch to family and friends. Once motorization accelerated, we assumed 20 % of the artisanal fishery to be subsistence, and by 2010 we assumed this had declined to 10 %. The ratio of subsistence was interpolated from 1980 to 2010. The catch composition of artisanal fishery was used to calculate the composition of subsistence fishery, except sharks and rays were excluded because they are mainly for export markets.

Industrial Fishery

The earliest report of industrial fishery was documented by Reed (1962) for exploratory trawling for shrimp during his assignment in Sudan from 1958 to 1961. The trials were sporadic and aimed to identify trawling grounds. He reported best catches of up to 100 individual shrimp per hour at night, with lower values during daytime. The next trawling surveys were conducted by ODA in 1976 and 1981 (ODA 1983), which we used as data anchor points. The 1976 shrimp catch was 14.7 t, with 114.5 t of discards, while in 1981, the shrimp catch was 26 t and retained fish catch was 13 t. Although the main target of trawling at the time was shrimp, some fish were kept for the local market. The fish catch for 1976 was not available and was estimated using the shrimp-fish ratio for 1981, while the discard for 1981 was calculated using the shrimp discard ratio of 1976. From 1977 to 1980, data were not available for trawling. However, Sanders and Morgan (1989) reported that a company called Ross Sea Food International was doing exploratory trawling in 1978 and 1979, and estimated the potential catch of shrimp from the Sudanese coast to be 30 t·year^{-1}; however, their actual catch during the operation was not reported. Thus, the catches for 1978 and 1979 were interpolated using the data for 1976 and 1981 as anchors. Ross Sea Food International abandoned their operation when they concluded that the 30 t·year^{-1} shrimp potential was not worthwhile for a commercial venture. The catches in 1977 and 1980 were set to zero as there were no trawling activities reported for those years (ODA 1983; Sanders and Morgan 1989). Starting in the early 1980s, Sudan started issuing trawling permits to foreign vessels, mainly from Egypt (MEPI 1993). The annual catch from 1982 to 1984 was estimated based on Elawad (2002), who reported the minimum trawling total catch, including discards, from 1979 to 2000 was 90 t·year^{-1}. This value was assumed until 1984. The next reference available was for 1990, the last year the Egyptian trawlers operated before they were stopped for a few years. There were, however, a few other trawlers operating in Sudan. The total catch for 1990 was 544.2 t (MEPI 1993; Elawad 2002). The total catches from 1985 to 1989 were interpolated to depict the increase in the trawling activity during those years. The catch declined to 137 t·year^{-1} in 1991 (MEPI 1993). The next data reference was for 1998, being 290 t (Elawad 2002), when a special permit was given for trawling for Egyptian vessels. The catch from 1992 to 1997 was kept constant at the level of catch in 1991. Catches for 1999, 2002–2010 were available from the files of Fisheries Administration and Fisheries Research Centre (FA 2007, 2012). Catches for 2000 and 2001 were interpolated.

The catch composition of the trawl catch in 2004 was taken from files of the Fisheries Administration (FA 2007), i.e., 5 % shrimp (Penaeidae), 27 % lizard fish (*Saurida* spp.), 7 % threadfin bream (*Nemipterus* spp.), 4 % goatfish (Mullidae) and 57 % discards. These ratios were used from 1982 to 2006. From 1976 to 1981, the catches of shrimp, all fish and discard were calculated as mentioned in the previous paragraph. The shrimp catch and discard were used directly in the catch composition, while the fish total was divided according to the fish ratios from 2004. The main shrimp species were deep-water species, i.e., *Penaeus semisulcatus*, *Metapenaeus monoceros* and *Melicertus latisulcatus*, which contributed more than 90 % of the catch. For coastal lagoon, which contributed less to the total catch, *P. monodon* and *P. indicus* were the only species reported (Elnaiem 2002). The composition of the discarded catch was calculated according to data from Yemen (see Chap. 5), since trawl operations are similar between the two countries.

The purse seine fishery in Sudan started in 2002 by vessels from Egypt. Data were available for the total catch and catch composition from 2002 to 2005 (FA 2007) and from 2006 to 2010 (FA 2012) from the Fishery Administration in Port Sudan. This fishery is assumed to not generate any discards. In addition, as a relatively limited and controlled fishery, there is good data recording system; hence the values were used as they were.

Comparing Reconstructed Catches with FAO Statistics

The reconstructed catches were compared to the composition of Sudanese catch in the FAO data. Annual catches are reported for Sudan in the FAO database. From 1950 to 1997, all the catch data are given as 'Marine fishes nei' without compositional breakdown. From 1998 to 2010, the two taxa 'narrow-barred Spanish mackerel' and 'sharks, rays, skates etc. nei' are represented separately and the rest lumped as 'Marine fishes nei'. The FAO data for Sudan were not refined enough to provide information about any of the fisheries.

The taxonomic group 'others' in the reconstructed catch, which includes the miscellaneous taxa not reported separately, was very high for some years. It was first reduced to 10 % of the total catch, and the difference was distributed to the previously identified taxa in proportion to their reported percentage except 'sharks'. The procedure was needed because when the unidentified group 'others' assumes high proportion of the total catch, the resolution of the catch is not very informative. The group 'sharks' was excluded because it was reported separately in the FAO data. Then the FAO category 'marine fishes nei' was disaggregated further using the ratios in the reconstructed catch, again after excluding 'narrow-barred Spanish mackerel' and 'sharks'. Finally, each taxon in the reconstructed catch was compared to the FAO data to calculate misreporting. The part of the reconstructed catch that is accounted in the FAO data is referred to as 'reported catch' in our result. If the value of a taxon in the reconstructed catch was higher than its value in the FAO data, then the difference was labeled as 'unreported catch', and if the FAO value for a taxon was higher than the reconstructed catch, the difference is 'over-reported' catch.

Acknowledgements We would like to thank the personnel of the Fisheries Administration and Fisheries Research Centre in Port Sudan. We would like also to extend our thanks to the personnel of the Ministry of Animal Resources and Fisheries Research Centre in Khartoum. Our special thanks goes to Kalid and Mohammed from the Fisheries Administration, Port Sudan for their help in compiling the catch data from scattered records and their assistance during interviews with fishers; their friendliness to the first author during his field work in Sudan was remarkable. We are grateful to Ahmed Khalil for reviewing the chapter. This research was supported by the *Sea Around Us*, a scientific collaboration between the University of British Columbia and the Pew Charitable Trusts.

References

Barrania AA (1979) The exploratory socio-economic survey of the Sudanese Red Sea fisheries. FAO/UNDP Project for Development of Fisheries in Areas of the Red Sea and Gulf of Aden, Rome, 36 p

Barrania A (1985) The socio-economic impact of project's activities in Mohamed Qol, Sudan. FAO/UNDP Project for Development of Fisheries in Areas of the Red Sea and Gulf of Aden, Rome, 10 p

Chakraborty D (1983) Marine fisheries statistics in Sudan – an expanded plan of development. FAO, Rome, 39 p

Chakraborty D (1984) Fish landings on the Red Sea and Gulf of Aden coast of the member countries: a preliminary estimate. Development of Fisheries in Areas of the Red Sea and Gulf of Aden, United Nations Development Programme (UNDP) and Food and Agriculture Organization (FAO) of the United Nations, Suez, 29 p

Elamin SEM, Mohamed AB, Mahmoud ZN (2014) Butterflyfishes (Chaetodontidae) from Abu Hashish fringing reefs, Port Sudan, Red Sea. J Biol Agric Healthcare 4(12):40–47

Elawad AN (2002) Some biological aspects of fish by-catch from Sudanese Red Sea Shrimp Trawling. MSc thesis, University of Khartoum, Department of Zoology, Khartoum, 133 p

Elnaiem AG (2002) Aspects of shrimp fisheries on Sudanese Red Sea coast. PhD thesis, University of Khartoum, Department of Zoology, Khartoum, 70 p

Eltayeb MM (1999) Studies on the biology, ecology and fisheries of *Tectus dentatus* (Kokian) in the Sudanese Red Sea. MSc thesis, University of London, Department of Geography, 92 p

Eltayeb MM (2004) Review of the trochus fishery in Sudan. SPC Trochus Inf Bull 11:5–7

FA (2007) Fishery statistics of Sudan: based on the files of Fisheries Administration office in Port Sudan. Fisheries Administration, Khartoum

FA (2010) Fishery statistics of Sudan: based on the files of Fisheries Administration office in Port Sudan. Fisheries Administration, Khartoum

FA (2012) Fishery statistics of Sudan: based on the files of Fisheries Administration office in Port Sudan. Fisheries Administration, Khartoum

FHAS (1984) Fish production in Sudan – a study on the methodology and the data collected during 1979 and 1980. FAO/UNDP, Suez 61 p

Garibaldi L (2012) The FAO global capture production database: a six-decade effort to catch the trend. Mar Policy 36(3):760–768

Kedidi SM (1984) The Red Sea reef associated fishery of the Sudan catches, efforts and catches per fishing effort survey conducted during 1982–1984. Project for the development of fishries in the areas of the Red Sea and Gulf of Aden, FAO/UNDP, Cairo, 18 p

Kotb MMA, Hanafy MH, Rirache H, Matsumura S, Al-Sofyani AA, Ahmed AG, Bawazir G, Al-Horani FA (2008) Status of coral reefs in the Red Sea and Gulf of Aden region. In: Wilkinson C (ed) Status of coral reefs of the world: 2008. Global Coral Reef Monitoring Network and Reef and Rainforest Research Centre, Townsville, pp 67–78

Kristjonsson H (1956) A brief survey of the Sudanese Red Sea fisheries – report to the government. FAO Report no. 510, FAO, Rome, 38 p

MEPI (1993) Identification study for Sudan Red Sea fisheries. Ministry of Economic Planning and Investment, Khartoum, 213 p

Morcos SA (1970) Physical and chemical oceanography of the Red Sea. Oceanogr Mar Biol 8:73–202

ODA (1983) Report on project activities in Sudan. Overseas Development Agency, London

Oswald E (1958) Sudan – Red Sea fisheries – report to the government. FAO Report no. 934, FAO, Rome, 27 p

Pauly D, Froese R (2012) Comments on FAO's state of fisheries and aquaculture, or 'SOFIA 2010'. Mar Policy 36(3):746–752

Pauly D, Zeller D (2003) The global fisheries crisis as a rationale for improving the FAO's database of fisheries statistics. In: Zeller D, Booth S, Mohammed E, Pauly D (eds) From Mexico to Brazil: Central Atlantic fisheries catch trends and ecosystem models. Fisheries Centre research reports, vol 11(6), Vancouver, pp. 1–9

Reed W (1962) Sudan – the sudanese shell industry and Red Sea fisheries – report to the government. FAO Report no. 1489, FAO, Rome, 56 p

Sanders MJ, Kedidi SM (1981) Summary review of Red Sea commercial fisheries catches and stock assessments including maps of actual and potential fishing grounds. FAO, Rome, 52 p

Sanders MJ, Morgan GR (1989) Review of the fisheries resources of the Red Sea and Gulf of Aden. FAO, Rome, 138 p

Tesfamichael D (2012) Assessment of the Red Sea ecosystem with emphasis on fisheries. PhD thesis, University of British Columbia, Resource Management and Environmental Studies, Vancouver, 241 p

Tesfamichael D, Pauly D (2011) Learning from the past for future policy: approaches to time-series catch data reconstruction. West Indian Ocean J Mar Sci 10(2):99–106

Tesfamichael D, Pitcher TJ (2006) Multidisciplinary evaluation of the sustainability of Red Sea fisheries using rapfish. Fish Res 78(2–3):227–235

Tesfamichael D, Pitcher TJ, Pauly D (2014) Assessing changes in fisheries using Fishers' knowledge to generate long time series of catch rates: a case study from the Red Sea. Ecol Soc 19(1):18

Vine PJ, Vine MP (1980) Ecology of sudanese coral reefs with particular reference to reef morphology and distribution of fishes. In: Symposium on coastal and marine environment of the Red Sea, Gulf of Aden and Tropical Western Indian Ocean, Khartoum, vol 2, pp 87–140

Watson R, Pauly D (2001) Systematic distortions in world fisheries catch trends. Nature 414(6863):534–536

Eritrea

4

Dawit Tesfamichael and Sammy Mohamud

Abstract

The fisheries catches in the Eritrean Red Sea Exclusive Economic Zone (EEZ) are presented from 1950 to 2010. Six major fisheries, different in terms of their operations, the fish they target and their market, were identified. Overall, the fisheries went through major shifts from the state of high catches dominated by small pelagic beach seining in the 1950s and 1960s (slightly under 30,000 t·year^{-1}) to the domination by bottom trawling, prevailing since the 1990s. The catches started to decline to less than 2,500 t·year^{-1}, a level which lasted from the mid-1970s to the first few years after independence (1991), before recovering and reaching a new peak of about 20,000 t·year^{-1} at the beginning of 2000s. The artisanal fisheries, which target mainly fresh fish for direct human consumption, have exhibited a relatively steady upward trend since independence. Major findings are (1) the total catch for the period from 1950 to 2010 was 2.2 times the data reported by Eritrea to the Food and Agriculture Organization of the United Nations (FAO); and (2) that political events strongly impacted the fisheries of Eritrea, notably the struggle for independence.

Keywords

Catch time series • Fisheries • Catch composition • Mis-reported catch • Political instability

Introduction

Eritrea borders the Red Sea (Fig. 4.1), and its continental shelf, i.e., waters shallower than 200 m, where most of the fishing occurs, cover about 50,000 km^2. Due to its harsh environmental conditions, the coast is not densely populated. However, archaeological studies of middle stone-age middens from the Eritrean coast indicate that humans were exploiting near-shore marine organisms, such as giant clams and other molluscs about 125,000 years ago (Walter et al. 2000). In the recent past, the most common fishing activity along the Eritrean coast has been diving for top shell (*Trochus*) and pearl oyster by local people, and handlining mainly by Yemeni fishers. In the mid-1940s, beach seining for schools of sardines and anchovies started and it flourished as the main fishing activity for the next few decades (Ben-Yami 1964). In terms of the total catch and the number of people involved either directly or indirectly, the Eritrean fishery reached its peak in the 1950s and 1960s. Starting in the 1970s, the fishery went through rapid decline and almost completely disappeared at the end of the 1980s. After the independence of Eritrea in 1991, the fishery started to recover and there has been an increase in fishing activities. The newly recovered fishery does not rely on beach seining. Rather it is

D. Tesfamichael (✉)
Fisheries Centre, University of British Columbia, Vancouver, BC, Canada

Department of Marine Sciences, University of Asmara, Asmara, Eritrea
e-mail: dawittes@gmail.com

S. Mohamud
Research Division, Ministry of Fisheries, Massawa, Eritrea

D. Tesfamichael, D. Pauly (eds.), *The Red Sea Ecosystem and Fisheries*, Coral Reefs of the World 7,
DOI 10.1007/978-94-017-7435-2_4

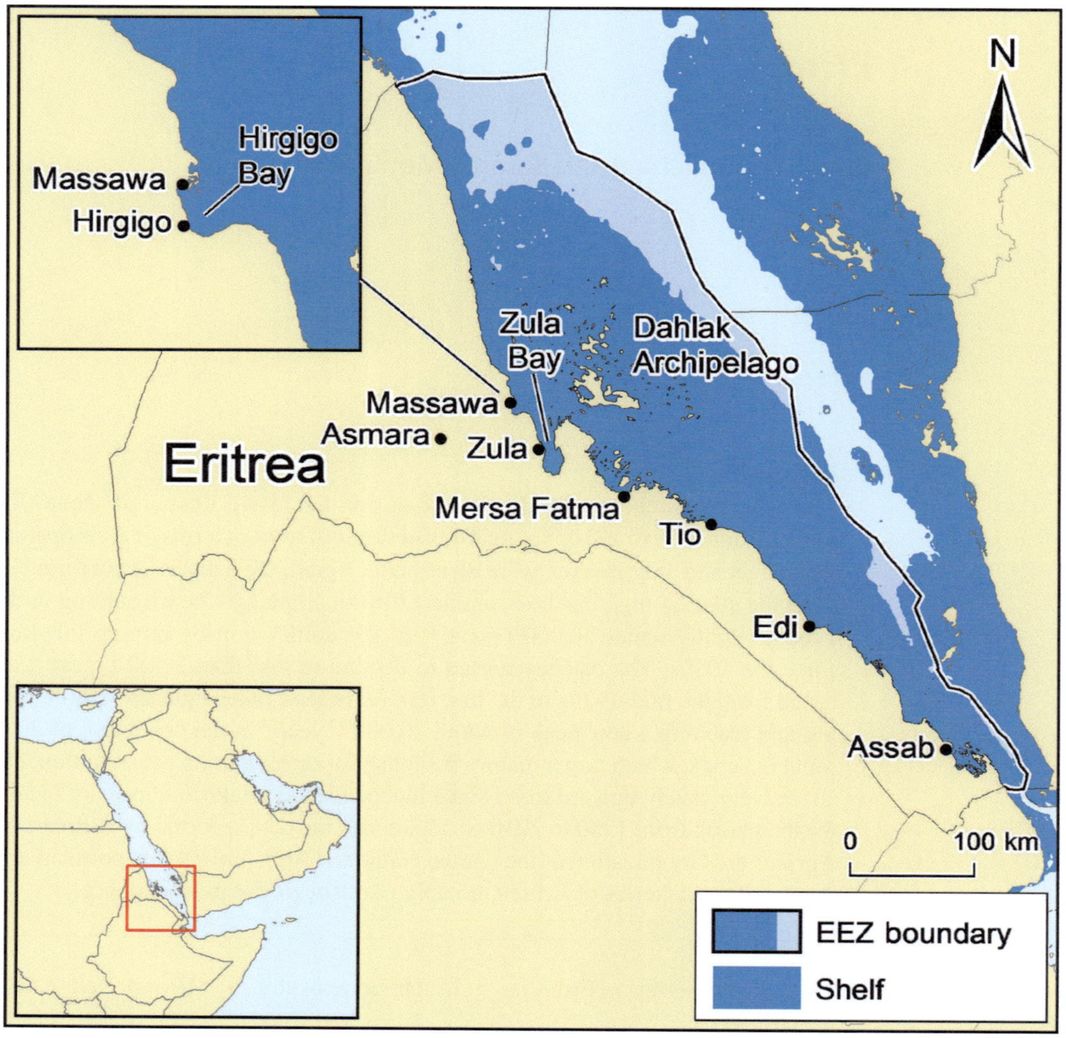

Fig. 4.1 Map of Eritrea and its Red Sea coast, shelf area and Exclusive Economic Zone (EEZ)

dominated by hook and line and gillnet fisheries in the artisanal sector, and bottom trawling in the industrial sector (Tesfamichael 2001). Eritrea was part of Ethiopia until 1991 and it is not uncommon to find pre 1991 reports referring the fisheries as 'Ethiopian'. In 1991, Eritrea gained its *de facto* independence and international recognition followed in 1993. Some of the older reports also frequently use the Ge'ez Calendar, used by the Orthodox church of Eritrea and Ethiopia, which starts in September and is seven and half years behind the Gregorian Calendar. All dates in this report refer to the Gregorian Calendar unless specified otherwise.

In this chapter, the catch of fisheries in Eritrean Red Sea EEZ, both by domestic and foreign fleets, are presented from 1950 to 2010. Overall, the fisheries can be divided into three sectors: artisanal, subsistence and industrial. Six major fisheries, categorized by the gear they use and the resources they target, were identified and treated separately. The catch of each gear was further divided to the species or other higher taxonomic levels that make up the total catch. We will briefly describe the coral reef ecosystems of Eritrea in the Red Sea, followed by an introduction to each fishery type based on its development over time, operation, species targeted and market, followed by the catches (amount and composition). The sources and methods to estimate the catches and composition are presented at the end.

The Coral Reef Ecosystems

Coral reefs in the Eritrean Red Sea are mainly of the fringing type, and they are found mainly in the central part of the coast around the Dahlak Archipelago. Coral reefs are also found in the north close to the border with Sudan and in the south, south of Tio all the way to Assab and the border with Djibouti (Fig. 4.1) (Kotb et al. 2008). The reefs along the continental coast are relatively less developed and they cover about 18 % of the coast. On the other hand, the reefs around the islands are well developed, relatively pristine and have global significance as reservoirs of marine biodiversity (Pilcher and Alsuhaibany 2000). The coral reef ecosystems along the Eritrean coast are not well studied, i.e., their distribution, coverage and composition are not well documented. However, there are a few studies looking at specific aspects, e.g. the dynamics of macroalgal communities (Ateweberhan 2004), the ecology and population dynamics of butterflyfishes (Zekaria 2003) and parrotfishes (Afeworki 2014). The diversity of coral species is quite high; recent surveys estimate the total number of species to be 220 in 38 genera (John. E.N. Veron, unpublished report). Coral coverage can reach up to 100 % especially in deeper waters around southern Dahlak islands and islands off the port city of Assab (Tilot et al. 2008), but less so around major settlements such as the port cities of Massawa and Assab, where mean live coral cover was 16–37 % and dead coral cover 16–29 %, at the end of the 1990s. These live cover percentages are less than what they were in earlier years, indicating deteriorating health of the reefs (Pilcher and Alsuhaibany 2000).

The coral reef ecosystems in Eritrean waters support a multitude of fish and invertebrates targeted, especially, by the artisanal fishery (Zekaria 2003; Tesfamichael 2012). Overall, fishing pressure on coral reefs is relatively low, as the hot and dry coastal areas of Eritrea are not densely populated. There are only few major settlements on the coast (Massawa and Assab being the two biggest coastal cities) and only 7 of the more than 350 islands are inhabited. However, some areas which are frequently visited by fishers have exhibited local depletion and deterioration (Tesfamichael 2001). Recently, though, development of coastal areas is increasing, and the threats to coral reef health is growing. Two marine protected areas (at Sheikh Seid Island near Massawa and Dissie-Madote Islands) have been recently proposed, but it is too early to assess their effectiveness.

Fisheries

Artisanal Fishery

The artisanal fishery is the most prominent fishery in the Eritrean Red Sea. It is a small-scale fishery with less capital investment (compared to industrial), locally owned and operated, and its main market is domestic. There are few gear types used by the artisanal fishery.

Beach Seine Fishery

Beach seining started in the mid-1940s (Ben-Yami 1964) and grew until it became the dominant fishing method in the country in the late 1950s and early 1960s, then declined and faded away in the mid-1970s. The beach seine fishery targeted mainly small pelagic fishes (sardines and anchovies) when they approach the coastal waters during the colder months, i.e. starting in October and extending until May. Through many years of practice, the fishers developed a great deal of experience and knowledge in spotting fish schools over clear bottoms, encircling them, and hauling their catch. The fishing operation included a mother *dhow* carrying the seine net and other miscellaneous supplies, and escorted by one or two small canoes. The mother *dhow* would anchor near a spot where the school of fish was heading, and the canoe(s) would encircle the fish, putting one end of the rope on the shore (Fig. 4.2). The catch was hauled to shore manually and left to dry above the high tide mark for 1–2 days. Occasionally the catch would be damaged by rain. Women and children from nearby villages used to scare off birds, thus earning the right to collect and sell what was left on the beach by the fishers. The dried fish would be transported to either Massawa or Assab, the major ports in Eritrea. Sardines and small anchovies were used almost exclusively for fishmeal (reduction fisheries) which was exported to Italy, Greece, Spain, Switzerland and other European countries, while large anchovies were sun-dried, handpicked and mechanically cleaned of impurities to be exported for human consumption to Sri Lanka and other Asian countries (Grofit 1971).

Sardines were found only around Massawa and the Dahlak archipelago, while large anchovies were found only in the southern part around Assab. When this fishery was at its peak in the 1950s and 1960s, there were up to 6 factories in Massawa for fishmeal and cleaning of large anchovies for human consumption, while there were a couple of them in Assab, devoted to cleaning anchovies only. The landing in Massawa was about double that of Assab (Ben-Yami 1964). This fishery accounted for up to 90 % of the total landed catch and it played a large role in the economic and social life of the local coastal communities.

Beach seining was also used, although to a lesser extent, to exploit stocks of mackerels (*Scomberomorus* spp.) and mullets for direct human consumption. Later, starting in the 1970s, when the export of dry fish and fishmeal stopped, beach seining was used to catch fresh fish for a few years and the accidentally caught anchovies and sardines were discarded (Ben-Yami 1964; Grofit 1971). At present, although the potential still exists, commercial beach seining for small pelagics does not occur.

Handline Fishery

Handlining (hook and line fishing) is one of the oldest forms of fishing practiced along the Eritrean coast, and is a technique that has been used continuously and almost exclusively by small-scale fishers. In the 1940s, fishers of Yemeni origin, of which some later settled in Eritrea, started to use handlining to supply markets in Eritrea and Yemen

Fig. 4.2 Fishers pulling a beach seine net, Eritrea; the person at the forefront on the left is the co-author of the chapter (Photo: Steffan Howe)

(Ben-Yami 1964). Since 1991, the handlining catch is sold both locally, or exported to markets in the Middle East and Europe. Fishers target mainly reef-associated fishes, such as snappers (Lutjanidae), groupers (Serranidae) and emperors (Lethrinidae). Handline fishery occurs around the Dahlak Archipelago and along the northern part of the Eritrean coast, where coral reefs are more developed. Handlining is performed while on foot from the shore, or from small canoes. Also wooden boats of about 10 m with inboard or outboard engine are used in this fishery.

Usually, only one hook is tied to one end of the line, which is lowered into the water. When the fishers feel that a fish is biting, they start pulling the line back to the boat with their bare hands; indeed, many fishers have cuts on their hands, sustained when they fight to haul the fish into their boats. Fish caught in previous trips are cut into smaller pieces to be used as bait at the beginning of a given trip. Later, low-grade fish are used as bait. Fishing starts in the late afternoon and is carried out until dawn. Overall trip length may range from a few to 10 days, usually as determined by either fish storage capacity (ice boxes), or the ice supply. Because the technique is very selective, this fishery does not have a discarding problem. Recently handlining and gillnetting represent the major artisanal fisheries in Eritrea.

Gillnet Fishery

Gillnets, which have been used for large pelagic species, started to be common only in the 1980s, according to interviews with old fishers, and Giudicelli (1984). Once gillnet fishing started, however, it became an important part of the artisanal fisheries. Their main target is Spanish mackerel (*Scomberomorus commerson*), but also barracudas (*Sphyraena* spp.), jacks (Carangidae) and sharks. The fishers set their nets at night mainly in the central and southern part of the coast. The net is immersed for about 2–3 h, to reduce spoilage of entangled fish, and post-capture damage by predators, notably sharks. Gillnet fishing is highly affected by moon phase. The new moon is the best time to catch fish using gillnet, where trips last only up to 2 days to fill the boat's storage. However, during full moon, trips are longer and some fishers switch to handlining. This fishery is less selective than handline and has discards, mainly small tunas such as kawakawa (*Euthynnus affinis*), which fetch a relatively low price.

Shark Fishery

Shark fishing has been common along the Eritrean coast for a long time (Ben-Yami 1964). Sharks were caught by handline, bottom longline and inshore gillnet mainly in the central and southern part of the coast. Shark fishing trips are very long. Ice is not needed in this fishery, and as a result, fishers stay up to a month at sea with short stops at villages near their fishing to refill water, drying and collecting their catches (fins and meat) until they come back to land their catches. During the 1950s and 1960s, shark fishing occurred during the hot months of summer, i.e., it alternated with the beach seine fishery that took place only in the cold season. Whenever the beach seining fishery faced problems, fishers used to switch to shark fishing. For example in 1967 when export of fishmeal was disrupted by the closure of the Suez Canal, most fishers switched to shark fishing, resulting in the highest catch of sharks. Shark flesh is not favored in Eritrea; thus shark fishing has been mainly for export. In the past, the flesh used to be salted and sun dried on the beach after gutting and fining. The dried meat would be tightly stacked and sewn in straw mats locally called '*ferasila*' (Campbell 1993), which weighed about 15 kg. About 5–6 kg of wet shark flesh were needed to produce 1 kg of dried shark meat (Grofit 1971). Both the dried fins and meat were exported to Aden, Yemen, where they were re-exported to East Asian markets. Because the flesh was dried on the beach, it had sand and other impurities; hence, it was not the best quality in the Aden market and did not fetch good price. This eventually decreased the demand of shark meat from Eritrea, and together with political instability, it resulted in the decline of the fishery starting in the early 1970s. After 1991, dried shark fins were exported, but not continuously. However, the flesh was not used and it is common to see rotting shark carcasses on the beaches. The fishers have a tradition of not throwing away the unwanted catch back into the water because they believe it will contaminate the sea and scare off the fish they target.

Sea Cucumber Fishery

Sea cucumber gathering has been done in shallow waters by women and young kids for many years at low level (Ben-Yami 1964). A major sea cucumber fishery started in 2000 exclusively for export due to the high demand for the product in China and other East Asian countries. Since its start in 2000, the reported local catch (mainly of *Holothuria* spp. and *Actinopyga* spp.) increased rapidly, peaked in 2002 and then started to decline (Tewelde and Woldai 2007), despite an increasing demand. The decline is a sign of overexploitation of the resource. Some of the catch is known to be sold illegally in Yemen, where prices are higher, and fishing supplies cheaper (Tewelde and Woldai 2007). Gathering is mainly done by skin diving, but as the stocks in shallow areas are being depleted, those in the deeper parts are increasingly being exploited as well. Collectors use air compressor and hookah diving to reach deeper waters. As they operate with little or no training and inadequate knowledge of equipment use, accidents are common, with serious health problems and deaths among divers. Proper SCUBA equipment is used very rarely. The collectors dive with a sack to fill with hand-picked sea cucumbers. The catch is processed first by boiling and then drying the boiled sea cucumbers on the beach (Fig. 4.3). When this fishery started in 2000, the catch was small and used to be sold in markets in Yemen, but later as the catch increased, it was exported directly to East Asian markets.

Other Artisanal Fisheries

The fisheries included in this category are all for invertebrates: shell fish, lobster, snail 'nail' and pearl fisheries. Shell collection in shallow waters by women and young children and skin diving by men was one of the oldest fishing activities in the Eritrean Red Sea. Most of the shell collection occurred in the summer months, when the water is warm enough for skin diving, and there was no beach seining. The main targets were top shell, mother of pearl shells, pearl oyster, and ornamental conchs (Ben-Yami 1964). Similar to those in the shark fishery, the trips were very long, up to 30 days, with 20–40 crews and supplies replenished from the settlements near the fishing grounds. No equipment was used except sometimes home-made masks called '*nadur*', which limited operations to shallow waters although gradually, proper dive masks were introduced. The main product of the shells was semi-finished buttons, exported mainly to Italy.

Fig. 4.3 Sea cucumber catch drying on sandy beach (Photo: Ministry of Fisheries, Eritrea)

The industry was very active in the 1950s when six button factories were established in Massawa. Most fishers preferred to sell their catch in Suakin, Sudan, where they received better prices. Based on interviews with old fishers, about 40 sailing boats used to go to Sudan to sell shells until 1978, before it was prohibited by the then Ethiopian Government for the fishers to sell outside the country. Most of the fishers switched to other fishing activities and the shell fishery was abandoned.

Snail 'nails' (operculum) were and still are collected by women and children in shallow waters. The nails are dried and sold for use in the cosmetic industry. Lobster fishing occurs mainly in the Assab area. So far, the main bottleneck of this fishery is lack of market.

Subsistence Fishery

The majority of the subsistence catch is associated with the artisanal fishery. It includes part of the catch that is consumed by the crew, and freely given to family members, friends and people in the community who need help. There is a well-established social tradition locally called '*kusar*' throughout the Red Sea, where part of the catch is put aside to be given freely before the remainder is sold. In some smaller communities, it is a taboo to sell the whole catch before part of the catch is given to people who need support. It is a local social support system, which is enforced by social sanctions, and ostracising those who do not conform to the tradition.

Industrial Fishery

The industrial fishery, has higher capital investment, uses more advanced technology (e.g., GPS, winch) and uses larger vessels compared to the artisanal sector. It includes two gear types, typically bottom trawl and longline. The former is almost exclusively owned and operated by foreign companies, while the latter is a local operation. The industrial fishery is mainly export oriented.

Trawl Fishery

The trawl fishery in the Eritrean Red Sea, which is operated mainly by foreign vessels, was divided, especially in the early years, into inshore shrimp trawling, which was done mainly around Hirgigo Bay, and offshore trawling targeting fish. The operations were different and data reported separately. The protected shallow bays (less than 45 m) along the Eritrean coast were continuously fished for shrimp since the 1940s (Grofit 1971) by small inshore trawlers owned by locals of Italian ancestry (from colonial times). Mediterranean trawls with a cod end of 10–15 mm (stretched) were used to catch shrimp, which were sold to foreigners living in Eritrea and Ethiopia. About 90 % of the catches were discarded, as only shrimps were retained. The shrimp fishery restarted again in the 1990s after being inactive in the 1970s and 1980s, with landings an order of magnitude higher than the 1950s and 1960s. The recent shrimp fishery is geared toward the export market.

The first exploratory offshore trawling survey was done in 1957 by Israeli trawlers, and later a commercial trawl fishery started in 1959 (Ben-Yami 1964). They used Mediterranean trawl at depths of 45 m to more than 100 m around and south of Massawa. The main catch included lizard fish (*Saurida* spp.) and threadfin bream (*Nemipterus* spp.), which were refrigerated in the trawlers and landed at Eilat, Israel. This fishery had a lot of discard, consisting of fish families such as Leiognathidae, Platycephalidae, Fistulariidae, Trichiuridae and others, and smaller sizes of the targeted species. The retention size for lizard fish and threadfin bream were 15 cm and 22 cm, respectively. The fishery was active until the mid-1970s, before it was disrupted by war and instability. It started again after 1991 by trawlers mainly from Egypt and rarely from Saudi Arabia.

Longline Fishery

Although longlines were used for shark fishing in the past, a major longline fishery was started in 1999 by an Australian/Eritrean joint venture. It uses mainly fibreglass boats with inboard or outboard engine with a winch to pull in the line. The bait used is mostly Indian mackerel (*Rastrelliger kanagurta*), caught either by a dedicated seining boat, or sometimes imported. Most of the boat operators reported that bait had been the main bottleneck for their operation. They target coral reef-associated carnivorous fishes, but they do not fish in the coral reef area itself because of gear entanglement problems. The fishers mark the spots (using GPS) where they had the best catch before and return to the same site again and again until the catch declines to a level such that they deem it not worth their while to return. This can cause localized overfishing of some species, especially when they have limited ranges. As the line is hauled in, it is common to see retrieved fish that are half eaten, usually by sharks. It is also common to see sharks caught in the line; they are usually thrown back to the sea either dead or alive. The catch of this fishery was very rewarding; about 2 t were caught in 2 days by a fast boat that had 4 crew members, as noted by the first author while onboard in 2000.

Fisheries Catches

The total fisheries catch in the Eritrean Red Sea Exclusive Economic Zone (EEZ) by the sectors (both domestic and foreign vessels) is shown in Fig. 4.4, together with the total reported catch by FAO for the country. The fishery went through major changes. The total catch was quite high in the 1950s (about 29,000 t) and the late 1960s (about 26,000 t). It started to decline in the early 1970s and remained very low

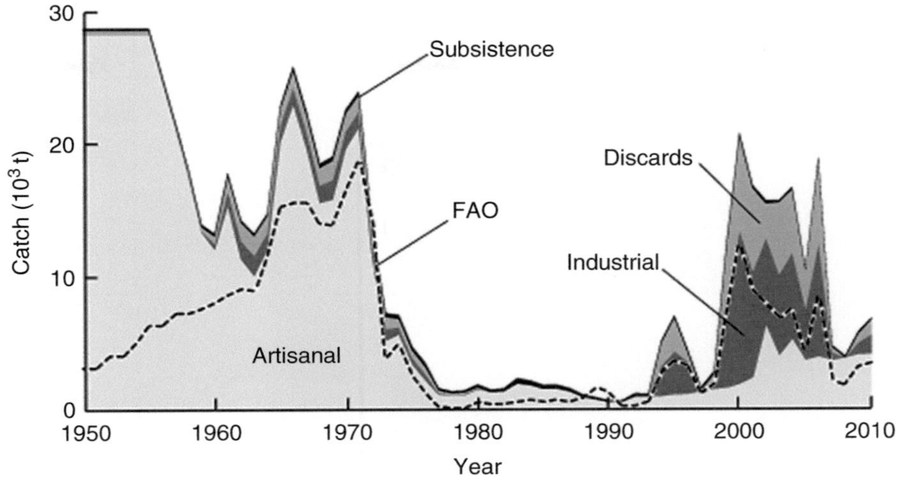

Fig. 4.4 Fisheries catches in the Red Sea EEZ of Eritrea by sector, and the catches reported to FAO from 1950 to 2010

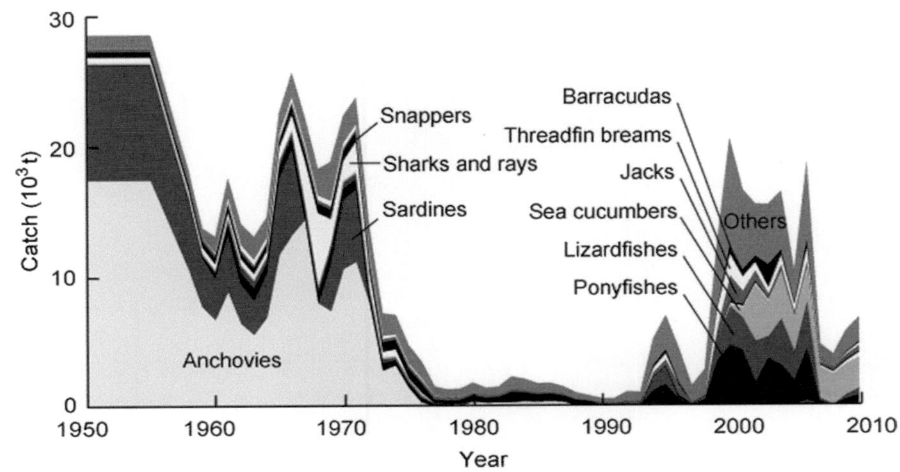

Fig. 4.5 Taxonomic composition of the total catch of the fisheries in the Red Sea EEZ of Eritrea from 1950 to 2010

in the 1970s and 1980s, but started to increase again in the mid-1990s. The artisanal fisheries were the dominant by far (76 %), followed by the industrial fisheries (22 % total, 12 % retained and 10 % discards) and subsistence 2 %. The industrial fishery had the highest contribution from 1994 to 2006, where it contributed more than 50 % of the total catch (except for 1997 and 1998). The trawl fisheries' contribution (Fig. 4.4) varied depending on whether they received permits or not.

The reported catch (part of the reconstructed catch accounted in the FAO data) represented 38 % of the total catch, which obviously brings the accuracy of the reported data into question. The unreported landed catch had the highest contribution to the total catch (52 %). The major part of the reported catch was from the beach-seine fishery in the 1950s and 1960s, which was under-reported in the FAO data. The discards, which were not reported in the FAO data, contributed 10 %, and they became more visible starting in the late 1990s, with the increase of the trawl fishery.

Taking into consideration individual gear types, the main contributor to the relatively high catch in the 1950s and 1960s was the beach seine fishery for small pelagic fishes. The increase in total catch after the mid-1990s was mainly due to bottom trawling. The small-scale artisanal fisheries, which consisted mainly of handlining and gillnetting, were the most steady in their operation. The shark and sea cucumber fisheries made relatively high contributions to the total catch in the late 1960s and early 2000s, respectively. Once these fisheries declined, they did not recover, despite a strong demand. The post 2000 high does not match the highest catch in the early period, mainly because of the absence of the small pelagic beach seine fishery.

The taxonomic composition of the total catch is shown in Fig. 4.5, while the compositions for each gear types are given in Fig. 4.6. The small pelagic fishes (anchovies and sardines) dominated the 1950s and 1960s, while demersal fishes dominated after 1990. The changes in species composition over time are explained largely by the changes in the fishing gear prevailing during the periods in question. The catch compo-

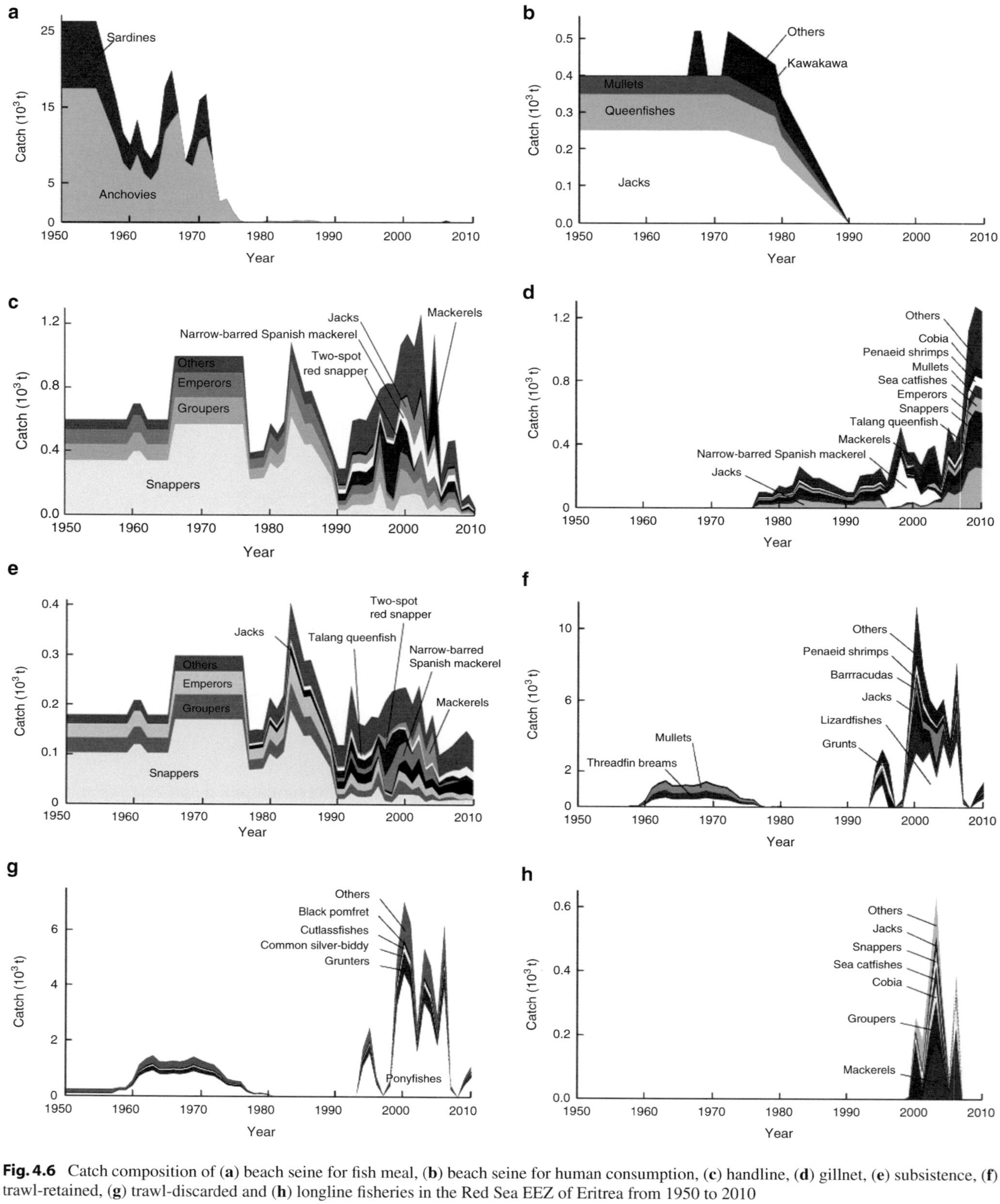

Fig. 4.6 Catch composition of (**a**) beach seine for fish meal, (**b**) beach seine for human consumption, (**c**) handline, (**d**) gillnet, (**e**) subsistence, (**f**) trawl-retained, (**g**) trawl-discarded and (**h**) longline fisheries in the Red Sea EEZ of Eritrea from 1950 to 2010

sition after 1990 shows more diverse groups with sizable catches, unlike in the early years which were dominated by a few taxa. Discards, which are usually omitted from official reports, were high, especially in the latter years. This is due to the fact that demersal trawling became prominent in recent years, and the absence of an industry to make use of the bycatch they generate, e.g., through processing it to fishmeal. An attempt was made to convert bycatch to shrimp feed for an aquaculture operation, but this did not last.

The fisheries in Eritrea's EEZ went through some major fluctuations, with a relatively high total catch during the early 1950s, followed by a decline in the late 1950s and a recovery starting in the mid-1960s. Then it went through a rapid decline starting in the early 1970s and stayed very low until the beginning of the 1990s. Towards the end of the 1990s, the catch increased again, but not to the high levels of the early years. These fluctuations are explained by major geopolitical changes and market availability for the fishery products. This is in contrast to fisheries that have more stable political and market environments, where changes are explained in terms of the changes in the fishery resources. In the latter case, the changes are not usually as abrupt as those seen in Eritrea.

The first decline, which occurred in 1955, was due to the fact that the small pelagic beach seine fishery, which was dominant at the time, used to be conducted by fishers from Yemen, who were banned from Eritrean waters in late 1954 (Ben-Yami 1964). The (small) next peak, which occurred in 1961, was due to the Yemeni fishers being allowed back into Eritrean waters, but only for 1 year. Then the fishery recovered, and attained a new peak in 1966, as a result of motorization (Grofit 1971). The decline after 1966, especially in the beach seining fishery, was mainly due to the closure of the Suez Canal, associated with the 6-Day War between Israel and the neighbouring Arab countries. The Suez Canal was the main route for the export of the beach seine fisheries product, i.e., fishmeal, which was exclusively export-oriented. After 1966, some of the decline was compensated by the growth of the shark fishery. The next peak occurred in 1971, and is attributed to the creation of alternative markets in neighbouring countries and Asia for the beach seine fishery catch (Grofit 1971).

The contribution of foreign trawl fishery was high from 1961 to 1971. The major reason for the strong decline after 1971 was political instability. As the Eritrean independence war gained momentum, it expanded to the coastal area. As a result, the Ethiopian Government, which was in control over Eritrea at the time, declared martial law. The fishery infrastructure was destroyed and most of the foreign companies involved in the fishing industry, such as trawling and fishmeal processing plants, left. Many local fishers also went into exile in neighbouring countries. Eritrea became independent in 1991, and soon after, the industry started to recuperate. The first peak of the post-independence period was in 1995, as some foreign trawlers resumed operation in Eritrean waters. They stopped in 1997, however, causing a decline in total catch. But they resumed their operations in 1998 and the catch increased rapidly, reaching a new peak in 2000. After 2000, the decline in the trawl fishery catch was partly compensated by the rapidly increasing catch of sea cucumber.

Overall, the most stable fishery throughout the whole period was the artisanal fishery, which targets fish for direct human consumption, and mainly deploys handlining and gillnet. Most of the fresh fish caught by the artisanal fisheries is sold in local markets. A small proportion of the catch is distributed freely, and this sustenance catch goes straight to the households. Fish is the main staple food for some of the coastal people, especially in the southern part, and that is why handline and gillnet fisheries were active even during the war. The fisheries that showed rapid fluctuations (even when the demand was still high) were those that targeted a small number of vulnerable species, i.e., the shark and sea cucumber fisheries. These two fisheries are similar in many ways. First, they both yield a dry product, which allows fishers to remain fishing at sea for longer periods, unlike fisheries that yield fresh product, which forces the fishers to land their catch before it is spoiled. Second, their products are exclusively for export and the demand for shark fin and sea cucumber has been increasing globally. As these products have high market prices, they are preferred by fishers , except in the 1950s and 1960s, when the high catch of beach seining made this operation very profitable. The rapid decline of these two fisheries, despite a high effort level, suggests much diminished populations of the target groups, which demands immediate research and action.

This is, to our knowledge, the first attempt to document and standardize the fisheries catches in the Eritrean Red Sea EEZ (both by local and foreign vessels) comprehensively over a long time series (1950–2010), with each gear and type of fishery covered and their catch composition established. We believe this is a crucial step for the assessment of the existing fisheries, and the potential development of new ones (Tesfamichael 2012). The fisheries in Eritrea's waters went through several shifts due to geopolitical changes that have affected how data have been recorded – or not. For those years where data were not available, it was necessary to infer catches from other years, combined with the best knowledge available to us regarding what kind of changes had happened during those years. We used all the reports we could find to understand the fisheries in Eritrea and also interviewed more than 200 fishers and managers (Tesfamichael et al. 2014). Our procedures and assumptions are transparent and open to objective criticism. We considered it unacceptable to simply ignore certain fisheries or years from consideration because detailed data were not readily available (Tesfamichael and Pitcher 2006).

Sources and Methods

An extensive search was made for published papers, reports and other documentations to be used as the basis for the reconstruction of the Eritrean Red Sea fisheries. Most of the materials for the early years were government reports, and expert technical and survey reports. A good description of the fisheries by resource type, gear, operation, total catch, composition and some effort data for the past fisheries was given by Ben-Yami (1964), who was in Eritrea as an expert working for the then Ethiopian Government as a master fisher from 1960 to 1963. Grofit (1971), who was also an advisor to the Ethiopian government, from 1966 to 1969, gave a follow-up account of the fisheries based on Ben-Yami (1964) and his own experiences. Most of the information in the 1980s and early 1990s was based on FAO technical reports written by experts who made short visits to the area, notably Giudicelli (1984). For the period after Eritrea became independent in 1991, most of the data were obtained from the Ministry of Fisheries (MOF), which keeps relatively good records of the fisheries activities in the country. Interviews were also done by the first author with fishers of a wide age range to obtain a general understanding of the fisheries at different periods, which were also used as supplementary information when quantitative data were missing (Tesfamichael et al. 2014). For example, the qualitative information given by the fishers was used for interpolations where there were data gaps (see below).

For some years, catch data were given in some reports and those were used as anchor points for the reconstruction procedure. For years for which data were unavailable, interpolations or extrapolations were performed using the anchor points and information as to what happened in the fisheries during those years. Different approaches were used for different fisheries based on the information available and the nature of their operations; the reconstruction method is given below for each fishery. A spatial distribution of the catch is given at www.seaaroundus.org.

Artisanal Fishery

Beach Seine Fishery

The beach seine fishery in Eritrea can be divided into two categories: one targeting export markets for fishmeal and dried fish of small pelagic species such as sardines and anchovies. The other category is beach seining of relatively larger fishes for direct consumption through the local market. The records on the export fishery were thorough and informative, including files from customs office, which provided information on how much of the final products, fish meal powder and dried fish, were exported (Tesfamichael and Pauly 2011). The earliest reported catch of the beach seine fishery for export was 25,000 t for the early 1950s (Johnson 1956). Because this fishery was well established by the mid-1940s (Ben-Yami 1964), 25,000 t is taken as the total annual catch from 1950 to 1955. Catch estimates from 1958 to 1963 were available from Ben-Yami (1964), 1964–1967 from Grofit (1971) and 1968–1975 from Sanders and Morgan (1989), which also presented data for 1979 and 1980. The next data for the small pelagic beach seine fishery was from FAO, reported under Ethiopia for 1976 and 1980–1987 (FAO 2010). Following that period, the fishery declined, and collapsed in 1990. In 2000, an exploratory project was introduced to exploit the small pelagic resource. The catch was generally very low except for 2006, when 293 t was reported (MOF 2007); there were no catches between 2007 and 2010 (MOF 2012).

From exported dry sardines and anchovies and fishmeal, Ben-Yami (1964) calculated the total catch using the ratio 4:1 of gross catch to final processed product. For the reconstruction, first the data were assigned to the corresponding Gregorian Calendar based on the months the fishing occurred, i.e., from October to May, e.g., 3/8 of the catch reported for Ge'ez Calendar year 1951 was allocated to Gregorian calendar year 1958 and 5/8 to 1959. Based on the yearly catch calculated for 1959–1962, the ratio of catch for October – December to January – May was computed to be 1:1.9; the catch for the few months in 1958 and 1963 were scaled up to the whole year using this ratio. The landings in Assab were estimated to be half that of Massawa (Ben-Yami 1964). The national total was calculated by adding the Massawa and Assab landings, which were the only two places where the beach seine catch was landed. To the total landed catch, 5 % was added to account for unreported catch from 1950 to 1975 when the fishery was active. Because of fishmeal production, there was no discard in this fishery; however, part of the catch was spoiled by rain during drying, some eaten by birds and some spoiled due to bad handling. Based on interviews with fishers, the unreported catch was estimated to be 5 % of the landed catch, a conservative estimate (Tesfamichael and Pitcher 2007). The same procedure was followed for 1964–1967 from Grofit (1971) and 1968–1975 from Sanders and Morgan (1989). For the years 1956 and 1957, interpolations were used to match the decline of the fishery in those years.

The species composition of the total catch was calculated based on the information that the landings in Massawa were generally equal amounts of anchovy and sardine; however, for Assab, it was all anchovy (Ben-Yami 1964). As the Massawa landings were double those of Assab, a ratio of 1:2 sardine to anchovies was used for 1950–1971, except for 1967 and 1968, where all the catch was allocated to anchovy. According to Grofit (1971), export of fishmeal stopped in

1967 due to the closure of the Suez Canal, which was the main route to the market in Europe; then, all the fisheries switched to anchovy and shark fishing, and handlining. Fishmeal export started again in 1969, as alternative markets were found in neighbouring countries and Asia (Grofit 1971), resulting in increase in catch. Thus, the ratio 1:2 for sardine and anchovies was used. Although the exact year when the fishmeal industry stopped again was not given, we assume that it was in 1972, based on the qualitative information that Giudicelli (1984) gave about the decline in the industry and the fact that the catch was low, back to the same level as in 1968 when fishmeal export stopped. Thus, the catch from 1972 to 1976 and 1980 to 1987 was allocated only to anchovies. There were two common species of anchovies, *Encrasicholina heteroloba* and *Thryssa baelama*, and one species of sardine, *Herklotsichthys quadrimaculatus*. After 1987, neither anchovy nor sardine was landed as the whole industry was shut down and whatever was reported for beach seine was assumed to be for fresh fish consumption.

The average annual catch of beach seine for fresh fish for direct human consumption was estimated to be 400 t in the early years (Ben-Yami 1964). This value was used from 1950 to 1972, as the fishery was more or less stable during that period. Two data points were available for 1979 and 1980, 331 and 269 t, respectively (Sanders and Morgan 1989). These data were recorded at the main landing sites where fish were sold through formal market channels, which would usually be delivered to the fish market in Asmara, the capital city of Eritrea. These data points were given in the Ge'ez Calendar, and were converted to the Gregorian calendar; they pertained only to fresh food fish, as fishmeal and dry fish were not being produced at that time, according to interviews conducted by the first author, and Giudicelli (1984). For the years 1973–1979, interpolation was applied to estimate the catch, which suggested a slight decline in the fishery during those years. The fishery stopped in 1990, as was the case with some of the other fisheries as well, due to civil unrest. From 1981 to 1989, the catch was estimated by interpolation. Not all fish that was caught by the beach seine fishery for fresh fish went through the proper market channels where data recordings were possible. The low grade fish were either sold in bulk to the fish meal processing plants, when they were operating, or sold through the informal markets to the local people on the coast. Sometimes not all the low grade fish could be sold to the locals; the rest would be thrown away. For 1967, 1968 and since 1972, when there was no production of fishmeal, the unrecorded catch (consisting mainly of kawakawa, *Euthynnus affinis*, which did not fetch a good price in the market) was estimated to amount to 30 % of the landed catch (Giudicelli 1984). This was added to the landings of food fish by beach seine to calculate the total catch.

The species composition of fresh fish from beach seining used for direct consumption was calculated based on data given in Ben-Yami (1964), who reported the catch to consist of jacks (*Carangidae;* 62.5 %), queen fish (*Scomberoides* spp.; 25 %), and mullets (Mugillidae; 12.5 %). The unrecorded catches were assumed to consist of 67 % kawakawa (*Euthynnus affinis*) and the rest, a mix of many taxa with minor contributions, based on the qualitative information given in Giudicelli (1984).

Handline Fishery

Ben-Yami (1964) estimated the annual landings of the handline fishery to be 300 t, which was used from 1950 to 1965, except for 1960 and 1961, where 100 t were added to reflect the extra effort in those years. In 1966, some fishers, who used to do beach seining, switched to handlining and the catch increased to 1,000 t annually (Grofit 1971). This value was used from 1966 to 1976, a period when the fishery did not change much. In 1977, war broke out in the coastal area and disrupted the artisanal fisheries. After 1977, estimates were given for 2 years, 1981 and 1983, by Giudicelli (1984), while Sanders and Morgan (1989) compiled total landings from 1977 to 1986. Those landings were assigned only to the artisanal fisheries, which were predominantly handline and gillnet fisheries, as they were the only fisheries operating mainly to supply local communities. Gillnet started to be prominent only later, in the 1970s (Giudicelli 1984). Total artisanal landing for 1992–1995 was obtained from Tesfamichael and Zeremariam (1998) and for 1996–2010 from the Eritrean Ministry of Fisheries database (MOF 2007, 2012), which separated the artisanal catch by gear type.

This fishery uses very selective gear, thus it hardly has any discard. However, not all the catch is reported in the formal channels. Some of the fish is directly sold in areas where data recording does not occur. This is in addition to part of the catch being given freely to family and friends, and which is treated separately, as part of subsistence fishery, in this chapter. Based on interviews with old fishers who were active in the 1950s and 1960s, we found out that up to half of their catch did not go to formal market channels (landing sites). Thus, we assumed that the total catch to be double of what was reported, as was also done in a survey of artisanal fisheries in neighbouring Sudan (Chakraborty 1983). Thus, the unreported catch was estimated to be 300 $t \cdot year^{-1}$ from 1950 to 1976. From 1966 to 1976, the catches given by Grofit (1971) were taken as an estimate of the total, including unreported. Given that Ben-Yami (1964) estimated a maximum of 700 t, including unreported catch and that, after 1991, when the fishery again was in full swing, an annual maximum of 1,300 t was being caught, the 1,000 $t \cdot year^{-1}$ estimate by Grofit (1971) from 1966 to 1976 seemed reasonable to be the total catch. For the period 1977–2010, unreported catch was estimated based on Giudicelli (1984), who estimated the landings to be 328 t for 1981. Based on the assumption that about half of the catch did not go through proper channels, i.e., remained unrecorded, the unreported catch was estimated to be 328 t for 1981. For the rest of the years from 1977 to 2010, population size was used as a proxy to calculate the total unreported catch, here assumed to go only to local consumption. Hence, the unreported catch was calculated by multiplying 328 t by the ratio of population size of the respective year to 1981. In 1990 and 1991, the peak of the independence war, fishing was much reduced; however, the coastal people were still fishing, and supplied some of the local demand. Thus, we assumed the amount caught to be equal to the unreported catch estimated from demographics.

The artisanal catch from 1977 to 1995 was not divided between handline and gillnet. The first distinction appeared in the Ministry of Fisheries database starting in 1996 (MOF 2007), and the average ratio of handline to gillnet was 3.13:1. This ratio was used to divide the reconstructed total artisanal landing from 1990 to 1995. From 1977 to 1986, a ratio of 4.5:1 was used as gillnetting was just starting and more catch came from handlining. No data were available from 1987 to 1989, hence they were interpolated.

For 1983, Giudicelli (1984) estimated the total artisanal catch to be around 2,000 t, based on the number of boats during his visit to the Eritrean coast and catch rates from Yemen calculated by Walczak (1977). In his report, Giudicelli (1984) stated that the fishery showed some sign of renewal, following its decline at the end of the 1970s, with the fishers using mainly handline and gillnet, and targeting prime fresh fish species such as snappers, groupers and emperors. However, we think this estimate is too high: first, the total catch of prime fresh fish, prior to the decline, was 1,000 t according to Grofit (1971). Secondly, the maximum catch was 1,319 t in 2002 when the fishery flourished again after Eritrean independence. Hence, we assume the 2,000 t reported by Giudicelli (1984) was an over estimate, and we reduced it to 1,000 t, with an estimated unreported catch of 346 t.

The only taxonomic composition record for the early catches was given to be 57.2 % snappers (Lutjanidae), 16.9 % groupers (Serranidae), 15.6 % scavengers or emperors (Lethrinidae) and 10.3 % other taxa (Ben-Yami 1964). These ratios were used from 1950 to 1989. From 1996 to 2004 a detailed catch composition of the handline fishery was available (MOF 2007), and its weighted average was used from 1990 to 1995 and from 2005 to 2010.

Gillnet Fishery

The gillnet fishery for shark is reported separately (see below); here only the gillnet fishery for fresh food fish is reported. According to Giudicelli (1984) and interviews by the first author with fishers, gillnets for non-shark fishery, although used for a long time, started to be important to artisanal fishers around 1977. The gillnet catch was calculated from the total, including unreported catch, of the artisanal fishery (see the handline fishery section for the calculations) by dividing it by 5.5 from 1977 to 1986 and by 4.13 from 1990 to 1995. Gillnet landings are given separately in the MOF database for 1996–2010 (MOF 2007, 2012). From 1996 to 2010, the unreported catch ratio of gillnet fishery from the total unreported catch of artisanal fishery as described above was taken to be proportional to the gillnet to handling reported catch. For the period from 1987 to 1989, the gillnet fishery catch was interpolated. The catch composition was calculated based on data from the database of the Ministry of Fisheries of Eritrea from 1996 to 2004 (MOF 2007). For the rest of the years, the weighted average of 1996–2004 was used.

Shark Fishery

The shark fishery was active for a long time; however, the earliest available catch estimates were from 1963 to 1968 (Grofit 1971); these catches were aggregated with landings of small pelagics. However, from 1966 to 1968, they were disaggregated, and the ratio of shark to small pelagics for 1966 and 1967 were used to disaggregate the landings of 1963–1965. The 1968 ratio was not included because many fishers switched from beach seining for small pelagics to shark fishery by the sudden closure of the Suez Canal, which resulted in a spike in shark catch in 1968. From 1969 to 1977, landings were taken from Sanders and Morgan (1989). All catch reports were given in Ge'ez Calendar and were converted to Gregorian Calendar based on the fact that in the past, the shark fishery operated mainly during the hot months, i.e., alternating with the beach seine fishery.

Catches from 1950 to 1962 were assumed to be the same as in 1963, because the shark fishery had been active for a long time in the area and was relatively stable during that period. Shark catches from 1996 to 2010 were obtained from the MOF database (MOF 2007, 2012). However, the reported catch did not include all sharks caught; a conservative estimate of unreported catch to be 10 % of the reported catch was used. The data gap between 1977 and 1996 was filled by estimating the catch to be 14 $t \cdot year^{-1}$, which is the amount for both 1977 and 1996. This is a reasonable estimate as interviews with fishers revealed that the shark fishery continued during those years.

Sea Cucumber Fishery

Although the main and latest fishery for sea cucumber started mainly around 2000 (Tewelde and Woldai 2007), catches of sea cucumber together with snail nail and pearl were reported from 1962 to 1965 (Grofit 1971). The average of the total catch over those years was 16 t and was taken to be the annual catch from 1950 to 1976. Out of the 16, 15 t were assumed to be sea cucumber (mainly *Holothuria* spp. and

Actinopyga spp.) while the rest is assumed, based on interviews, to be snail nail and pearl. From 1998 to 2006, data were available from Tewelde and Woldai (2007), who reported on an extensive socio-economic survey of the Eritrean sea cucumber fishery. Their catch estimates are likely to be reliable, because the fishery is for export and only few companies were involved. Data were not available from 2007 to 2010, and we used the value of 2006. All sea cucumber data were expressed as dry weight, after the boiling and drying processes and were converted to wet weight based on data from Purcell et al. (2009). Here, we used the mean of their conversion data for boiled and dried *Holothuria* spp. and *Actinopyga* spp. (i.e., 10.2 % of wet weight).

Other Artisanal Fisheries

The fisheries included in this category are trochus shells (top shell), lobsters, shell fish nail and pearls. Earliest trochus shell collection data available were from Reynolds et al. (1993), who reported 300 t for 1955; the same amount was assumed for 1950–1954 during which the fishery was relatively stable. Data from 1958 to 1962 and 1963 to 1968 were available from Ben-Yami (1964) and Grofit (1971), respectively. From 1969 to 1974 and 1977, data were obtained from Reynolds et al. (1993). From 1978 to 2010, no data were available on shell collection; however, the collection of shell by women and young children in shallow water was still occurring. Thus, a minimum of 1 t·year^{-1} was assumed for that period. Interpolation was used to estimate the catches for 1956, 1957, 1975 and 1976.

Grofit (1971) reported an annual catch of 5 t of lobster, and this value is used from 1950 to 1977. For the period from 1978 to 2010, a minimum catch of 1 t·year^{-1} was estimated, which is close to the reports from MOF (2007). Based on observations, interviews and field notes of Eritrean fishery officers, the snail nails and pearl catch was estimated at a conservative minimum value of 1 t·year^{-1} from 1950 to 2010.

Subsistence Fishery

As the source of the subsistence catch is the artisanal fishery, its magnitude and composition was estimated as a proportion of the artisanal catch. However, not all the fisheries categorized in the artisanal fishery are represented in the subsistence fishery. The catch or product of the artisanal fishery which are targeted for export – such as shark fin, sea cucumber, shell fishes – are not given freely to family and friends. These products do not contribute to the local food supply. Hence, only the gillnet and handline fisheries were considered for the subsistence fishery reconstruction. Two separate estimations, corresponding two periods were used. The first period was from 1950 to 1992, when the gillnet and handline fisheries were less commercialized and a good proportion of the catch was given freely to family members and friends. Although motorization of boats, which triggers the commercialization of catches, started in the 1960s, we set the less-commercialized period until 1992, because before that period the independence war affected the commercialization of the fishery and kept it very local. We estimated 30 % of the reconstructed catch of gillnet and handline fisheries to represent subsistence catch from 1950 to 1992. This is a very conservative value, as interviews with fishers and managers indicated that about half of the catch used to be given for free. The percentage was reduced to 20 % for 1993, when fishery infrastructures and markets started to flourish after independence, and for 2010, we assumed the subsistence catch to be only 10 % of gillnet and handline fisheries catch. The percentages for the years between 1993 and 2010 were interpolated.

Industrial Fishery

Trawl Fishery

Since the shrimp fishery was reported separately for most of the period, we estimated it separately from the finfish trawl fishery. A total of 30 t of shrimp were landed from inshore waters around Massawa during the early years (Grofit 1971), and this value was assumed to apply from 1950 to 1970. Grofit (1971) also reported that shrimp accounted for only 10 % of the total catch, which was used to calculate the discard amount to be 270 t·year^{-1}, which is similar to the estimate by Ben-Yami (1964) for 1960–1963. The main shrimp species caught were *Penaeus semisulcatus, Marsupenaeus japonicus, Melicertus latisulcatus* and *Metapenaeus monoceros*, while the discard was composed of small fishes belonging mainly to the families Leiognathidae, Fistulariidae and Trichiuridae. Sanders and Morgan (1989) reported landings of 20 t for 1976, from which the discard was calculated to be 180 t. Giudicelli (1984) reported that trawling totally disappeared by the beginning of the 1980s; as a result, a catch of zero was assigned from 1981 to 1995. The catch from 1971 to 1975 and 1977 to 1980 were interpolated. The first shrimp trawl catch after Eritrea's independence occurred in 1996 and it was not reported separately as 'shrimp catch' as it was in the past. Rather, it was a component of the trawlers' catch; thus, for 1996–2010, the shrimp catch was calculated from the total catch composition of trawlers (MOF 2007, 2012).

The highest catch of shrimp was in 2001, more than 700 t. This is higher than the generally cited 500 t maximum sustainable yield (MSY) of shrimp for Eritrea (Giudicelli 1984), which was made in the 1970s and 1980s and based on little research. In addition, the fishery has expanded spatially, i.e., new fishing grounds are fished now, which were not known during the limited surveys of the 1970s and 1980s.

Before the 1970s, the trawlers operating along the Eritrean coast came from Israel. The first report of offshore trawling for fish, as opposed to inshore for shrimp, was for the first experimental Israeli trawling off the coast of Eritrea in 1958 (Ben-Yami 1964). The report also contained data of landings until 1962. The landings for 1963–1967 were obtained from Grofit (1971). Data from 1968 to 1980 were obtained from Sanders and Morgan (1989), who reported the landings for 1969/1970 to be only 100 t. This data point was not used as there was no justification for the sudden drop. It is possible this is a typographic error, and the average of the landings of the years before and after was taken instead. All records were adjusted from the Ge'ez Calendar to the Gregorian Calendar. Giudicelli (1984) reported that trawlers were not operating in the 1980s; as a result, the catch was zero until 1994, when trawler landings appeared again. Data for 1994 and 1995 were taken from Tesfamichael and Zeremariam (1998) and those for 1996–2010 from the MOF database (MOF 2007, 2012).

In addition to the shrimp trawl catch, as described above, the taxonomic composition of the retained catch of non-shrimp trawling was calculated from 1958 to 1980 using the ratios given in Ben-Yami (1964). MOF (2007) reported catch composition data from 1996 to 2006. They were used here as presented, while for 1994 and 1995, and 2007–2010 the weighted average ratios of 1996–2006 were used. The dominant taxa were lizard fish and threadfin bream, which accounted for more than 50 % of the total retained catch. There are two clearly separate phases in the trawl fishery in the Eritrea Red Sea based on the origin of the trawlers. The first, from late the 1950s to the end of the 1970s was when trawling was done by Israeli trawlers; the second from the mid-1990s to 2010 involved Egyptian trawlers (Tesfamichael and Mehanna 2012).

The trawl fishery has large amount of discards, and hence they were estimated separately. For the earlier period (1950–1980), the discards were calculated from two separate datasets: trawling for fish and shrimp trawling. For the former, the discarded catch amounted to 44 % of the total catch (Ben-Yami 1964), while for the latter it was 90 % of total catch (Ben-Yami 1964; Grofit 1971). For the period from 1981 to 2010, unlike the earlier period, there was not separate shrimp dataset, hence discards were calculated all together. Discard data were available from the MOF database from 1996 to 2003 (MOF 2007), and the average of those years, i.e., 43.5 % (similar to the value from 1950 to 1980) was used for 1994, 1995 and 2004–2010. As there was no trawling from 1981 to 1993, discards were zero for that period. Once the total discarded catch amounts were established, the next step was to disaggregate them to their taxonomic components. There was no information on the composition of discarded catch from Eritrea during our analysis of Eritrean catch. Thus, it was calculated using data from Yemen (Tesfamichael et al. 2012), which has similar ecosystems and trawling practices. Two separate periods were considered in calculating the composition of discarded catches: 1950–1957 and 1958–2010. In the earlier period, the sole target of trawling was shrimps and everything else was discarded. In the second period, however, some fish were retained; hence they were eliminated from the calculation and the ratios of the remaining taxa were scaled up to the total.

Longline Fishery

Before the 1970s, different reports mentioned that longline was used in the shark fishery. The recent longline fishery started in 1999 and targeted non-shark fishes that dwell in and near coral reefs. Since the shark fishery is presented separately in this report, we describe here only the recent longline fishery. Catch data were available from 1999 to 2010 (MOF 2007, 2012). Based on the first author's observation onboard longline fishing boats, about 10 % of the total catch was discarded, which included mainly top predators such as sharks. The composition of the landed catch, i.e., excluding discard, was available from the MOF database for 2000, 2001, 2004 and 2006 (MOF 2007). The 10 % discard was added to the unidentified group 'others'. For years for which catch composition data were not available, the weighted mean of the reported composition was used.

Comparing Reconstructed Catches with FAO Statistics

The reconstructed catch data were compared to what Eritrea (Ethiopia pre-1993) reported to the FAO. After the total catch and composition of each gear were reconstructed, whenever the contribution of 'others' was more than 10 %, it was reduced to 10 % and the difference distributed to the rest of the taxa proportional to their values. Comparison was made for the different taxonomic groups. There were few taxa that were in the FAO data but not in the reconstructed catch. They were allocated to the appropriate sector in the reconstruction. Their amount in the reconstructed catch was taken to be the same proportion as in the total FAO data. The amounts were later deducted from the 'others' of the sector to which they were placed. Since the FAO data were only a total by taxon, they were divided into the different gears based on the proportion of the taxa for each gear in the reconstructed catch. Then each taxon in the reconstructed data was compared with its corresponding value in the FAO data. The part of the reconstructed catch that is accounted in the FAO data is referred as 'reported catch' in our result. When the catch of any taxon was higher in the reconstructed dataset compared to the FAO data, which was in most cases, the difference was taken to be 'unreported catch'. In a few cases, the FAO data were higher than the reconstructed catch, and were taken as 'over-reporting'.

Acknowledgements We would like to express our thanks to the staff of the Ministry of Fisheries of Eritrea for their cooperation and in providing access to fisheries data, reports and other documents. Heartfelt thanks also go to assistants during the field work: Ahmed, Aron, Bokeretsion and Yonathan. This research was supported by *Sea Around Us*, a scientific collaboration between the University of British Columbia and The Pew Charitable Trusts. Additional supported was obtained from Eritrea's Coastal, Marine & Island Biodiversity Conservation project (ECMIB). Finally, we thank Mebrahtu Ateweberhan for reviewing the chapter.

References

Afeworki Y (2014) Population ecology of the rusty parrotfish *Scarus ferrugineus*, a dominant grazer on a seasonal coral reef. PhD thesis, University of Groningen, Groningen, 219 p

Ateweberhan M (2004) Seasonal dynamics of coral reef algae in the southern Red Sea: functional group and population ecology. PhD thesis, University of Groningen, Groningen, 187 p.

Ben-Yami M (1964) Report on the fisheries in Ethiopia. Minister of Foreign Affairs Department for International Cooperation, Jerusalem, 106 p

Campbell G (1993) The structure of trade in Madagascar, 1750–1810. Int J Afr Hist Stud 26(1):111–148

Chakraborty D (1983) Marine fisheries statistics in Sudan – an expanded plan of development. FAO, Rome, 39 p

FAO (2010) FishStatJ – software for fishery statistical time series. Food and Agriculture Organization of the United Nations (FAO), Rome. Available at. http://www.fao.org/fishery/statistics/software/fishstatj/en

Giudicelli M (1984) The Ethiopian fisheries: situation, development needs and opportunities. A report prepared for the Fishing Planning and Development Project, FAO, Rome, 157 p

Grofit E (1971) The Red-Sea fisheries of Ethiopia (1966–1969). Minister of Foreign Affairs, Department for International Cooperation, Jerusalem, 82 p

Johnson R (1956) Preliminary survey of Ethiopian fisheries (USOM, Ethiopia). Ethiopian Fisheries Division, Addis Ababa

Kotb MMA, Hanafy MH, Rirache H, Matsumura S, Al-Sofyani AA, Ahmed AG, Bawazir G, Al-Horani FA (2008) 4. Status of coral reefs in the Red Sea and Gulf of Aden region. In: Wilkinson C (ed) Status of coral reefs of the world: 2008. Global Coral Reef Monitoring Network and Reef and Rainforest Research Centre, Townsville, pp 67–78

MOF (2007) Eritrean fisheries database system. Ministry of Fisheries, Massawa

MOF (2012) Eritrean fisheries database system. Ministry of Fisheries, Massawa

Pilcher N, Alsuhaibary A (2000) Regional status of coral reefs in the Red Sea and the Gulf of Aden. In: Wilknson C (ed) Status of coral reefs of the world: 2000. Australian Institute of Marine Science, Dampier, pp 35–64

Purcell SW, Gossuin H, Agudo NS (2009) Changes in weight and length of sea cucumbers during conversion to processed bêche-de-mer: filling gaps for some exploited tropical species. SPC Bêche-de-mer Inf Bull 29:3–6

Reynolds JE, Appleton J, Bellemans MS, Bozon A, Christy LC (1993) The fisheries of Eritrea, pt. 1: sector review. pt. 2: proposed national fisheries strategy and development programme. FAO, Rome

Sanders MJ, Morgan GR (1989) Review of the fisheries resources of the Red Sea and Gulf of Aden. FAO, Rome, 138 p

Tesfamichael D (2001) Characterising the Eritrean artisanal fisheries using catch and effort data analysis. MSc thesis, Wageningen University, Fish Culture and Fisheries Group, Wageningen, 42 p

Tesfamichael D (2012) Assessment of the Red Sea ecosystem with emphasis on fisheries. PhD thesis, University of British Columbia, Resource Management and Environmental Studies, Vancouver, 241 p

Tesfamichael D, Mehanna SF (2012) Reconstructing Red Sea fisheries of Egypt: heavy investment and fisheries. In: Tesfamichael D, Pauly D (eds) Catch reconstruction for the Red Sea large marine ecosystem by countries (1950–2010). Fisheries Centre Research Reports, Vol 20(1). Fisheries Centre, University of British Columbia, Vancouver, pp 23–50

Tesfamichael D, Pauly D (2011) Learning from the past for future policy: approaches to time-series catch data reconstruction. West Indian Ocean J Mar Sci 10(2):99–106

Tesfamichael D, Pitcher TJ (2006) Multidisciplinary evaluation of the sustainability of Red Sea fisheries using Rapfish. Fish Res 78(2–3):227–235

Tesfamichael D, Pitcher TJ (2007) Estimating the unreported catch of Eritrean Red Sea fisheries. Afr J Mar Sci 29(1):55–63

Tesfamichael D, Zeremariam S (1998) The status of fisheries in Eritrea. In Coetzee L, Gon J, Kulongowski C (eds) African fishes and fisheries diversity and utilisation. Poissons et Pêches Africains: Diversité et Utilisation, International conference for the Paradi Association and the fisheries society of Africa, Grahamstown (South Africa), 13–18 Sept 1998. FISA/PARADI, Grahamstown (South Africa), p 174

Tesfamichael D, Rossing P, Awadh H (2012) The marine fisheries of Yemen with emphasis on the Red Sea and cooperatives. In: Tesfamichael D, Pauly D (eds) Catch reconstruction for the Red Sea large marine ecosystem by countries (1950–2010). Fisheries Centre Research Reports, Vol 20(1). Fisheries Centre, University of British Columbia, Vancouver, pp 105–152

Tesfamichael D, Pitcher TJ, Pauly D (2014) Assessing changes in fisheries using fishers' knowledge to generate long time series of catch rates: a case study from the Red Sea. Ecol Soc 19(1):18

Tewelde M, Woldai M (2007) Socioeconomic study of the sea cucumbers fisheries in Eritrea. Eritrea's Coastal Marine & Island Biodiversity Conservation Project (ECMIB), Massawa, 35 p

Tilot V, Veron C, Grissac AJ (2008) The coral reefs of Eritrea: little known gems. In: Wilkinson C (ed) Status of coral reefs of the world: 2008. Global Coral Reef Monitoring Network and Reef and Rainforest Research Centre, Townsville, p 78

Walczak P (1977) The Yemen Arab Republic. A study of the marine resources of the Yemen ArabRepublic. A report prepared for the fisheries development project, FAO, Rome, 67 p

Walter RC, Buffler RT, Bruggemann JH, Guillaume MMM, Berhe SM, Negassi B, Libsekal Y, Cheng H, Edwards RL, von Cosel R, Neraudeau D, Gagnon M (2000) Early human occupation of the Red Sea coast of Eritrea during the last interglacial. Nature 405(6782):65–69

Zekaria ZA (2003) Butterflyfishes of the southern Red Sea: ecology and population dynamics. PhD thesis, University of Groningen, Groningen, 120 p

5 Yemen

Dawit Tesfamichael, Peter Rossing, and Hesham Saeed

Abstract

Following a brief description of the Yemen's Red Sea coast and its coral reefs, the marine fisheries catches in the Exclusive Economic Zone (EEZ) of Yemen in the Red Sea are presented from 1950 to 2010. Reported catches were separated into different sectors, mainly artisanal, subsistence and industrial, and further into taxonomic composition. In the Red Sea waters of Yemen, the only active fisheries were the artisanal and subsistence until 1970; then, the industrial fishery started. The total catch remained low (around 10,000 $t \cdot year^{-1}$ in the 1950s) until the formation of fishery cooperatives and the availability of loans from the Agricultural Credit Bank in the mid-1970s, which allowed for the motorization of many vessels. The peak catch of about 90,000 $t \cdot year^{-1}$ was achieved at the end of the 1990s and then it declined to about 44,000 $t \cdot year^{-1}$ by the end of the 2000s. The industrial fishery picked up only in the mid-1990s, but its catches began to decline around 2003. The reconstructed catches were 1.9 times the Food and Agriculture Organization (FAO) catch data for the Red Sea part of Yemen.

Keywords

Time series • Fisheries • Catch • Catch composition • Mis-reported catch • Fishers cooperative • Credit Bank

Introduction

The Republic of Yemen is situated on the southwest corner of the Arabian Peninsula, and is bordered by Saudi Arabia to the North, the Red Sea to the West, the Gulf of Aden and the Arabian Sea to the South and Oman to the East (Fig. 5.1).

D. Tesfamichael (✉)
Fisheries Centre, University of British Columbia, Vancouver, BC, Canada

Department of Marine Sciences, University of Asmara, Asmara, Eritrea
e-mail: dawittes@gmail.com

P. Rossing
Fisheries Centre, University of British Columbia, Vancouver, BC, Canada

H. Saeed
Marine Research and Resource Center, Hodeidah, Yemen

Hence, Yemen has access to both the Red Sea and the Gulf of Aden. Prior to its (re)unification, in 1990, Yemen consisted of two entities, the Yemen Arabic Republic (YAR; or North Yemen) and the People's Democratic Republic of Yemen (PDRY, or South Yemen). The border between the two entities was where the Red Sea opens to the Gulf of Aden; hence, the Yemeni Red Sea coast was entirely part of the North Yemen (YAR) and the Gulf of Aden under South Yemen (PDRY). This research emphasizes the fisheries along the Red Sea coast of Yemen. All the fisheries, both by domestic fleets and foreign, are included.

Yemen is divided into governorates and three of these border the Red Sea, i.e., Hajja, Al Hodeidah and Ta'izz. Fish are landed at 31 locations along the Red Sea coast with the largest proportion taken from the Al Hodeidah governorate (Akester 2007). The main ports where fishing is concentrated are Hodeidah, Al Khauka, Al Khoba and Mocha. Fishery is an important aspect of the coastal communities and artisanal

D. Tesfamichael, D. Pauly (eds.), *The Red Sea Ecosystem and Fisheries*, Coral Reefs of the World 7,
DOI 10.1007/978-94-017-7435-2_5

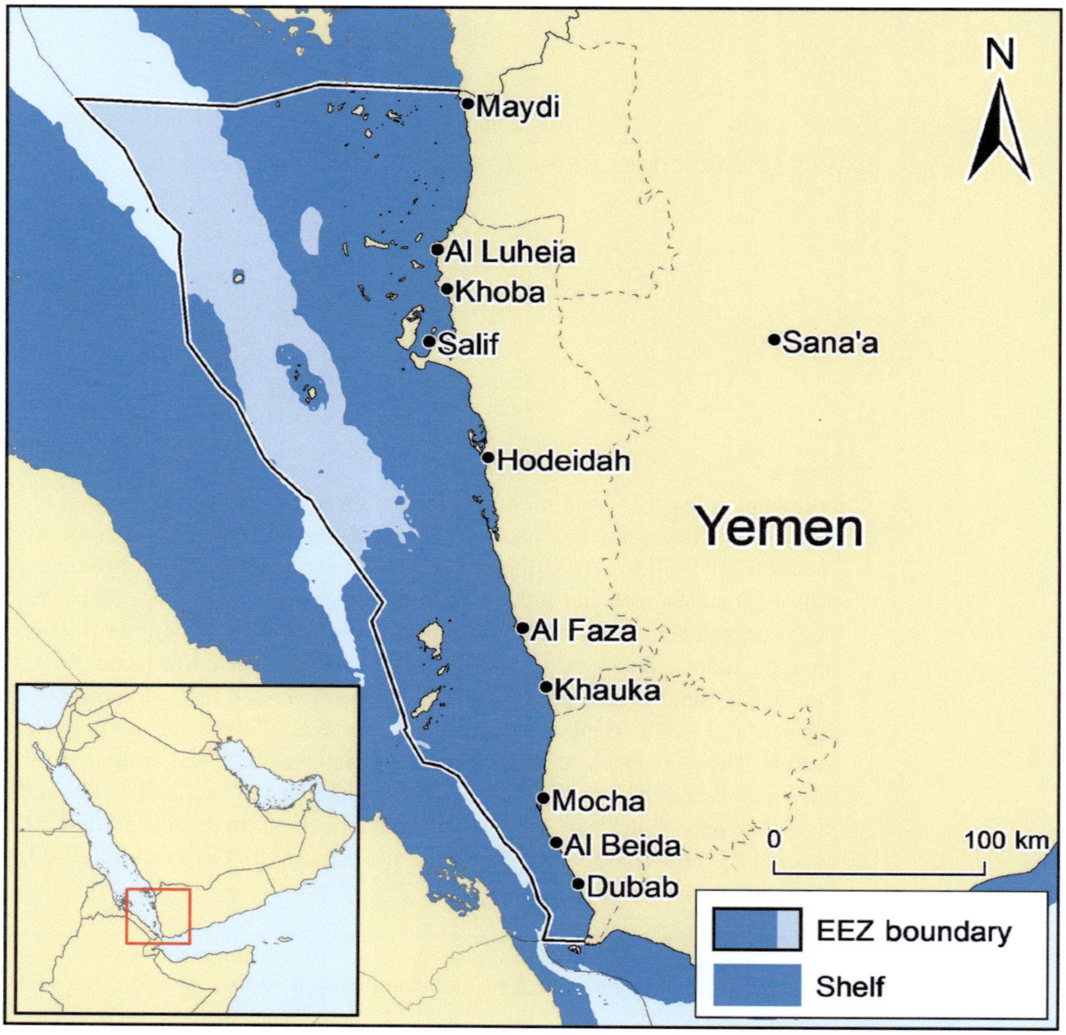

Fig. 5.1 The Red Sea coast of Yemen with its shelf area and Exclusive Economic Zone (EEZ)

fishery is very prominent in Yemen. It has long history and is relatively well organized through cooperatives.

While Yemen has introduced recent measures to better manage its fisheries, its limited capacity to effectively monitor existing regulations has limited the impact of these measures (Pramod et al. 2008). Compounding the overfishing issue, official catch statistics greatly misreport what is caught (Morgan 2006). Reported landings have historically not accounted for all species and have failed to include the recent drastic increase in the number of artisanal boats (Herrera and Lepere 2005). Official landing statistics have overlooked significant illegal, unregulated and unreported (IUU) catches, known to be substantial in Yemen's industrial fishing sector (Koehn and Aklilu 1999). In this chapter we present the catches in the Yemeni Red Sea waters from 1950 to 2010. The catches are segregated by sectors and also by taxonomic groups. A brief account of the coral reef ecosystem of Yemen is first presented, followed by introduction to the fisheries and the catches. For those interested in more details how the catches were obtained, the sources and methods used to estimate the catches are given at the end.

Coral Reef Ecosystems

The coral reefs along the Red Sea coast of Yemen are mainly accreting fringing reefs along the coast and islands, with some submerged patch reefs and non-accreting coral assemblages associated with red algae reefs, relic Pleistocene to Holocene reefs, and lava flow terraces and volcanic rock pinnacles. These structures host over 220 species of stony (Scleractinian) coral fauna, some of which are endemic to the Red Sea (Turak et al. 2007), and occur along about 25 % of the coastline. The fringing reefs are present both in the limestone islands and the volcanic oceans islands. The latter have clear water, which is conducive to coral growth. The central and northern parts of the coast have better coral coverage than the southern part, which has a shallow muddy shelf and strong southerly winds, which stir up sediments. These factors reduce water visibility and stress the corals, hence limit coral growth and distribution. The reef structures are flat without true crests and slopes (Kotb et al. 2008). In terms of coral reef ecology, the southern part of the Red Sea coast of Yemen shares similarity with the Gulf of Aden (Turak et al. 2007).

While there are recent and detailed coral reef studies for Yemen in the Gulf of Aden, around the developments of the Yemen Liquefied Natural Gas (Kotb et al. 2008), there are few recent studies in the Red Sea part of Yemen. What little exists suggests that, in 2002, average coral cover was 53 %, with a maximum of 70 %. In limited survey conducted in 2004, the coverage ranged from 28 % to 63 %. The recovery from the coral reef bleaching of 1998 has been good, especially in areas around Tigfash and Kamaran islands. The dominant coral species are *Stylophora* spp. (Kotb et al. 2008).

As many coral reefs in the Red Sea are near the temperature tolerance limit for coral growth, additional stress may hinder their ability to maintain themselves (Turak et al. 2007). Crown of thorn starfish (*Acanthaster* spp.) outbreak, sea snail (*Drupella* spp.) outbreaks and fishing also add to the stress of the coral reefs, at times (for example in the 1990s) causing the loss of up to 90 % of coral coverage in some areas (Kotb et al. 2008). Fishing is so important to the coastal communities of Yemen and their dependence on the reef ecosystems is so high that any conservation work has to face this barrier (Marshal et al. 2010). There are no marine protected areas (MPA) in the Red Sea coast of Yemen. The only MPA in Yemen is Dihamri Marine Reserve in Socotra Island, in the Gulf of Aden.

Fisheries

To date, over 600 commercial species of fish and invertebrates have been recorded from Yemen (see www.fishbase.org and www.sealifebase.org). Of these, 40 fish species (mainly pelagic) contribute the bulk of the Red Sea catch, notably sharks, jacks and tunas (Brodie et al. 1999). The fisheries are an important source of foreign exchange, generating an estimated 2–3 % of Yemen's GDP in the mid-2000s (Koehn and Aklilu 1999; FAO 2002a).

The Yemeni Ministry of Fish Wealth (MoFW) is mandated with the management of fishery resources, and controls the licensing of boats and the collection of data, while the Marine Research and Resource Center (MRRC) is a government institute active in fisheries and environmental research. Both institutions collaborate when conducting fisheries research, which faces numerous human and financial constraints that have affected the data collection system. The branch offices of MoFW and MRRC in Hodeidah are responsible for the Red Sea coast of Yemen. The fishing activities of Yemen can be broadly divided into (a) artisanal (small scale), (b) subsistence and (c) industrial (large scale). Recreational fishing does not exist in Yemen (Morgan 2006). The book 'Salmon fishing in the Yemen' (Torday 2008) and successful film suggesting the opposite were regrettably fictional.

Artisanal

Yemen has a long history and tradition of artisanal fisheries. Yemeni fishers are the most experienced in the wider Red Sea, where they operate, legally or not, along the coasts of several countries. Yemeni fishers also venture into the Indian Ocean, and in their wide range of operation, they are similar to the Fante people of West Africa (Atta-Mills et al. 2004). They are also innovative, and were often the first to introduce new fishing practices and gears. Yemeni fishers also benefit from well-organized cooperatives, a financing system and an infrastructure unique among countries in the region that includes a logistics system providing with all the basic necessities for their fishing trips, such as fuel, food and gear.

Many coastal people of Yemen depend on fishing for their livelihood. In the Red Sea, the number of Yemeni fishers has increased from approximately 3,000–4,500 in the mid-1970s to over 37,000 in 2007, while the number of artisanal boats (Fig. 5.2) grew from approximately 1,000 to 7,600 (Walczak and Gudmundsson 1975; Agger 1976; MoFW 2008). The reported artisanal Red Sea landing, however, declined from its peak 51,247 t in 1993 to 28,641 t in 2007. In 2010, it declined further, to 20,751 t (MoFW 2008, 2012), as many resources have become over-exploited (Morgan 2006).

The artisanal or traditional fisheries are mainly restricted inshore, with fishing taking place close to the landing areas and targeting mostly pelagic species (PERSGA 2001). Catches are landed directly on the beach, or are brought beyond the surf line, on the back of porters (Bonfiglioli and Hariri 2004). The primary vessels used in artisanal fisheries are *sambuk*, a wooden vessel which can range from 12 to 20 m, with an inboard diesel engine, and *huri*, a smaller canoe-like vessel of 7–12 m, fitted with outboard engine and/or sails (Sanders and Morgan 1989). In the 1970s, Yemen's artisanal fleet underwent massive technological changes; notably, sailing *huris* were motorized as fishers got access to loans from the Yemeni government and foreign aid agencies (Barraniya 1979). In the mid-1970s, the Agricultural Credit Bank was established and, together with the flourishing of fishery cooperatives, facilitated the growth of fisheries. This trend has continued and engines have become more powerful, enabling skippers to operate further away from shore with larger crew (Walczak and Gudmundsson 1975; Brodie et al. 1999). This, combined with the availability of ice, has

Fig. 5.2 Artisanal boats docked near the fish landing site in Hodeidah, Yemen, which has the highest concentration of artisanal fishing boats in the Red Sea (Photo: Dawit Tesfamichael)

Fig. 5.3 Catch of Yemeni artisanal fishers, landed and marketed in bundles of 5–6 fish (Photo: Dawit Tesfamichael)

enabled fishing trips to last for several days (PERSGA 2001). Hook and line fishing (hand or troll); gillnetting (drifting or set) and purse seine are the most important types of gears used (Barraniya 1979). A small number of *sambuks* have since the 1980s also been equipped to trawl for shrimp (Sanders and Morgan 1989). Besides those gears, there are some minor fisheries in which artisanal fishers get involved, such as crabs, cuttlefish and sea cucumber fisheries for which diverse gears such as trap, skin and hookah diving, spear-guns, etc., are used.

Most of the artisanal catch (Fig. 5.3) is consumed locally; however, export of the higher value component of the catch of artisanal fisheries is growing. The low grade fish, such as Indian mackerel, are dried and sold all the way to the interior of Yemen. Yemeni society, especially along the coast, has a long tradition of seafood consumption. Fisheries are an important part of the socio-economic and cultural part of the community, and command strong political and financial support, partly due to the strength and success of the fishery unions, important in organizing and financing the fishery.

Subsistence

Subsistence fishing takes place at beaches free of rocks and coral reefs using beach seines to catch sardines, anchovy and other small pelagic species, locally referred to as '*wasif*' fishery (Walczak 1977). This is the least capital intensive fishery and solely for consumption by local communities; hence it is here categorized as a subsistence fishery. In addition to the small pelagic fishery, the subsistence fishery includes the catch of the artisanal fishery that is consumed by the crew and the catch that is freely given to family, friends and part of the communities who need such help. This portion can be as much as half of the total catch in the earlier years. Hence, it is important to explicitly represent this component in the catch reconstruction, even though it declined as the marketing of seafood grew.

Industrial

Prior to reunification, the PDRY and YAR sought to develop their fisheries in different ways. The government of South Yemen focused on the development of industrial foreign and state-owned fisheries in the 1970s and the early 1980s, while that of North Yemen aimed to develop its artisanal small scale fisheries, notably by offering loans to fishers to improve their boats and equipment (Koehn and Aklilu 1999). Thus, industrial fishing was extremely rare in the YAR, one of the exceptions being shrimp trawling by a Kuwaiti company operating along the Red Sea coast of Yemen from 1970 to 1978 (Walczak and Gudmundsson 1975; Sanders and Morgan 1989; Morgan 2006). However, following reunification, industrial vessels that had been operating in the Gulf of Aden were granted access to the Red Sea as well. From 1993 to 2010, most of the trawlers were from Egypt and a few from Lebanon. In 1998, there were 63 licensed boats in the Red Sea, catching a total of 4,200 t (FAO 2005). At present, the landings from industrial vessels consists mostly of shrimp, cuttlefish, emperor, snappers, lizardfish and threadfin bream. Estimates from Brodie et al. (1999) suggested that industrial vessels were typically 20–40 m long, with motors of 500–800 hp, and caught 600–800 kg/day of shrimp, which is much higher than the 25 kg of shrimp caught daily by *sambuks* in the same period. The rapid growth of effort and the changes it induced in the last few decades have led to dramatic decline in catch rates, which could be a sign of overfishing in many areas and leading, among other things, to conflict between artisanal and industrial fisheries (Bonfiglioli and Hariri 2004).

Fisheries Catches

The total catch in Yemen's EEZ in the Red Sea (both by domestic and foreign fleets) was low, started to increase in the early 1970s, and reached its peak at the end of 1990s (Fig. 5.4). The bulk of Yemen's catch in its Red Sea EEZ is generated by the artisanal fisheries (66 %), and went through different phases: relatively low level until the beginning of the 1970s, a slow increase until the mid-1980s, followed by a rapid increase until a peak was reached in 1993 and a phase of rapid decline since. The rapid increase was due to motorization of boats. The decline may have been caused – at least in part – by a conflict with Eritrea over the Hanish Islands, in the southern part of the Red Sea, which prevented Yemeni from accessing some major fishing grounds. The unreported catch of the artisanal fishery accounted for 40 % of the total artisanal catch from 1950 to 2010.

For the period 1993–2007, when the industrial fishery was active, the artisanal catch was on average 58 % of the total catch. The contribution of industrial fisheries increased only after 1990, the year of Yemen's re-unification, when industrial fishing was encouraged, and permission was given to foreign fleets to operate in Yemen's Red Sea waters. Overall, the industrial sector accounted for 18 % of the reconstructed catch, with only 6 % being retained. The subsistence catch, third in its contribution (16 %), follows the trend of the artisanal fishery, because its main component is

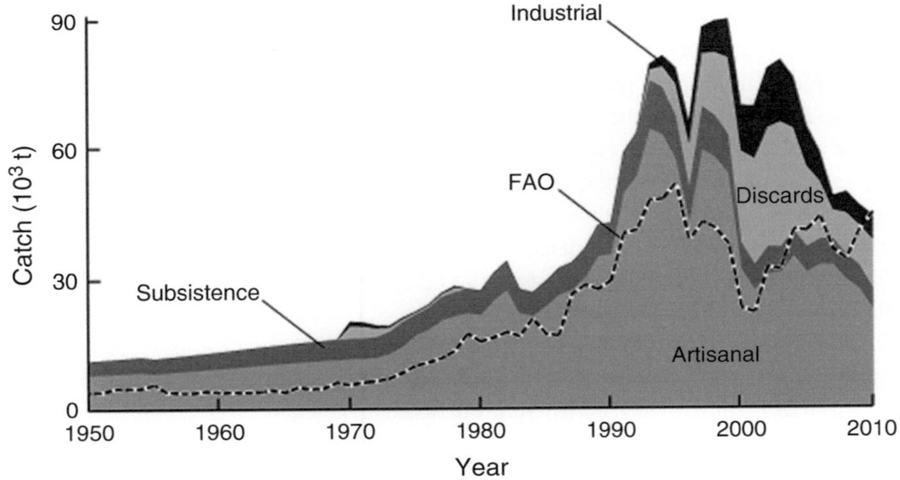

Fig. 5.4 Reconstructed catch in Yemeni Red Sea waters from 1950 to 2010 by sectors compared to the total supplied to FAO

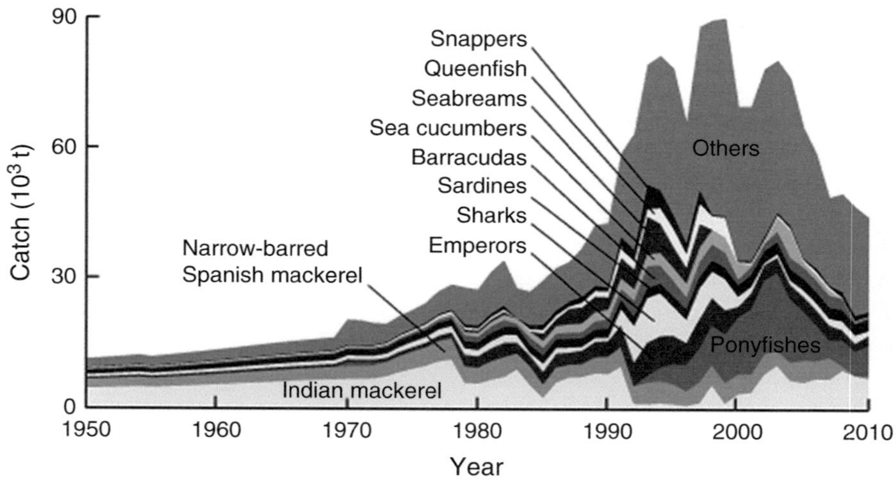

Fig. 5.5 Reconstructed total catch by major taxa in the Red Sea EEZ of Yemen from 1950 to 2010

computed as a proportion of the artisanal catch. Discards accounted for 12 % of the total catch (all from the industrial sector) and appeared mainly after the 1990s with the resumption of industrial fishery.

The 'reported' catch, which is the amount of the reconstructed catch reported in the FAO data for Yemen for its EEZ in the Red Sea, and the unreported landings, i.e., the difference between our reconstructed catch and the FAO data, each accounted for 44 % of the total catch. Unreported landed catches existed throughout the whole period and was more stable than the other components. The discarded bycatch appeared in the latter period when the industrial fishery was active (Fig. 5.4). Both the unreported catch and discards are not recorded in the official statistics, the main difference being that the unreported catches are landed while discards are not. The artisanal fisheries use selective gear; hence, almost all the catch is retained. However, not all their catch is reported.

With regard to the taxonomic composition of the total catch in Yemeni Red Sea EEZ, Indian mackerel (*Rastrelliger kanagurta*) and kingfish (*Scomberomorus commerson*) are dominant at the species level (Fig. 5.5), with 17 % and 9 %, respectively. They are so dominant that they are represented individually in Yemeni fisheries statistics, while other taxa are usually lumped together at family level. A total of 43 taxa make up 94 % of the total catch in the Red Sea, while the rest are taxa with minor contributions. Only the major taxa are shown in Fig. (5.5), the rest were added to the group 'others'. Ponyfish (Leiognathidae), a discarded group in the bottom trawl fishery, started to increase since the early 1990s, when the industrial fishery increased its effort.

The number of taxa in the catch of the artisanal fishery is quite high. The dominant taxa are Indian mackerel (*Rastrelliger kanagurta*), narrow-barred Spanish mackerel (*Scomberomorus commerson*), emperors (Lethrinidae), and sharks (Carcharhinidae) (Fig. 5.6a). Most of the dominant species are caught by gillnet, the major gear of artisanal fishers of Yemen in the Red Sea. The number of taxa increased in later years, because, based on interviews with fishers, the species that were very important in the earlier years started to decrease and fishing effort switched to previously non-targeted taxa.

The total catch of the minor artisanal fisheries for sea cucumber, cuttlefish and crabs is very low, compared to the major artisanal fishery or other sectors. Sea cucumbers were the largest component of the total catch, and hence, their catch time series shapes the pattern of the total catch of the minor artisanal fisheries. The sea cucumber fishery started only in the 1970s. Before 1970, the catch of cuttlefish and crabs was very low, but it then increased.

The subsistence beach seine fishery for small pelagic fish (mainly sardines) is not detailed in the fishery statistics of Yemen. Although its economic value may not be as important compared to the other fisheries, the amounts caught are quite large and their contribution to food security are very important, as beach seining is a fishery whose catch is fully consumed and distributed within all of Yemen. In the absence of other data, it is reasonable to assume that the total catch of this fishery changed with population size. Accordingly, the catch was low, around 1,200 $t \cdot year^{-1}$, until the 1980s and then started to increase continuously even after we have assumed the catch per capita in 2010 to be only half of that of 1975, which we believe is a very conservative estimate (Fig. 5.6b).

The other subsistence fishery catch (i.e., the fraction of the artisanal catch that is given away) has relatively less difference in its amount from 1950 to 2010 compared to the other fisheries, with the exception of a sharp increase in the 1990s (Fig. 5.6b), during which time the artisanal fishery – the main source of subsistence catch – reached its peak. Even for the 1990s, however, the change is less pronounced in the subsistence than in artisanal fishery. This is a likely more

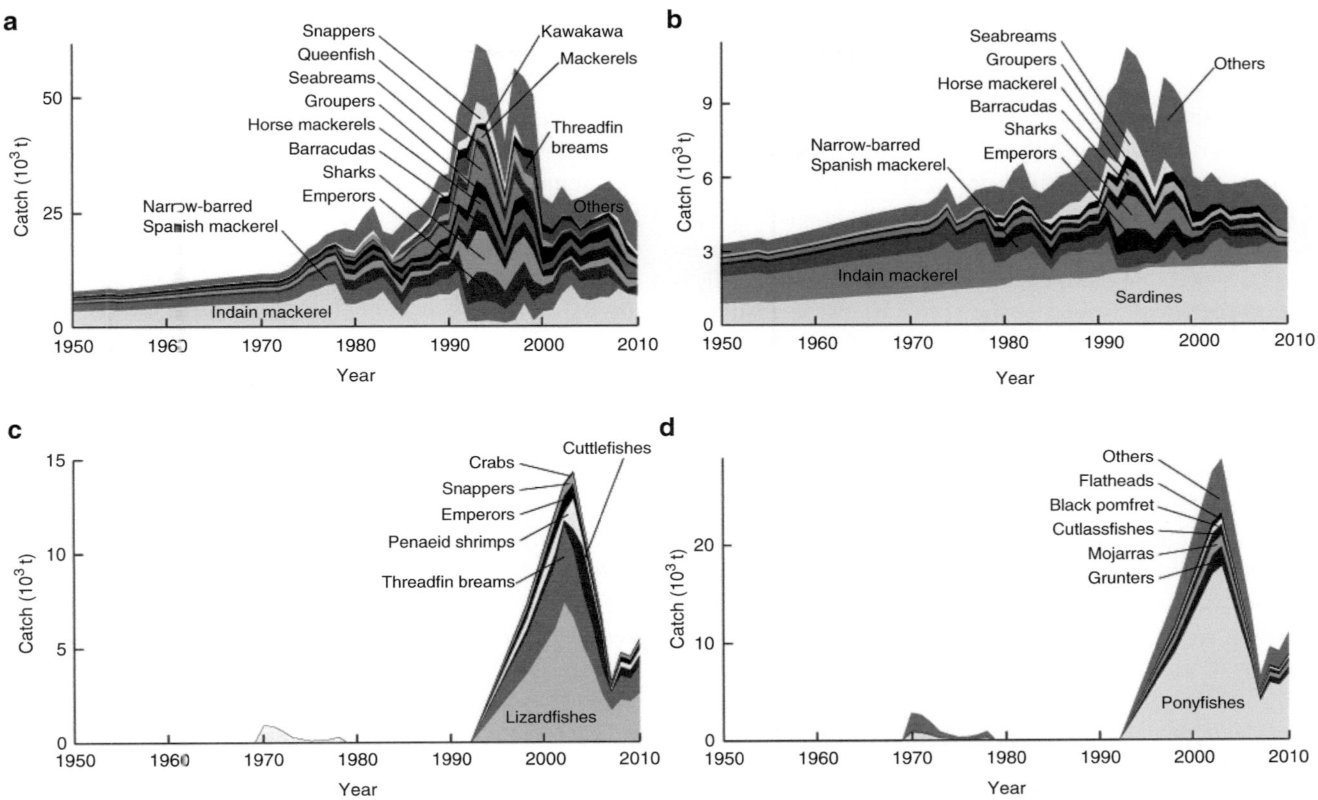

Fig. 5.6 Taxonomic composition of (**a**) artisanal, (**b**) subsistence, (**c**) industrial-retained and (**d**) industrial-discards of Yemen in Red Sea from 1950 to 2010

realistic, as a subsistence fishery it not likely to be strongly affected by market fluctuations.

Industrial fisheries played an important role in the Red Sea part of Yemen only since the mid-1990s. Prior to this, in the 1970s, it was relatively small, and caught only shrimps. After the reunification in 1992, the industrial fishery became more important, targeting shrimp and a wide variety of finfish. Since the fishery uses non-selective trawling gear, a huge proportion of the catch is thrown overboard. The highest peak of the industrial fishery occurred in 2003, when a large number of foreign vessels were given permits to operate along Yemen's Red Sea coast. However, the number of licences was reduced after 2003, because the government wanted to restrict this fishery (Akester 2007), hence the decline in total catch (Fig. 5.6c). According to our catch reconstruction, from 1950 to 2010, reported catch, unreported catch and discards accounted for 22 %, 11 % and 67 % of the total catch, respectively.

Lizard fish and threadfin bream, with a combined 73 %, are the dominant taxa that are retained by the industrial fishery. There are a number of other taxa that appear in the retained catch, but their contribution is limited (Fig. 5.6c). The main target of the industrial fishery was and still is shrimp, which accounts for less than 10 % of the total hauled catch. Some of the taxa that are not prime targets are retained, but the majority, mainly demersal species, are discarded. The number of taxa in the discarded catch is quite large and they are usually discarded because of their extremely low market value (Fig. 5.6d). The dominant group in the discards is ponyfishes (Leiognathidae) with 61 %.

Lack of a structured data recording system is a serious hindrance for assessing the catch of Yemen in the Red Sea by gears and in terms of its taxonomic compositions. Although such data do not exist continuously for the whole period from 1950 to 2010, pieces of information exist from different periods reported by various authors and institutions in Yemen. In this report, the total catch in the Red Sea EEZ of Yemen is reconstructed from 1950 to 2010 and is divided into artisanal, subsistence and industrial sectors. It is divided, as well, into the taxa that compose the catch. Assumptions were made to fill in some of the data gaps, based on the best knowledge available to us about the fisheries.

Yemen's artisanal fishery is one of the best established in the Red Sea. It has a large number of boats and fishers, a strong cooperative system, a relatively smoothly working financing system, and an effective marketing system that meets a high demand. Yemen has a long and strong fishing tradition and Yemeni's fishers are found throughout the Red

Sea and Indian Ocean, actively spreading their fishing skills to neighboring countries (Tesfamichael and Pitcher 2006). For example, Yemeni fisheries were active in the small pelagic fishing industry in Eritrea, where they fished in the EEZ of Eritrea and sold their catches to the fish meal processing plants within Eritrea in the 1950s and 1960s (see Chap. 4). The increase in total catch of the artisanal fishery in the mid-1970s, from its near constant value of around 10,000 $t \cdot year^{-1}$ to more than 20,000 $t \cdot year^{-1}$ in the 1980s is due to the strengthening of the sector by the formation of fishery cooperatives and the establishment of the Agricultural Credit Bank of Yemen. These two institutes allowed the availability of loans for the fishers to buy new boats and more importantly motorization of the boats. The cooperatives became a collective bargaining body to negotiate with the government in issues such as securing subsidies to the fishers. Although, the Agricultural Credit Bank was established in 1976 and opened the door for fishers requiring loans for motorization, it took several years for the fishers to adopt the new technology. The momentum increased and major motorization occurred at the end of the 1970s (Barraniya 1979), which further increased the number of artisanal fishers. The number of boats and fishers increased significantly at the end of the 1980s resulting in a rapid increase of total catch (Brodie et al. 1999). Although fishers reported a decline in their catch rates, the larger number of boats combined with more powerful engines and availability of ice, which allowed the fishers to stay longer at sea and go to further fishing grounds, resulting in higher total catch.

As far as the industrial fishery is concerned, its contribution to the total catch was considerable during the years it was given permission to operate. The official reports of the industrial fishery do not account for the discards that can make up to 90 % of the total catch. This omission can be misleading in any fishery management decision-making process. It is helpful to present the discards clearly so that by-catch mitigation strategies can be investigated. This can involve the designation of areas suitable for trawling or not, the use of suitable mesh size or by-catch excluding devices. On the other hand, if the discard information is not presented at all, as is currently the case, there will not be any urgency to deal with the real problem that it represents.

Pelagic fish contributed the largest fraction of the catch throughout the period from 1950 to 2010, and did not decline, at least not strongly, in contrast to demersal and reef-associated fish. The high fluctuation of the trawl fishery is due to whether or not permits were given to foreign vessels. An interesting case is the difference between the catch of pelagic and reef fish. They are both generated by artisanal fisheries; however, reef fishes exhibit a sharper increase and later decrease than pelagic fishes, which may be due to the extensive migrations which pelagic fish undertake. The southern part of the Red Sea gets replenishment of migratory pelagic species from the more productive Gulf of Aden; hence, it can sustain a larger fishery. On the other hand, reef-associated fish are territorial and hence their biomass will be strongly affected by localized effort increases.

Overall, the reconstructed catch is higher than the catch that Yemen reported to the FAO. This is critical information for future plans in Yemen. Due attention to the reconstructed catch can prevent some serious mistakes in the assumption on the status of the resources (Tesfamichael 2012). The fact that the total catch is declining should alert the decision makers to initiate ways of managing effort before the resources get too depleted.

Sources and Methods

Separating the FAO Data into Red Sea and Gulf of Aden

Yemen reports its catch annually to the FAO through one data set for the country which combines both the Red Sea and Gulf Aden, and thus needs to be split, as the two areas are here treated separately. For this, we used sources which reported the catch of the two seas separately, i.e., the catch ratios they provided were used to divide the total Yemen catch in the FAO records into their respective seas. Sanders and Morgan (1989), PERSGA (2001), FAO (2002b), MoFW (2004) and MoFW (2008) provided reported total landings for both the Red Sea and the Gulf of Aden/Arabian Sea for the periods 1976–1986, 1986–1994, 1998, 2002–2003 and 2006–2007, respectively. For 1986, data were obtained from Sanders and Morgan (1989) for the Red Sea and from (PERSGA 2001) for the Gulf of Aden and Arabian Sea. We used the reported proportions to divide the FAO data into Gulf of Aden and Red Sea components for these years. Also, in absence of a better alternative, the estimated catch proportion for 1976 was used for the period 1950–1975. All other missing years were estimated using interpolation except for 2009 and 2010, where the average ratio from 2006 to 2008 was used. This split was used as an initial value to compare the reconstructed values with the FAO data. The final FAO data for each sea, however, were calculated later by taking into account the taxonomic breakdown information which caused the proportions to change slightly, so this first split is not necessarily representative of the final data set. A spatial distribution of the catch is given at www.seaaroundus.org.

Catch Reconstruction for the Red Sea Waters of Yemen, 1950–2010

There is no catch statistics that presents Yemen's landings in the Red Sea as a complete time series from 1950 to 2010.

Similarly, no published information was found that accounts for the unreported catches in Yemen. Catch time series were therefore estimated using a reconstruction method based on assumption-driven inferences (Pauly 1998; Zeller and Pauly 2007). Such an approach is justified, despite data uncertainties, given the less acceptable alternatives that users of official data will interpret non-reported or missing data components as zero catches (Pauly 1998; Pitcher et al. 2002; Tesfamichael and Pitcher 2007). Using data from published, gray literature and interview sources as anchor points, time series data can be reconstructed using interpolation and extrapolation (Tesfamichael and Pauly 2011). Anchor points included fishery surveys, national reported catch data, peer-reviewed literature, and field trip interviews and data collected in Yemen by the first author in 2007. We worked in close collaboration with local experts from Yemen's Ministry of Fish Wealth (MoFW) and Marine Research and Resource Center (MRRC) to get feedback and update our results.

The Yemeni artisanal and industrial catches in the Red Sea (i.e., of pre-unification North Yemen) were estimated separately. For the artisanal fisheries, this involved reconstructing landings and unreported catch for the major fisheries targeting large pelagics and coral reef-associated fishes using *huri* and *sambuk*, which constitute the 'artisanal fisheries'. The same procedure was needed for the beach seine fishery (a subsistence fishery). Also reconstructed were the catches of the minor artisanal fisheries for sea cucumber, crabs, and cuttlefish. For the industrial fishery, the elements involved in the reconstruction included estimating reported landings, discards and unreported landings. The estimation of catch composition was mostly based on reports with reliable data and the database of the Ministry of Fish Wealth (MoFW), with interpolations for missing years.

Artisanal Fisheries

Reported Landings

The first catch estimates for Yemen in the Red Sea were made in the 1970s by the staff of FAO field projects, based on surveys of the artisanal fisheries for the number of boats, number of fishing days and catch rates to derive yearly catches for the various boat types (Agger 1976; Campleman 1977; Walczak 1977; Barraniya 1979). They provided reliable estimates for Yemen's *huri* and *sambuk* catches for the 1970s, which are used as anchor points.

Barraniya (1979) presented two different sets of catch statistics covering 1970–1978: one based on the General Directorate of Hodeidah and another based on the Central Planning Agency of Yemen. The former data set was used because its data matched the results of an independent extensive survey for 1973 by Agger (1976). As these data did not pertain to a calendar year, but applied from June to May of the next year (Walczak 1977), they were adjusted to the calendar year using monthly catch ratios for 1975 (Barraniya 1979). This approach reconstructed only 6 months of the catches for 1970 and 1978. For those 2 years, total annual catch was calculated using the average proportions from 1971 to 1977. Since most of the catch by the artisanal fisheries was for local consumption, we assumed, to estimate the total catch from 1950 to 1969, that catches grew proportionally with Yemen's population. Hence, catches from 1950 to 1969 were estimated using 1970 as an anchor point and population sizes given in www.populstat.info/.

Brodie et al. (1999) reported landings were used for 1979–1982 and 1987–1997. Landings provided by Sanders and Morgan (1989) were used, instead of the ones reported by Brodie et al. (1999), for the years 1983–1986, as these data had a better species breakdown. A dataset of the Ministry of Fish Wealth was used for the years 1998–2007 (MoFW 2004, 2008) and 2008–2010 (MoFW 2012).

Unreported Catch

Reported landings for Yemen's artisanal fisheries are believed to be severely underestimated (Chakraborty 1984; Koehn and Aklilu 1999; PERSGA 2001; Herrera and Lepere 2005; Morgan 2006; Pramod et al. 2008). In Yemen, fishers do not necessarily land their catches at landing sites where landings are recorded. Based on interviews with fishers, a sizeable fraction of the total catch is landed in areas remote from major landing sites, where it goes unreported (Tesfamichael et al. 2014).

In the 1970s, reported landings were based on the number of boats multiplied by observed landings, based on observations for some landing places and some days of the year (Chakraborty 1984). Interviews with fishers who were active since the 1950s suggest that they graded their catch. Usually the high grade fish and the large individuals of selected species were sold in the formal market where fishery statistics were recorded, and the rest was sold in the informal markets. This is supported by the relatively small number of taxa reported in early statistics. Sanders and Morgan (1989) noted that grading was an issue in the artisanal fisheries. Species now landed, notably catfish, sharks and the smaller grunts, were not preferred in the past. So, we assumed, conservatively, the level of under-reporting, to have been 30 % of the reported landing from 1950 to 1975.

In the mid-1970s, Yemen's catch statistics for the Red Sea improved considerably as a more comprehensive system of fisheries data collection was put in place (Sanders and Morgan 1989). The catch data we used from 1983 to 1986, provided by Sanders and Morgan (1989), were therefore considered quite reliable (Herrera and Lepere 2005). The level of underreporting was assumed to be 20 % and 10 % for 1976–1982 and 1983–1990, respectively. For about a decade after 1991, no reliable analysis of landings statistics, resource surveys or stock assessments were undertaken, due to the system for collecting fisheries statistics having broken down after the re-unification of Yemen. This resulted in gross underestimation of artisanal catches. Thus, for example, statistics were not adjusted for the growth of the artisanal fishery, including the number of boats, failed to account for fish sold outside auctions and relied on historical prices to convert auction sales to weight (Koehn and Aklilu 1999; PERSGA 2001; Herrera and Lepere 2005; Morgan 2006). We assumed the level of unreported catch to be 20 % for the period 1991–2001. Catch statistics after 2002 are believed to have improved, as Yemen took a number of steps to improve its fisheries management with external financial and technical assistance (Morgan 2006). Thus, 10 % was assumed to be the level of unreported catch from 2001 to 2010.

Catch Composition

Catch composition data for 1974–1976 were available in Walczak (1977), which were used to disaggregate our reconstructed total catch (reported and unreported) for the period 1950–1978. However, the components of a few groups were modified. Walczak (1977) reported catch ratio for 'jacks', a group we split into two, i.e., 'jacks' (Carangidae) and 'queen fish' (*Scomberoides* spp.), contributing 72 % and 28 %, respectively, based on the catch ratio of these species for the years 1979–1986. This was done as interviewed fishers indicated that they always historically caught queen fish. Similarly, snappers (Lutjanidae) and emperors (Lethrinidae), which were reported together, were split in a similar way using data from 1979 to 1986, with relative contributions of 25 % and 75 %, respectively.

The average of the 1983–1986 catch composition reported by Sanders and Morgan (1989) was used to disaggregate total landings from 1979 to 1982. From 1983 to 2010, the data were already reported by taxonomic composition. We used Sanders and Morgan (1989) for 1983–1986, Brodie et al. (1999) for 1987–1997, MoFW (2004, 2008) for 1998–2007 and MoFW (2012) for 2008–2010 who reported catch compositions for the periods indicated.

Other Minor Artisanal Fisheries

The relatively minor artisanal fisheries included in this group target sea cucumber (Holothuroidea), crabs (Brachyura) and cuttlefish (Sepiidae). Sea cucumbers have been harvested for many years, but have traditionally been omitted from the catch statistics (Bonfiglioli and Hariri 2004). Catch data for this fishery are therefore very sporadic. Walczak (1977) indicated that 20 t of sea cucumber were exported in 1975. Sea cucumber landings were also reported from 2000 to 2010 (MoFW 2012). As sea cucumber catches are reported in dry weight units we converted the reported catch data to wet weight by multiplying them by a factor of 9.54 based on Purcell et al. (2009). Based on qualitative information from interviews, we assumed that catches began in 1970 and used interpolations between anchor points to derive the missing years 1971–1974 and 1976–1999.

Records of the artisanal catches for crab and cuttlefish were only available for 2002–2003 (MoFW 2004) and for 2006–2010 (MoFW 2012). Based on interviews, these taxa are likely to have been caught in the past as well, but remained unrecorded, as is the case for the sea cucumber fishery. Catches for 1950–2001 were estimated by adjusting the catch from 2002 relative to estimated population size. The catches for 2004–2005 were interpolated between the landings of 2003 and 2006.

Subsistence Fishery

The subsistence fishery includes beach seining for small pelagic fishes, which is considered an entirely subsistence fishery, and the catch of the artisanal fishery that is consumed by the crew and/or given to family, friends and people in the fishing communities who need help.

Beach Seine Fishery

The beach seine fishery for sardines and anchovies was estimated to be 1,500 t in 1976 (Walczak 1977). Unfortunately, no other data point was available. However, based on interviews with fishers, it appeared that the fishery has been active for a long time at subsistence levels. The catch for the whole period 1950–2010 was calculated by assuming that it was directly proportional to Yemen's population size. This is reasonable, because the beach seine fishery catches were mainly for subsistence and strongly affected by population size. However, an adjustment was applied to the population-related catches from 1976 to 2010, where a multiplier of 1 was used for 1976 and 0.5 for 2010 and the multipliers between the two points were interpolated. This was done to reflect the likely decrease in catch per unit effort of the fishery over its long period of operation and arrive at a conservative catch estimate.

The subsistence fishery catch that comes from the artisanal fishery was calculated from a ratio of the artisanal fishery catch (excluding the minor artisanal fishery described above). We assumed the subsistence catch to be 30 % of the artisanal catch from 1950 to 1974. The ratio was reduced to 20 % for 1975, when the motorization of the artisanal fleet was in full swing and the artisanal fishery started to become more commercialized, hence likely decreasing the proportion of the catch freely given to the community. The ratio was assumed to be 10 % in 2010 and the ratios between 1975 and 2010 were interpolated. Based on interviews with fishers and the fishery administration in Yemen, fishers used to give up to 50 % of their catch away, so these ratios are quite conservative.

Industrial Fisheries

The industrial fisheries in the Yemeni Red Sea waters are operated by foreign vessels. Reported, unreported catches and discards were reconstructed as three separate components of industrial fisheries from 1950 to 2010 in Yemen's Red Sea. Industrial fishing in Yemen's Red Sea went through three distinct phases. First, there was a period of limited shrimp trawling in the 1970s (Agger 1976; Walczak 1977), followed by a period of no industrial fishing lasting from the 1980s to the early 1990s (Chakraborty 1984; PERSGA 2001), and lastly a period characterized by an in-flux of foreign bottom trawlers from 1992 to 2010 MoFW (2004, 2008, 2012).

Reported Catch

Industrial fisheries did not begin in Yemen's Red Sea until the late 1960s. Agger (1976) indicated that the Greek Achilles Frangistas Co. had gained permission to trawl in the waters of Yemen's Red Sea, Saudi Arabia and Eritrea with at least two 86 m 2,000 hp factory trawlers from the period 1966/1967. As this operation was a form of exploratory fishing with very limited catches, it is not considered here.

Shrimp catches reported by the General Directorate of the Hodeidah from 1970/1971 to 1977/1978 were used as basis to reconstruct industrial catches for 1970–1978 (Barraniya 1979). The catches from 1970/1971 to 1973/1974 were identical to those reported for the United Fishing Company of Kuwait, or UFCK, formerly Gulf Fisheries. The company operated a total of 43 trawlers and two mother-ships until it stopped in May 1974 due to declining catches (Walczak 1977). As the records were for the period lasting from June to May of the next year (Walczak 1977), the catches were recomputed for calendar year based on the assumption that monthly catches had the same distribution as observed in 1975 (Barraniya 1979). This gave an estimate of the industrial shrimp catches for 1970–1978. For 1970 and 1978, this approach only allocates half a year of catch data. These 2 years where raised to annual catches using seasonal average for 1971–1977.

We assumed industrial catches to be nil between 1979 and 1992, as no industrial fishing occurred in Yemen's Red Sea in the 1980s and the early 1990s (Chakraborty 1984; PERSGA 2001). It recommenced when Yemen, after reunification, changed its policy to allow foreign fishing fleets access to the Red Sea, beginning with two Lithuanian double-rigged shrimpers in 1993 (PERSGA 2001). The fishing was carried out mainly by Egyptian trawlers and to a less extent by Lebanese from 1993 to 2000 and from 2001 to 2010 by Egyptian only. In 1998, it was reported that 40–44 foreign industrial bottom trawlers caught 4,186 t of shrimp and fin fish (FAO 2002a). Total reported industrial landings for the Red Sea were also available for 2002 and 2003 (MoFW 2004) and 2006 and 2007 (MoFW 2008). These years were used as anchor points, and for the missing years 1993–1997, 1999–2001 and 2004–2005, catches were estimated by interpolation between these anchor points. The total catch for 2008–2010 was estimated using the average ratio between reconstructed catch and FAO data for 2006 and 2007. The average total industrial reconstructed catch, including retained, unreported and discarded, for 2006 and 2007 was 47 % of Yemen's FAO data for the Red Sea. The total was divided between retained and discard using the ratio 1:2, the ratio used from 1993 to 2007 (see discard calculation below).

Unreported Catch

The unreported catch refers to the catch that is landed but not recorded in the official statistics, while the discards (see below) refer to the catch that is not recorded nor landed. We added a conservative 10 % to the reported landings for the industrial fleet from 1970 to 1978, as the catches of the United Fishing Company of Kuwait (UFCK) fleet were thought to be underreported (Walczak 1977). Yemen now requires observers onboard industrial vessels, but this has not solved the problem, because observer coverage is partial (Pramod et al. 2008), and the effectiveness of the limited onboard observation is also very questionable. Local experts estimate the unreported catch, in the Red Sea, to be 75 % of the total catch from 1993 to 2007. The unreported catch from 2008 to 2010 is calculated as part of the total industrial catch using the reconstructed catch to FAO ratio as described above.

Discards

Discards are catch that are not landed, hence not recorded as well. Discards from trawling, especially from shrimp, are substantial and must therefore be added as a component of industrial fisheries catch. However, data regarding Yemen's discard levels in the Red Sea were limited. Lisac (1971), while onboard a United Fishing Company of Kuwait trawler, observed discards from shrimping to be up to three times that of shrimp caught. Losse (1973) found that boats fishing for shrimp discarded approximately 4.4–5.6 t of fish for every tonne of shrimp caught. For 1950–1969, discards were zero as industrial fishing did not occur during these years. For the period 1970–1978, we assumed discards to be 75 % of total catch (reported + unreported catch + discards) based on Lisac (1971). A reduced discard rate of 67 % was assumed from 1993 to 2007, as shrimp were no longer the only species retained. The discard amount from 2008 to 2010 was calculated as part of the total catch, using the average ratio of reconstructed catch (including retained, unreported and discard) to FAO data for 2006 and 2007, i.e., reconstructed catch was 47 % of FAO data for the Red Sea. The discarded catch was 67 % of the total industrial catch, the ratio used from 1993 to 2007.

Species Composition

The species composition of industrial catches was estimated separately for total retained catches and for discards. For the retained catch, total catches from 1970 to 1978 were assumed to be shrimp only (Barraniya 1979), reflecting the nature of the fishery then. The earliest catch composition data available for Yemen since the industrial fishery started to retain non-shrimp taxa in 1993 were for 2002 and 2003 (MoFW 2004). For the years 1993–2001 we used the weighted average ratios of 2002 and 2003 to calculate the catch composition. Catch composition data were available for 2006 and 2007 (MoFW 2008) whose average values were used for 2004 and 2005. From 2008 to 2010, for which the total retained catch was calculated using the FAO to reconstructed catch ratio of the previous years, the composition was calculated using the weighted average of the ratios from 1998 to 2007. The data for 2002 and 2003 had more generalized categories for which 'deep water fishes' accounted for more than 85 %. This was later subdivided to taxa using ratios from 2006 to 2007.

Surveys estimating the species composition of trawlers' discards were available from the FAO research vessel *R/V Orion* from 1974 to 1977 (Walczak 1977). These percentages were used in unaltered form to breakdown discard totals by species from 1970 to 1978. However, in recent years, many species previously discarded are retained, as the most sought-after species are getting scarcer. Thus, these were removed from the survey, and the total of the ratio of the remaining discarded species was scaled up to 100 % to divide the discard into its composition from 1993 to 2010.

Comparing Reconstructed Catches with FAO Statistics

The catch composition of each sector was compared with the taxonomic composition of the data Yemen reported to FAO. Only the artisanal and industrial sectors were compared with the FAO data, as the subsistence and discarded catches are not reported to FAO. A few taxa were reported to the FAO, but were not in our reconstructed catch composition. They were allocated to the appropriate sector in the reconstruction. Their amount in the reconstructed catch was taken to be the same proportion they had in the total FAO catch. The amounts were later deducted from the 'others' of the sector to which they were allocated. For each sector, for the years the group 'others' was higher than 10 % it was reduced to 10 % and the rest distributed to the taxa already identified according to their proportion in each sector.

After the reconstructed and the corresponding FAO catch by taxa were tabulated, comparison was done at the taxonomic level. The part of the reconstructed catch that is accounted in the FAO data is referred as 'reported catch' in our result. If the value of a taxon in the reconstructed catch was higher than its value in the FAO data, then the difference was labeled as 'unreported catch'. If the FAO value for a taxon was higher than the reconstructed catch, the difference is 'over-reported' catch in the EEZ of Yemen into the Red Sea. At this stage, each sector had a more detailed catch composition than it started with in the catch composition methods presented above. Since, the comparison of the reconstructed catch with the FAO data has modified the catch composition of the reconstructed catch, the final ratios are not exactly what is reported in the methodology in the above.

Acknowledgments We would like to thank the personnel of the Marine Research and Resource Center in Hodeidah, especially to Fahad Shooi Zabaan for his help with interviews and data acquisition, and Aref for helping in the arrangement of this work and for hospitality. We also thank the personnel of the Ministry of Fish Wealth, especially Khalid Hezaam Al Makrami. Margaret North's inputs after reading the first draft are appreciated. This research was supported by *Sea Around Us*, a scientific collaboration between the University of British Columbia and The Pew Charitable Trusts.

References

Agger P (1976) Yemen Arab Republic – stock assessment FI: DP YEM/74/003/3. FAO, Rome, 42 p

Akester S (2007) Yemen's fisheries: the need for management. British-Yemeni Soc, Vol. 15

Atta-Mills J, Alder J, Sumaila UR (2004) The decline of a regional fishing nation: the case of Ghana and West Africa. Nat Res Forum 28(1):13–21

Barraniya AA (1979) Socioeconomic aspects of the Red Sea fisheries in the Yemen Arab Republic. FAO/UNDP Project for Development of Fisheries in Areas of the Red Sea and Gulf of Aden, Rome

Bonfiglioli A, Hariri KI (2004) Small-scale fisheries in Yemen: social assessment and development prospects. FAO/The World Bank, Washington, DC, 111 p

Brodie J, Al-Sorimi M, Turak E (1999) Fish and fisheries of Yemen's Red Sea. In: DouAbul A, Rouphael TS, Marchant R (eds) Ecosystems of the Red Sea coast of Yemen. Protection of marine ecosystems of the Red Sea Coast of Yemen. Hassell & Assoc., AMSAT and UNOPS

Campleman G (1977) Fisheries and marketing in YAR. FAO, Rome

Chakraborty D (1984) Fishery statistics in Yemen Arab Republic – an expanded plan of development. FAO/UNDP, Suez, 48 p

FAO (2002a) Fishery country profile – the Republic of Yemen. www.fao.org

FAO (2002b) Information on Fisheries Management in the Republic of Yemen, February 2002. www.fao.org

FAO (2005) The FAO yearbook of fishery statistics – capture production 2003. FAO Fishery Information/Data and Statistics Unit, Rome

Herrera M, Lepere L (2005) Revised catch estimates for tuna and tuna-like species caught by artisanal boats in Yemen Indian Ocean Tuna Commitee (IOTC), Victoria, Seychelles, 15 p

Koehn D, Aklilu P (1999) Republic of Yemen – fisheries sector strategy note. World Bank, pp 1–23

Kotb MMA, Hanafy MH, Rirache H, Matsumura S, Al-Sofyani AA, Ahmed AG, Bawazir G, Al-Horani FA (2008) Status of coral reefs in the Red Sea and Gulf of Aden region. In: Wilkinson C (ed) Status of coral reefs of the world: 2008. Global Coral Reef Monitoring Network and Reef and Rainforest Research Centre, Townsville, pp 67–78

Lisac H (1971) Report to travel to Yemen Arabic and the People's Republic of Southern Yemen, (16 January–1 February 1972) FAO fish. Travel aide mem. FAO, Rome

Losse G (1973) Marine resources of the Yemen Arabic Republic. A report prepared for the the UNEP/FAO Food and Nutrition programme. FAO, Rome

Marshall NA, Marshall PA, Abdulla A, Rouphael T (2010) The links between resource dependency and attitude of commercial fishers to coral reef conservation in the Red Sea. AMBIO 39(4):305–313

MoFW (2004) Fisheries statistics book. Ministry of Fish Wealth, General Department for Planning and Statistic. 224 p

MoFW (2008) Catch statistics Yemen 1990–2007. Ministry of Fish Wealth, General Department for Planning and Statistic

MoFW (2012) Catch statistics Yemen 2008–2010. Ministry of Fish Wealth, General Department for Planning and Statistic

Morgan G (2006) Country review: Yemen. In: De Young C (ed) Review of the state of world marine capture fisheries management: Indian Ocean. FAO Fisheries Technical Paper. No. 488. Rome, pp 337–348

Pauly D (1998) Rationale for reconstructing catch time series. Bulletin. EC Fisheries Cooperation/Bulletin. CE Cooperation Pêche. Brussels 11(2):4–10

PERSGA (2001) Strategic action programme for the Red Sea and Gulf of Aden. Regional Organization for the conservation of the environment of the Red Sea and Gulf of Aden. PERSGA, Jeddah, pp 149–180

Pitcher TJ, Watson R, Forrest R, Valtýsson HP, Guénette S (2002) Estimating illegal and unreported catches from marine ecosystems: a basis for change. Fish Fish 3(4):317–339

Pramod G, Pitcher TJ, Pearce J, Agnew D (2008) Sources of information supporting estimates of unreported fishery catches (IUU) for 59 countries and the high seas fisheries. Fisheries Centre Research Reports, Vol 16(14), Fisheries Centre, University of British Columbia, Vancouver, 243 p

Purcell SW, Gossuin H, Agudo NS (2009) Changes in weight and length of sea cucumbers during conversion to processed bêche-de-mer: filling gaps for some exploited tropical species. SPC Bêche-de-mer Inf Bull 29:3–6

Sanders MJ, Morgan GR (1989) Review of the fisheries resources of the Red Sea and Gulf of Aden. FAO, Rome, 138 p

Tesfamichael D (2012) Assessment of the Red Sea ecosystem with emphasis on fisheries. PhD thesis, University of British Columbia, Resource Management and Environmental Studies, Vancouver, 241 p

Tesfamichael D, Pauly D (2011) Learning from the past for future policy: approaches to time-series catch data reconstruction. West Indian Ocean J Mar Sci 10(2):99–106

Tesfamichael D, Pitcher TJ (2006) Multidisciplinary evaluation of the sustainability of Red Sea fisheries using rapfish. Fish Res 78(2–3):227–235

Tesfamichael D, Pitcher TJ (2007) Estimating the unreported catch of Eritrean Red Sea fisheries. Afr J Mar Sci 29(1):55–63

Tesfamichael D, Pitcher TJ, Pauly D (2014) Assessing changes in fisheries using Fishers' knowledge to generate long time series of catch rates: a case study from the Red Sea. Ecol Soc 19(1):18

Torday P (2008) Salmon fishing in the Yemen. Harcourt, Orlando

Turak E, Brodie J, DeVantier L (2007) Reef-building corals and coral communities of the Yemen Red Sea. Fauna Arabia 23:1–40

Walczak P (1977) The Yemen Arab Republic. A study of the marine resources of the Yemen Arab Republic. A report prepared for the Fisheries Development Project. FAO, Rome, 67 p

Walczak P, Gudmundsson J (1975) Yemen Arab Republic – summary of fish stock assessment work. FAO, Rome, 14 p

Zeller D, Pauly D (eds) (2007) Reconstruction of marine fisheries catches for key countries and regions (1950–2005). Fisheries Centre Research Reports, Vol 15(2), Fisheries Centre, University of British Columbia, Vancouver, 163 p

Saudi Arabia

6

Dawit Tesfamichael and Peter Rossing

Abstract

Saudi Arabia is the largest country in the Arabian Peninsula with access to both the Persian Gulf and the Red Sea, of which it represents most of the east coast. Despite the Saudi Arabian coastline in the Red Sea being three times longer than its Gulf coast, Saudi catches from both coasts are similar. The catches of Saudi Arabian fisheries in the Red Sea are presented from 1950 to 2010, based on data from various sources. This reconstruction was conducted separately for each fishery sector: artisanal, subsistence, industrial and recreational. The total catch of each sector was further divided into its component species or groups of species. The catch was low at the beginning of 1950s, about 7,000 t·year^{-1}, and grew only slowly. The major change in total Saudi Arabian catch occurred at the beginning of the 1980s, with the massive motorization of artisanal boats and the beginning of industrial fisheries. Peak catch, i.e., about 50,000 t·year^{-1} occurred in the mid-1990s, after which catches decreased to about 40,000 t·year^{-1} at the end of the 2000s. The artisanal fishery had the highest contribution to the total catch (64 %), followed by the industrial (23 %), subsistence (10 %) and recreational fisheries (3 %). While a large number of taxa were identified in the catch, few groups were dominant. The reconstructed catches were compared with the data Saudi Arabia reported to the Food and Agriculture Organization of the United Nations (FAO), and was found to be 1.5 times the catch reported by FAO on behalf of Saudi Arabia from 1950 to 2010. The major discrepancy occurred following the mid-1980s, because of the industrial fishery, which became very active then, and whose substantial discards remain unreported. The procedures and assumptions used here are clearly stated, because they may be useful for further research on specific aspects of the fishery, and to improve the catch time series presented herein.

Keywords

Catch time series • Fisheries • Catch composition • Mis-reported catch • National wealth

Introduction

Saudi Arabia occupies 80 % of the Arabian Peninsula, and has coastlines on both the Red Sea and the Persian Gulf (also known as Arabian Gulf), with the former more than three times as long as the latter. People in the coastal areas of Saudi Arabia have been fishing in both seas since ancient times. In fact, in the past, almost all coastal communities derived their entire livelihood from fishing (Neve and Al-Aiidy 1973). However, the development of fishing was

D. Tesfamichael (✉)
Fisheries Centre, University of British Columbia, Vancouver, BC, Canada

Department of Marine Sciences, University of Asmara, Asmara, Eritrea
e-mail: dawittes@gmail.com

P. Rossing
Fisheries Centre, University of British Columbia, Vancouver, BC, Canada

D. Tesfamichael, D. Pauly (eds.), *The Red Sea Ecosystem and Fisheries*, Coral Reefs of the World 7,
DOI 10.1007/978-94-017-7435-2_6

Fig. 6.1 Map of the coast of Saudi Arabia in the Red Sea, shelf area and Exclusive Economic Zone (EEZ)

uneven, with the Red Sea providing only about half of the Saudi Arabian catch despite being about three times longer.

The aquaculture sector, both freshwater and marine, has grown from 2,700 t in 1995 to over 14,000 t in 2005; mariculture, which produced only 158 t in 1995 grew to over 12,000 t in 2005 (MAW 1996, 2006).[1] The freshwater sector emphasizes tilapia farming, while mariculture emphasizes shrimp. The total aquaculture output is still increasing and production in 2009 was about 26,000 t (Tim Huntington, Poseidon Aquatic Resource Management Ltd, pers. comm.). The rapid increase in aquaculture is in line with a government policy to supply the rapidly growing population with fish, given the stagnating yields of the capture fisheries (FAO 2003, and see also below).

Saudi Arabia has the largest shelf in the Red Sea, about 70,000 km^2. This shelf is narrow in the north, about 40 km, but broadens in the south (Fig. 6.1). In the past, most of the landings came from the productive southern grounds, particularly those adjacent to the Farasan Banks (Barrania et al. 1980; Sanders and Morgan 1989). More recently, about half of the catch originated from the southern part and the rest from the central and northern part (MAW 2006). Because the Saudi Arabian coast in the Red Sea has a narrow shelf and deep waters occur close to the shore, terrigenous nutrients are not recirculated and hence pelagic production and fish landings are relatively low. The Farasan Islands (not Farasan Banks) are probably the most productive grounds in the Saudi Arabian Red Sea because of the considerable run-off. In addition, the southern part (South of Jeddah) has well developed mangroves that serve as nursery areas for many reef-associated fish (Price et al. 1987).

The coastal area is hot and dry, with air temperatures reaching more than 45 °C in the hot season. There are as

[1]The reports with fishery statistics from the Saudi government that were available to us do not appear to have publication years. We have assigned them the year next to the latest data presented in the reports, e.g., data for 2005 would be published in 2006.

many as 88 settlements along the coast where fish are landed. Of these some of the largest are Yanbu, Jeddah and Jazan in the northern, central and southern part of the coast, respectively. The population density on the Red Sea coast of Saudi Arabia has traditionally been low, due to the dry and hot climate. The economy was based on agriculture, fishing (including pearling), nomadic herding, commerce, and since the ascent of Islam, catering to pilgrims to and from Mecca. The discovery (in 1949) and the exploitation of petroleum in Saudi Arabia profoundly modified the economy, with oil accounting for over 85 % of government's revenue in the early 1970s (Neve and Al-Aiidy 1973), reducing the traditional sectors to insignificance; thus, in 1998, fisheries accounted for only 0.3 % of GDP (Sakurai 1998). Easy access to air conditioners allowed major population centers to grow along the hot coast. The oil wealth also changed the demography of fisheries in that foreign fish workers, both on boats and in processing, mostly from India and Bangladesh, increased significantly starting in the mid-1980s and account for more than 60 % of the fishers (Sakurai 1998). The biggest fish market in Jeddah, the highly populated coastal city in the Red Sea, is locally called '*Bangala*', owing to the large number of Bangladeshi workers who are involved in fishing, trading, fish cleaning and other activities (Fig. 6.2). Most of the local people prefer the meat of land animals (goat, mutton, beef, chicken and camel) over seafood. On the other hand, the rather large expatriate community has a high seafood consumption, and thus Saudi Arabia imported 58,300 t, half of its seafood supply, in 1996 (Sakurai 1998) and 112,683 t, about two third of total supply, in 2003 (FAO 2003). Saudi Arabia exported a small amount of seafood, about 2,000 t in the 1990s and early 2000s, of which shrimp accounts for about 20 %.

Fig. 6.2 Auction at the fish market in Jeddah, Saudi Arabia (Photo: Dawit Tesfamichael)

Establishing a time series of catches is a crucial starting point to understand the impact of fishing in the ecosystem. In this chapter, the Saudi Arabian fishery catches in its Red Sea Exclusive Economic Zone (EEZ) are presented from 1950 to 2010, i.e., an attempt is made to account for all catches of all fisheries including recreational fisheries, the discards of industrial trawlers and other catches usually ignored in official fisheries statistics (see e.g., Zeller et al. 2011). The fisheries are reviewed and the catch standardized by the major fishery sectors in order to establish a long time series data. The fishery is divided into four sectors: artisanal, subsistence, industrial and recreational. The catches are further disaggregated into taxonomic groups. First the coral ecosystems of Saudi Arabian coast in the Red Sea are briefly reviewed. Then for each fishery sector, general introduction is given, followed by the catches. The sources of the catch reconstruction and methods used are presented at the end.

Coral Reef Ecosystems

Saudi Arabia has the largest coral reefs area in the Red Sea covering 6,660 km^2 and extending from the Gulf of Aqaba in the north to the Farasan Islands in the south (Bruckner et al. 2011b). There are diverse coral reef structures along the coast: mainland fringing, island fringing, platform patch, barrier and ridge reefs (Bruckner et al. 2011a; Al-Sofyani et al. 2014). The ridge reefs are characteristics of the Red Sea. For a long time, information and knowledge about the coral reefs of Saudi Arabia were sparse (Berumen et al. 2013); however, in the last decade, a lot of research and publications are coming out of Saudi Arabian universities and research organizations, especially the Prince Khalid bin Sultan Living Oceans Foundation and the Red Sea Research Centre, King Abdullah University of Science and Technology (KAUST).

The inner shelf is characterized by shallow fringing reefs, which run for few meters to more than half a kilometer. The reefs are interrupted at the mouths of valleys or *wadi* beds, which created natural deep water harbors (Neve and Al-Aiidy 1973). Bay (locally called *sharm*) areas also support extensive coral reef growth and are characteristic of the Red Sea. The reefs are more developed, with higher live coral cover, in the north and central region (Kotb et al. 2008; Bruckner et al. 2011a; Al-Sofyani et al. 2014), while in the south the water is more turbid due to run offs from the mountainous areas and influx of nutrient rich water from the Indian Ocean through the narrow strait of Bab-al-Mandeb. There is an almost continuous band of coral reefs along the Saudi coast from north of Jeddah to the Gulf of Aqaba. Also, there are well-developed off-shore reef systems around Al Wajh, Yanbu and the Farasan Islands (Bruckner et al. 2011a; Saleh and Hariri 2012). Beyond the narrow shelf, depth increases abruptly; however, deep water corals are found up to depths of 760 m off the coast of Saudi Arabia (Qurban et al. 2014). The ecosystem around Farasan islands is very important ecologically and contains some of the most diverse corals species (Gladstone 2000; Riegl et al. 2012).

The distribution and abundance of the coral reef-associated fishes follow a pattern similar to the coral reef distribution, with higher counts (density) in the northern than southern parts (Kotb et al. 2008). Environmental gradients are good predictors of the population structure of coral reef communities (Nanninga et al. 2014). The density of herbivorous fish was found to be higher in areas with less live coral cover than in reefs with higher live coral cover along the Saudi coast. However, the general density was low (Khalil et al. 2013). Annual aggregations of parrotfish (*Hipposcarus harid*) occur around Farasan islands, and a fishing event called 'Hareed Festival' is held every year (Gladstone 1996; Spaet 2013).

The overall health of the coral reefs off the coast of Saudi Arabia in the Red Sea is good, especially in areas that are not inhabited or with very low human density. Most impacted coral reefs are the ones near urban centers, where they are subject to land reclamation, dredging, urban run-off, pollution, anchorage and littering (Kotb et al. 2008; Saleh and Hariri 2012). The impact of climate change has been recorded in the coral reefs of Saudi Arabia (Baker et al. 2004). Increased sea surface temperature resulted in decreased coral reef growth (Cantin et al. 2010), decline of large corals in the last two decades (Riegl et al. 2012) and bleaching (Kotb et al. 2008; Furby et al. 2013). However, as temperatures subsequently decreased, most of the coral reefs recovered from the stress of higher temperature (Cantin et al. 2010). The corals in the Red Sea are unique in that they tolerate high temperature. Some claim that this makes the Red Sea a possible refuge for the survival of corals in case of mass mortality in other areas due to the rise in temperature from climate change (Riegl and Piller 2003), while others claim that Red Sea coral reefs are vulnerable to increasing pressure, as are coral reefs elsewhere (Furby et al. 2013). The main large scale management scheme to conserve coral reef ecosystem in Saudi Arabia is the marine protected area (MPA) in the Farasan islands (Gladstone et al. 2003). There are other smaller MPAs; however, they are not properly managed (Berumen et al. 2013).

Fisheries

In the 1950s, the fisheries along the Red Sea coast of Saudi Arabia were mainly artisanal and used small boats, which started to be motorized only at the end of the decade. In the past, most of the supplies for fishing such as ice-making machines, engines, spare parts, gears and workshops were found only in the major settlements (Neve and Al-Aiidy 1973). The industrial fishery, experimental at first, took off at the beginning of the 1980s (Ferrer 1958; Sanders and Morgan 1989). However, due to the narrow shelf area, and reefs, there are no large areas with soft bottom, able to support a large number of trawlers, except around Jazan where most of the trawling occurs.

Up to 1990, the administration of fisheries was placed under the Animal Husbandry Department of the Ministry of Agriculture and Water. In 1991, recognizing the importance of fishery in national food security policy, the government created a Deputy Minister for Fisheries Affairs under the Ministry of Agriculture and Water, tasked to manage the fishery resources of the country. The deputy minister, in 1993, created three departments each with specific responsibilities: Marine Fisheries Department, Aquaculture Department and Marine Protection Department (Sakurai 1998; FAO 2003). Although there are institutions and regulations to manage the resource, they are not usually enforced. For example, shark populations are declining from heavy fishing despite a royal decree prohibiting any shark fishery (Spaet and Berumen 2015) and spear-fishing continues, although it is banned (Oakley 1984). The lack of enforcement has compromised the sustainability of the fishery, especially since the early 1990s where all artisanal boats were motorized and the industrial fishery was in full swing (Jin et al. 2012; Spaet and Berumen 2015).

Artisanal Fishery

The artisanal fishery, sometimes referred to as 'traditional' fishery in Saudi Arabia, is conducted from small boats lacking sophisticated technologies such as winch, fish finder or communication equipment (MAW 2000). The boats range from 5 to 20 m, but most are 6–9 m, and only about 10 % exceed 9 m. In the past, traditional boats (locally called *sambuks* and *huris*), were exclusively made of wood and used to have sails (El-Saby and Farina 1954). The fishery was not highly developed, but the economy of coastal communities depended heavily on fishing (Tesfamichael and Pitcher 2006). Part of their catch went to local consumption, and another part was dried or salted to be sold or bartered in exchange for other necessities (El-Saby and Farina 1954). Dried and salted fish were the only fish product to be available inland (Neve and Al-Aiidy 1973). Fishers were part time, and often would work from September to January in the agriculture sector (Gilberg 1966). Fishing was done only during the day, because it was difficult to navigate the coral reefs at night. Small boats performed day trips, but larger boats performed trips of up to 7 days (Kedidi et al. 1984); in some cases, even trips of up to 14 days occurred (Neve and Al-Aiidy 1973). Except for a few large boats, ice was not carried onboard, and fish were sold fresh bundled in strings (Barrania et al. 1980). Crew size ranged 1–4 per boat (Kedidi et al. 1984). The artisanal fishery changed with the advent of oil wealth: in 1983, fiberglass boats were introduced (Sanders and Morgan 1989) and by the mid-1990s, they accounted for about 80 %, with the majority having outboard engines (Sakurai 1998). Motorization of boats started at the end of 1950s (Ferrer 1958) and was completed by the beginning of 1990s (Sakurai 1998). The dominant fishing gears are gillnet, and hook and line, which account for more than 90 % of the artisanal catch, while trolling and trap are rarely used (MAW 2000). Handlining is dominant in the northern coast, targeting coral reef fishes, while gillnets are used in the southern coast, targeting pelagic fishes (Fig. 6.3). Fishing licenses for the traditional fishery are issued by representatives of the Fisheries Out-Stations at the district level (MAW 2000). According to the fishery administration, traditional fishers are Saudi citizens who work on board one of their own traditional fishing boat; however, a traditional fisher can own up to four fishing boats (MAW 2000). Since the 1980s, most Saudi fishers are not directly involved in fishing, sometimes referred as 'investor fishers'. They own the boats, but the fishing is done by foreign workers through an arrangement where the owner gets half of the profit and the crew shares the other half. To limit effort, which had increased tremendously, and led to stagnating catches, the issuing of new licenses stopped in the mid-1990s (Sakurai 1998).

Subsistence Fishery

For cultural reasons, women are not involved in fishing in Saudi Arabia, and the overwhelming bulk of the subsistence catch consists of what the fishers consume or give to their families and friends to sustain their communities. These catches are not reported at all because the catch is given away before the official recording, if it exists. In the coastal areas, community members, mainly children, are also involved in catching fish for consumption by the family. The subsistence fishery catch can be a good proportion of the total catch: based on interviews with Red Sea fishers, it can amount up to half of the total catch. The trend is declining, however.

Industrial Fishery

The history of Saudi Arabian industrial fishery in Red Sea is tied to the history of the Saudi Fishing Company. Experimental fishing by chartered boats, mainly trawling for shrimps, started in 1952 (Ferrer 1958). The Saudi Fishing Company did not have its own boats until 1954, and it became inactive in 1961, due to engine failures, lack of maintenance and lack of profits (Gilberg 1966). The company re-established itself again in 1981, when shrimp trawling started with Thai vessels operating under contract (Sanders and Morgan 1989). Eventually, the Saudi Fishing Company grew, and owned and operated its own vessels. Currently, the company leased its operations to other companies. Its main base in the Red Sea is Jazan, in the south, close to the border with Yemen, and its main fishing grounds are located off the Farasan Islands, and near Al-Qunfudhah and Al-Khoriebah. Some industrial fishing vessels operate outside the Saudi EEZ, in the neighboring countries or interna-

Fig. 6.3 A fish trap on a coral reef, Saudi Arabia (Photo: Andrew Bruckner)

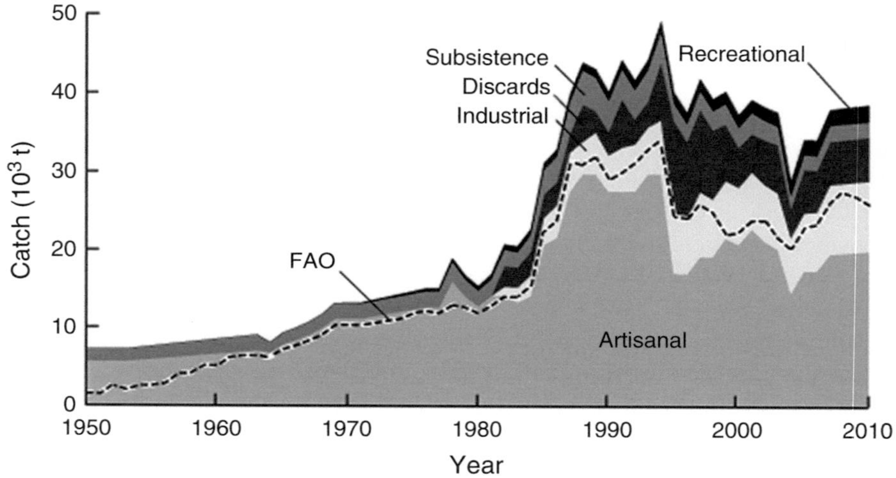

Fig. 6.4 Reconstructed catch of Saudi Arabia in the Red Sea by sectors and its comparison to the data reported by FAO on behalf of Saudi Arabia

tional waters (Sakurai 1998). Although the industrial fishery consists mainly of trawling (hence sometimes referred as 'trawl fishery'), the vessels are also involved in purse seining (MAW 2000). The issuance of fishing license for industrial fishing is a sole prerogative of the Minister of Agriculture and Water or the Deputy Minister for Fisheries Affairs in the same ministry (MAW 2000).

Recreational Fishery

Recreational fishing was not commonly practiced in Saudi Arabia. However, once the country acquired oil wealth, and saw numerous expatriates practice the sport, recreational fishery or 'pleasure fishing', as it is called locally, took off. In the mid to late 1990s, there were 2,126–2,445 boats registered as 'recreational' (Sakurai 1998; MAW 2000). Most of the fishing happens on the weekend (Thursday and Friday), using handline. There are reports of trolling, spear-fishing and longline fishing (MAW 2000). Recreational fishing with net is prohibited (Sakurai 1998). Spear-fishing is also prohibited by law; however, it is a common occurrence due to lack of enforcement (Oakley 1984).

Fisheries Catches

The Saudi fishery in its Red Sea EEZ had, for a long time, caught less than 10,000 t·year^{-1}. The fishery was dominated by traditional fishers who used small sailing boats and simple gears (El-Saby and Farina 1954; Neve and Al-Aiidy 1973; Tesfamichael and Pitcher 2006). The catches were largely for subsistence and very localized markets. This changed rapidly starting at the beginning of 1980s, when the catch increased drastically to a peak of 50,000 t in 1994, due to the motorization and the introduction of fast fibreglass boats in the artisanal fisheries (Sanders and Morgan 1989; Sakurai 1998). Since then, the total catch declined but remained higher than in earlier years (Fig. 6.4). The artisanal and subsistence fisheries were the only ones operating along the Saudi coast of the Red Sea until the 1980s, when the industrial and recreational fisheries started, and contributed to a drastic increase of the total catch (Sanders and Morgan 1989). The catch of the artisanal fishery decreased dramatically in 1995, but this was compensated for by an increase of the industrial catch. The slight decrease from the peak is taken as a sign of resource depletion by the fisheries administration and precautionary approaches are being considered; for example new licenses are not issued (MAW 2000). Overall, artisanal fishery contributed the highest to the total catch from 1950 to 2010 (64 %), followed by industrial fishery (12 %), discards (11 %), subsistence catch (10 %) and recreational fishery (3 %).

The reported catch (the reconstructed catch accounted in the FAO data) had the highest contribution to the total catch (54 %). Discarded catches (11 %), appeared in 1982, when the Saudi Red Sea trawl fishery started; unreported landed catch accounted for 35 %. The reconstructed catch and the catch reported by FAO on behalf of Saudi Arabia assigned to the Red Sea were close to each other from the 1960s to mid-1980s. The major differences are in the 1950s and from the mid-1980s on. In the 1950s, there was no recording system. The fishery was traditional and there was no regulatory body; hence its catches were not properly reported. The main difference after the mid-1980s is the discarded catch of the industrial fishery, which is included in the reconstructed catch, but missing in the data supplied to FAO. As in many other sub-tropical fisheries, the Saudi Red Sea fishery catch consisted a very large number of taxonomic groups. However, only 5 taxa made up 50 % of the total catch from 1950 to 2010: Spanish mackerel (17 %), emperors (9 %), jacks (9 %), groupers (8 %) and pony fish (7 %). Of these, 4 are caught by artisanal fisheries. The highest contribution of the industrial fishery were ponyfishes (Family Leiognathidae,

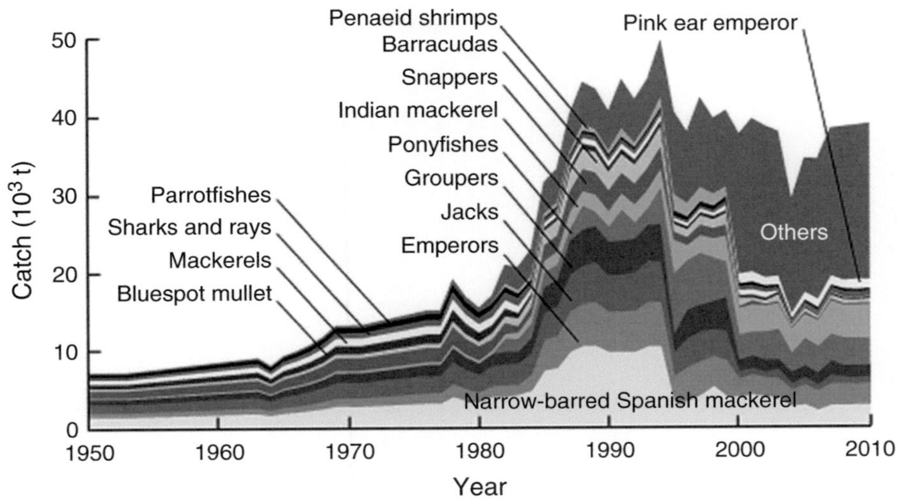

Fig. 6.5 Composition of the total catch of Saudi Arabia's fisheries in the Red Sea

which are systematically discarded) and which ranked fifth in the total catch (Fig. 6.5).

The artisanal fishery by its sheer size dominated the pattern of the total catch. Its catches increased rapidly in the early 1980s (Fig. 6.6a), which is thus reflected in the total catch. The arisanal fishery then declined from its peak in the mid-1990s. However, the magnitude of the decline is moderated to a lower level in the total catch by the increase in the catch of the industrial fishery, which started in 1982 (Fig. 6.6c). The highly sought-after narrow-barred Spanish mackerel has the highest contribution (22 %) in the catch composition of the artisanal fishery. The second highest is jacks (11 %), followed by emperors and groupers each with around 10 % of the total artisanal catch (Fig. 6.6a). These four taxonomic groups account for about 54 % of the total catch. The subsistance fishery should have a catch composition similar to that of the artisanal fishery, because the main source of subsistence catch is artisanal fishery, hence the former was calculated as a ratio of the latter (Fig. 6.6b). For the retained catch of the industrial fishery, Indian mackerel is the dominant taxon with 29 % Shrimps, the most important catch of the industrial fishery is second with 9 % and jacks is third with 8 %. The three, together with cuttlefish (6 %) account for 53 % of the retained catch of the industrial fishery (Fig. 6.6c). For the discarded catch of the industrial fishery, ponyfishes is by far the dominant taxon with 63 % (Fig. 6.6d). The recreational fishery exhibited a continuous increase from its start in 1970 and emperors have the highest contribution to the total catch at 40 % (Fig. 6.6e).

This long time series of reconstructed Saudi Arabian catches in the Red Sea, the first of its kind, is very informative, and should be useful in the assessment and management of the fisheries (Tesfamichael 2012). Unreported catches are estimated for the different sectors given explicit assumptions, on the basis of the best information available. Some of the procedures and assumptions used here will certainly require correcting when new information is available. In the meantime, this is provided in the hope that it will be found useful.

Sources and Methods

The different sectors of Saudi Arabian fisheries in the Red Sea are different in the fish they target, gear they operate, target market and overall structure. Hence, they are treated separately in the sources and methods we used to reconstruct the catch. The description given here is brief, as a way giving the reader an idea as to how it was done. A spatial distribution of the catch is given at www.seaaroundus.org.

Artisanal Fishery

The earliest catch estimate for Saudi Arabia's artisanal fisheries in the Red Sea was from a partial survey of the fishers from which the average catch was calculated and multiplied by the total number of fishers in order to estimate the total catch for 1953, i.e., 3,000–5,000 t (El-Saby and Farina 1954), whose mid-range was used here. The same amount was assumed from 1950 to 1952. After 1952, the next estimate was for 1963, from a survey which led to a total catch estimate of 5,000 t (Gilberg 1966). From 1965 to 1971, total catches were available for the whole of Saudi Arabia, i.e., for the Red Sea and Persian Gulf coasts, where it is stated that 'about half' of the catch came from the Red Sea (Neve and Al-Aiidy 1973). Another report gave a more precise estimate, 53 % (Barrania et al. 1980), which was used to calculate the Red Sea catch. From 1976 to 1986, catches from the Red Sea were estimated by halving the total Saudi Arabian reported catch (Sanders and Morgan 1989). Estimates for the Red Sea only were available for 1976 and 1977 (Peacock

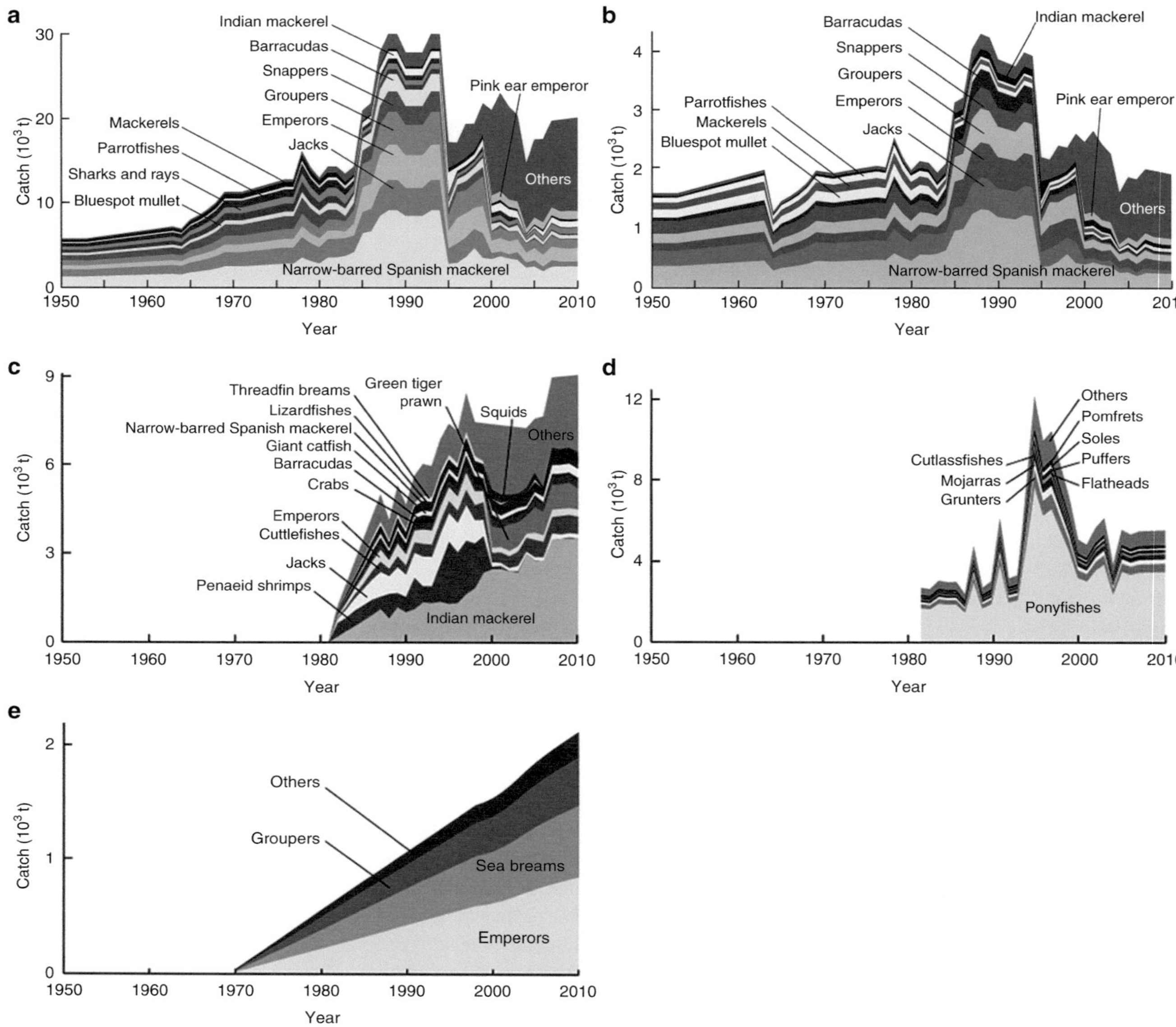

Fig. 6.6 Catch composition of the (**a**) artisanal (**b**) subsistence and (**c**) industrial retained (**d**) industrial discarded and (**e**) recreational fisheries of Saudi Arabia in the Red Sea from 1950 to 2010

1978), 1978 and 1980 (Barrania et al. 1980) and 1983 (Chakraborty 1984). The data from Sanders and Morgan (1989) were used for years for which there were no other estimates, after adjusting it using weighted mean for the years where there were estimates for the Red Sea only. This resulted in that the Red Sea catch was only 86 % of that reported by Sanders and Morgan (1989).

Starting 1987, data availability improved, and the following sources were used: 1987–1994 (Sakurai 1998), 1995–1998 (MAW 2000), 2004–2007 (MAW 2008) and 2009 (FSDP 2011). Also, the number of fishers data were available from 1991 to 2007 (MAW 2008), which was used to calculate the catch per fisher from 1991 to 2007. A linear regression was fitted to the data, and used to estimate the catch per fisher for 1999–2003, which was multiplied by the number of fishers to estimate the total catch. Interpolation of the total catch was used to fill in the data gaps in 1954–1962, 1964, 1972–1975, 2008 and 2010.[2]

Part of the artisanal fishery catch is not reported at all. The major contributions to unreported catch in artisanal fisheries are fish landed outside the major landing sites where data recording occurs, if any. The second contributions to unreported catch occurs when boats land their catch at the major landing sites, but before their catch is recorded, part of the

[2]After this chapter was drafted, Saudi Arabia published fishery statistics up to 2010. Our estimates are very similar to the official published values.

fish is sold in the non-formal market before recording. These catches are different from catch that is allocated for subsistence, which is treated separately below. It was difficult to estimate the unreported catch of the artisanal fishery, as we were unable to locate pertinent studies for Saudi Arabia. Thus, we used qualitative information that the first author collected during a field trip to the neighboring Red Sea countries of Sudan, Eritrea and Yemen (Tesfamichael et al. 2014). The market and data recording infrastructure improved with the motorization of artisanal boats starting the beginning of 1960s facilitated by the government. This knowledge, jointly with the general pattern of the fishery, was used to estimate the unreported catch. The reconstructed time series of reported Saudi artisanal catch can be roughly divided into three phases: low catch levels until 1963, a slight increase until 1984, and rapid increase after that, followed by decline. Thus, we assumed conservative unreported catch to be 30 % of the total catch from 1950 to 1963, 20 % from 1964 to 1984 and 10 % from 1985 to 2010. In addition, fishers from Jordan fished in Saudi waters from 1950 to 1984 (see Chap. 7). The catches were very small (maximum of 100 t) compared to the Saudi catch (minimum of 5,700 t); nevertheless, they were included in the unreported catch.

Catch composition data for Saudi Arabia's artisanal fishery were available for some years, the earliest being 1980. Data points were often only available for the major target species, reflecting the over-aggregated nature of the reported data. From 1950 to 1998, where the catch composition was highly aggregated, we created a comprehensive catch composition list consisting of 20 major taxa and the category 'others' for the species not identified. To fill the species composition matrix for all the years, broad taxonomic groups were disaggregated using data from other years. Indian mackerel and Spanish mackerel ('kingfish') were reported as 'Mackerel' for 1980 and 1983 (Barrania et al. 1980; Chakraborty 1984). This category was split based on disaggregated ratio reported for 1985, i.e., kingfish 91 % and Indian mackerel 9 % (MAW 1986). The same source was used to split parrotfish and surgeonfish, with contributions of 98 % and 2 %, respectively. Tunas were reported separately for most of the years, but not for all years. For 1985, their catch was calculated by splitting the reported 'others' using the proportion of tuna relative to 'others' for 1983. The catch ratio of tunas for 1995 (MAW 1996) was assumed to be similar to 1996 (MAW 1998; Sakurai 1998). Catch rates for cobia, wrasses, sea breams, rabbitfish, goatfish and cutlassfishes were not available for 1980 and 1983. We estimated their proportions from 'others' using ratios reported for 1985 (MAW 1986). The final result is a standardized catch composition.

The most detailed (disaggregated) catch composition available was from 2004 to 2007 (MAW 2008). The catch was divided into 40 taxonomic groups and 'others'. The 40 groups included all the 20 groups from 1950 to 1998 and 20 new ones. The 'others' from 1950 to 1998 were further divided to the 20 new groups in 2004–2007 using their average ratios. For the years without catch composition, it was estimated by interpolating between the closest years for which catch compositions data were available. From 1950 to 1979, the catch composition was estimated using the earliest available ratio (1980). From 1986 to 1994, the phase of the Saudi artisanal fishery characterized by high catches, the only composition available was for 1985, and it was used for the whole period. Species composition data were also missing for the period 1999–2003, the mean of 1996–1998 and 2004–2007 (i.e., similar phase of the fishery), was used. From 2008 to 2010, the catch composition of 2007 was used. The final catch composition of the artisanal fishery consisted of many taxa but for clarity of presentation of the result only the major taxa are included and the minor taxa are lumped together in the group 'others'.

Subsistence Fishery

The amount of the subsistence catch was estimated based on information obtained through interviews with fishers in neighboring countries of Yemen, Eritrea and Sudan (Tesfamichael et al. 2014). The amounts of fish allocated for subsistence changed over time; in the past: when fishing was mainly for subsistence, catch rates were high, and there was lack of extensive marketing, fishers reported to give about half of their catch. But later, as their catch became more valuable and their catch rates started to decrease, fishers gave less of their catch away. In order to capture the different phases of the fishery, we used the trend of the total reported catch of the artisanal fishery. The Saudi reconstructed reported artisanal catch can be divided into three phases: low catch level until 1963, i.e., the pre-motorization phase; a slight increase until 1984, and rapid increase after that, followed by decline. We assumed the subsistence fishery to be 30 % of the artisanal catch until 1963. This is a conservative estimate, because fishers told us in the years before motorization gained momentum, they used to give up to 50 % of their catch away. For 1964, the subsistence percentage was reduced to 20 %, and for 2010 only 10 %. From 1965 to 2009, the percentages were interpolated. Before these percentages were applied, some taxa not usually given away were eliminated. This included items usually targeted for the export market, i.e., shark fished for their fin and many invertebrates such as shrimp, crab and lobster. Traditionally most of these taxa were not consumed locally and their consumption was introduced by foreigner (mainly European) visitors to the region. Nowadays, they are consumed by the communities mainly in the affluent bigger urban centers, yet still not given freely to family and friends. The local names of most of these taxa are based on European names, rather than Arabic as is the case for most of the fish (see Chap. 10). The remaining taxa of the artisanal fishery constitute the composition of the subsistence fishery. Thus, our definition of subsistence fishery considers only the portion of the catch that is not sold by the artisanal fishers. We ignore classifications, wherein the entire artisanal fishery, especially before motorization, was considered a 'subsistence fishery'.

Industrial Fishery

An exploratory trawl fishing conducted for the Saudi Fishing Company in 1952 initiated industrial fishing in the Saudi Arabian waters of the Red Sea (El-Saby and Farina 1954; Ferrer 1958). The company became inactive and closed in 1961 (Gilberg 1966), but re-established itself in 1980 and carried out trial trawling until it started commercial operation in 1982 in the southern part of the Saudi Red Sea coast around Jazan (Sanders and Morgan 1989). The sporadic experimental trawling in the early 1950s, which yielded negligible catches, are not included in this reconstruction. Rather, it starts in 1982, with a first substantial catch, of which 466 t of shrimp and 703 t of fish were retained. The dataset is not continuous after 1982, but catch data were available for some years: 1987–1996 (Sakurai 1998), 1997–1998 (MAW 2000), 2004–2007 (MAW 2008), and 2009 (FSDP 2011) which were used as total reported catch for the respective years. For the periods where catch was not reported 1983–1986, 1999–2003, 2008 and 2010, interpolation was used to fill in the gaps. The interpolation for the earlier period reflects the rapid expansion of the industrial fishery in Saudi Arabian Red Sea (Sanders and Morgan 1989), while the latter period depicts a somewhat stable high catch.

For the years data were available, the sources reported total (both fish and shrimp) catch or usually the commercially lucrative shrimp only. To reconstruct the catch composition, first a continuous shrimp catch was established from 1982 to 2010. Shrimp data were available for 1982 (Sanders and Morgan 1989) and 1987–2007 (MAW 2008). The missing data from 1983 to 1986 were interpolated between the 1982 and 1987 data and for 2008–2010 the ratio of shrimp to the total catch for 2007 (11 %) was applied for the total reconstructed shrimp data. Then, the shrimp catches were subtracted from the total reconstructed to determine the total catch of fishes, i.e., non-shrimp retained catch. This procedure was used to get the best possible species distribution as available shrimp catches went back to the early 1980s, whereas the retained fish was available only for 1997 and 1998 (MAW 2000) and 2004–2007 (MAW 2008). The catch compositions of 1997 and 1998 were highly aggregated in 5 major groups and more than 30 % in the category 'others'. However, the data from 2004 to 2007 were more detailed with 31 groups and 'others'. The category 'others' for 1997 and 1998 was further disaggregated using the average of 2004–2007. From 1982 to 1996, the catch composition of 1997 was used, from 1999 to 2003, the average of 1998 and 2004 was used, and for 2008–2010 the composition of 2007 was used. The shrimp catch was added to the non-shrimp catch to establish the composition of the industrial total catch of Saudi Arabia in the Red Sea. *Penaeus semisulcatus* accounts for more than 90 % of the shrimp catch (Sakurai 1998).

Unlike the artisanal fishery, where the main source of unreported catch is the catch landed without being recorded, the main source of unreported catch for the industrial fishery of Saudi Arabia is discarding. Data of retained catch were collected using logbook (MAW 2000). Fish thrown overboard (i.e., non-target species, or the young of target species) usually have a low market value and are not recorded in the logbook at all. In the Saudi fishery recording system, the catches of industrial fisheries are not separated by gear, i.e., only the total is given, and usually, the same vessels use both trawling and purse seining gears (MAW 2000). Estimating the unreported catch from the total industrial catch would be misleading, because the source of discard is trawling and the proportion of trawl and purse seine catch is not constant. Using the percentage of shrimp and Indian mackerel, the main catch of trawl and purse seine, respectively, from the total catch for the years data were available for both; we found that they have inverse relationship, i.e., when the share of one increases that of the other decreases indicating the relative share of each gear. Because of their high value, shrimp are the main target of trawling and the main factor in discarding fish. Thus, given that shrimp catch data are available, they were used as a basis to estimate the amount of unreported catch. Estimates of retained and discarded catches were available from experimental fishing in 1952/53, where it was found that for 750 kg of retained catch, 1.5–2.0 t were discarded (Ferrer 1958). Taking the mid-range, discards were 2.3 times the retained catch or 70 % of the total catch. However, these values cannot be used directly as the retained catch (despite the main target being shrimp) contains other species and our basis for estimating unreported catch is shrimp. The earliest, which is also the highest, percentage of shrimp in the retained catch was 40 % for 1982 (Sanders and Morgan 1989), and is used to calculate the amount of shrimp in the retained catch of the experimental fishing. Using these ratios, the discard was calculated to be 5.8 times the shrimp catch. This is a conservative estimate because shrimp accounting 40 % of retained catch is high and for other countries in the Red Sea discard can be more than 6 times the amount of shrimp catch (see Chap. 4).

The species composition of discarded catch is different from that of the retained catch. Data are not available on the composition of discards from the Saudi Arabian Red Sea trawl fishery. However, such data were available from the Yemeni part of the Red Sea (Walczak 1977), which was used for Saudi Arabia with some modification. Jacks, lizard fish, breams, grunters, catfish and barracuda were reported as discards in Walczak (1977). They were removed from the discard list of Saudi Arabian industrial fishery, as these species were believed to have been retained (Sanders and Morgan 1989). The remaining taxonomic groups of the discard were scaled to 100 %.

Recreational Fishery

Very little data were available for Saudi Arabia's recreational fishery catches. They have not traditionally been accounted for in reported fisheries statistics, possibly because their catch is very small compared with artisanal or industrial fisheries. The only catch estimate available was 1,500 t for 1998 (MAW 2000). This tonnage was used as anchor point. We assumed, in lack of other data, that recreational catches had been ongoing since the beginning of 1970s, the time the oil wealth started to have effect in the fishing sector. Many Saudi citizens started to buy fiberglass boats and hired foreigners to do the fishing (Sanders and Morgan 1989; Sakurai 1998) while they would go fishing for pleasure usually in the weekend. Thus, the catch was assumed to be zero until 1969 and interpolated between 1969 and the anchor in 1998. The population size of Saudi Arabia was used as a proxy to calculate the recreational fishery catch from 1998 to 2010 using the 1998 catch as anchor.

Recreational catches were reported to consist mostly of emperors, then sea breams followed by groupers (MAW 2000). We transformed this qualitative information (using 10 % steps between the ranked groups) into percent contributions, which yielded: emperors (Lethrinidae) = 40 %; sea breams (Sparidae) = 30 %; groupers (Serranidae) = 20 % and 'others' = 10 %.

Comparing Reconstructed Catch with the FAO Statistics

The reconstructed catch was compared to the catch reported in the United Nations Food and Agriculture Organization (FAO) database on behalf of Saudi Arabia. FAO uses broad statistical areas to geographically subdivide catches, and in the case of Saudi Arabia, both coasts fall in the same statistical area ('Western Indian Ocean'). Thus, the Saudi Arabian catch reported to FAO needed to be divided into the two coasts. This was done using reports that allocated ratios to the Red Sea and the Persian Gulf.

The earliest available Red Sea-to-Gulf ratio was for 1979 (Barrania et al. 1980), 1987–1998 (Sakurai 1998; MAW 2000) and 2000 (FAO 2003). For 2002, the Regional Commission for Fisheries (RECOFI 2009), which is active in the Gulf, reported the Gulf catch of Saudi Arabia, which in turn was used to calculate the percentage for the Red Sea. In more recent years, Saudi Arabia published annual fishery statistical reports, separate for each coast, which we relied on for 2004–2007 (MAW 2008). The latest year with data disaggregated between the two coasts was for 2009, and statistical data were presented separately for the artisanal and industrial sectors in the Red Sea and the artisanal sector in the Gulf (FSDP 2011). The industrial catch in the Gulf was calculated using the ratio for the two sectors in 2007, where industrial catch was 0.38 % of the artisanal catch.

For 1950–1960 and 1975–1978, the earliest available data (for 1979) were used. Using the 1979 Red Sea-Gulf ratio for 1961–1974 resulted in unreasonably very high Red Sea catches, thus a different approach was used for this period. The closest period with data that separates the Red Sea and Gulf catches was for 1987–1994. Thus, an average ratio was calculated for the total reconstructed Red Sea catch without the industrial discard to the FAO catch of Red Sea from 1987 to 1994 (Note that discarded catches are not reported to FAO at all). The result, that FAO Red Sea data were on average 92 % of the reconstructed catch, was used to calculate the FAO Red Sea catch for 1961–1974. Then the FAO Gulf catch was obtained by subtracting the Red Sea amount from the total Saudi catch in the FAO database. The same ratio was also used for the period 1980–1986. Although ratios for 1995–1996 were available, they resulted in the FAO Red Sea catch being slightly higher than the reconstructed catch, which is unrealistic given the pattern for the other years. Thus, the reconstructed catch (without industrial discard) is assumed to be equal to the FAO Red Sea catch. For 1999, 2001, 2003, 2008 and 2010, the ratios were interpolated from the neighbouring years.[3]

Once the FAO data for Saudi Arabia in the Red Sea were separated, they were compared with our reconstructed catch. The FAO data have more taxa (127) than the reconstructed catch, which is strange given that we used the Saudi official national and technical reports for our catch reconstruction. The large number of taxa started in 2000 when the country introduced an extensive data recording and reporting system. Most of the taxa that were included starting in 2000 have very low catch amounts and they were aggregated as 'miscellaneous' in the national reports we used. To make full use

[3]After the completion of the report, we came to realize that data pertaining to the Gulf only were available at FAO/RECOFI. Our analysis gave similar results. Also note that the FAO data are here used only for comparative purposes, and do not have any impact on our reconstruction proper.

of the additional information on catch composition in the FAO data, it was used to further disaggregate the reconstructed catch composition. First the distribution of the taxa were verified using FishBase (Froese and Pauly 2012) to check if each taxon was to be included in both the Red Sea and the Gulf or only in one of these water bodies. Then, for the taxa included in the FAO data, but not in the reconstruction, the ratios of the taxa in the FAO data were used to disaggregate the catch composition of the reconstructed data.

After the reconstructed and the corresponding FAO catch by taxa were tabulated, comparison was done at taxonomic level. The part of the reconstructed catch that is accounted in the FAO data is referred as 'reported catch' in our result. If the value of a taxon in reconstructed catch was higher than its value in the FAO data, then the difference was labeled as 'unreported catch'. If the FAO value for a taxon was higher than the reconstructed catch, the difference is 'over-reported' catch in the EEZ of Saudi Arabia in the Red Sea. Since, the comparison of the reconstructed catch with the FAO data has modified the catch composition of the reconstructed catch, the final ratios are not exactly what are reported in the methodological tables given in the above.

Acknowledgments We thank Tim Huntington of Poseidon Aquatic Resource Management Ltd for providing us with the data of Saudi Arabian fisheries and also for his constructive comments on our first draft. Our gratitude goes to Andrew Bruckner, Julie Spaet and Mohamed Gabr for their help with reports of Saudi Arabian fisheries and reviewing the chapter. This research was supported by the *Sea Around Us*, a scientific collaboration between the University of British Columbia and The Pew Charitable Trusts.

References

Al-Sofyani A, Marimuthu N, Wilson JJ (2014) A rapid assessment of scleractinian and non-scleractinian coral growth forms along the Saudi Arabian coast, Red Sea. J Ocean Univ China 13(2):243–248

Baker AC, Starger CJ, McClanahan TR, Glynn PW (2004) Coral reefs: Corals' adaptive response to climate change. Nature 430(7001):741

Barrania A, Bringi MR, Saleh M (1980) Socio-economic aspects of the Saudi Arabian fisheries in the Red Sea. FAO/UNDP Project for Development of Fisheries in Areas of the Red Sea and Gulf of Aden, Suez, 54 p

Berumen ML, Hoey AS, Bass WH, Bouwmeester J, Catania D, Cochran JEM, Khalil MT, Miyake S, Mughal MR, Spaet JLY, Saenz-Agudelo P (2013) The status of coral reef ecology research in the Red Sea. Coral Reefs 32(3):737–748

Bruckner A, Rowlands G, Riegl B, Purkis S, Williams A, Renaud P (2011a) Khaled bin Sultan living oceans foundation Atlas of Saudi Arabian Red Sea marine habitats. Panoramic Press, Phoenix

Bruckner AW, Faisal M, Alnazry HH (2011b) A paradigm shift for fisheries management to enhance recovery, resilience, and sustainability of coral reef ecosystems in the Red Sea. Sustainable Fisheries: Multi-Level Approaches to a Global Problem. American Fisheries Society, Maryland, pp 85–111

Cantin NE, Cohen AL, Karnauskas KB, Tarrant AM, McCorkle DC (2010) Ocean warming slows coral growth in the central Red Sea. Science 329(5989):322–325

Chakraborty D (1984) Fish landings on the Red Sea and Gulf of Aden coast of the member countries: a preliminary estimate. Development of Fisheries in Areas of the Red Sea and Gulf of Aden, United Nations Development Programme (UNDP) and Food and Agriculture Organization (FAO) of the United Nations, Suez, 29 p

El-Saby MK, Farina L (1954) Report to the government of Saudi Arabia on possibility for development of marine fisheries. FAO, Rome, 29 p

FAO (2003) Kingdom of Saudi Arabia fishery country profile. FAO, Rome. Available at www.fao.org

Ferrer GG (1958) Report to the government of Saudi Arabia on exploration and commercial fishing operations in the Red Sea. FAO, Rome, 28 p

Froese R, Pauly D (2012) FishBase. Available at: www.fishbase.org

FSDP (2011) Fisheries Sector Development Plans (Phase 2): establishing entities to undertake marketing, processing and services for the aquaculture and fishery industries in the Kingdom of Saudi Arabia. KPMG and Poseidon, Aquatic Resource Management Ltd

Furby KA, Bouwmeester J, Berumen ML (2013) Susceptibility of central Red Sea corals during a major bleaching event. Coral Reefs 32(2):505–513

Gilberg YC (1966) Report to the government of the Kingdom of Saudi Arabia on a survey of marien fisheries development possibilities. FAO, Rome, 14 p

Gladstone W (1996) Unique annual aggregation of longnose parrotfish (*Hipposcarus harid*) at Farasan Island (Saudi Arabia, Red Sea). Copeia 2:483–485

Gladstone W (2000) The ecological and social basis for management of a Red Sea marine-protected area. Ocean Coast Manag 43(12):1015–1032

Gladstone W, Krupp F, Younis M (2003) Development and management of a network of marine protected areas in the Red Sea and Gulf of Aden region. Ocean Coast Manag 46(8):741–761

Jin D, Kite-Powell H, Hoagland P, Solow A (2012) A bioeconomic analysis of traditional fisheries in the Red Sea. Mar Resour Econ 27(2):137–148

Kedidi SM, Abushusha T, Allam K (1984) Description of the artisanal fishery at Tuwwal, Saudi Arabia: catches, efforts and catches per unit effort survey conducted during 1981–1982. FAO, Rome, 17 p

Khalil M, Cochran JM, Berumen M (2013) The abundance of herbivorous fish on an inshore Red Sea reef following a mass coral bleaching event. Environ Biol Fishes 96(9):1065–1072

Kotb MMA, Hanafy MH, Rirache H, Matsumura S, Al-Sofyani AA, Ahmed AG, Bawazir G, Al-Horani FA (2008) Status of coral reefs in the Red Sea and Gulf of Aden region. In: Wilkinson C (ed) Status of coral reefs of the world: 2008. Global Coral Reef Monitoring Network and Reef and Rainforest Research Centre, Townsville, pp 67–78

MAW (1986) Fisheries of Saudi Arabia, 1985. Ministry of Agriculture and water, Department of Marine Fisheries. Kingdom of Saudi Arabia

MAW (1996) Fisheries of Saudi Arabia, 1995. Ministry of Agriculture and water, Department of Marine Fisheries. Kingdom of Saudi Arabia

MAW (1998) Fisheries of Saudi Arabia, 1997. Ministry of Agriculture and water, Department of Marine Fisheries. Kingdom of Saudi Arabia

MAW (2000) Fisheries of Saudi Arabia, 1998. Ministry of Agriculture and water, Department of Marine Fisheries. Kingdom of Saudi Arabia

MAW (2006) Fisheries of Saudi Arabia, 2005. Ministry of Agriculture and water, Department of Marine Fisheries. Kingdom of Saudi Arabia

MAW (2008) Statistical indications about fisheries in the Kingdom of Saudi Arabia 2007. Ministry of Agriculture and water, Department of Marine Fisheries. Kingdom of Saudi Arabia

Nanninga GB, Saenz-Agudelo P, Manica A, Berumen ML (2014) Environmental gradients predict the genetic population structure of a coral reef fish in the Red Sea. Mol Ecol 23(3):591–602

Neve P, Al-Aiidy H (1973) The Red Sea fisheries of Saudi Arabia. Bull Mar Res Cent Saudi Arabia 3:32

Oakley SG (1984) The effects of spearfishing pressure on grouper (Serranidae) populations in the Eastern Red Sea. In: Saad MAH (ed) Proceedings symposium of the coral reef environment of the Red Sea, Jeddah, pp 341–359

Peacock NA (1978) Final report. The fishery resource survey of the Saudi Arabian Red Sea, February 1977–October 1979 Field rep. Fish. Dev. Proj. Kingdom of Saudi Arabia, 40: 28p

Price ARG, Medley PAH, McDowall RJ, Dawson-Shepherd AR, Hogarth PJ, Ormond RFG (1987) Aspects of mangal ecology along the Red Sea coast of Saudi Arabia. J Nat Hist 21(2):449–464

Qurban MA, Krishnakumar PK, Joydas TV, Manikandan KP, Ashraf TTM, Quadri SI, Wafar M, Qasem A, Cairns SD (2014) In-situ observation of deep water corals in the northern Red Sea waters of Saudi Arabia. Deep Sea Res Part I Oceanogr Res Pap 89:35–43

RECOFI (2009) Review of capture fishery statistics in the RECOFI area. Regional Commission for Fisheries (RECOFI), Dubai, 9 p

Riegl B, Piller WE (2003) Possible refugia for reefs in times of environmental stress. Int J Earth Sci 92(4):520–531

Riegl BM, Bruckner AW, Rowlands GP, Purkis SJ, Renaud P (2012) Red Sea coral reef trajectories over 2 decades suggest increasing community homogenization and decline in coral size. PLoS One 7(5), e38396

Sakurai T (1998) Fisheries of Saudi Arabia. Ministry of Agriculture and Water, Department of Marine Fisheries and Japan International Cooperation Agency, 41 p

Saleh M, Hariri B (2012) The present status of the Red Sea coral reefs between Haql and Yanbu, Saudi Aarabia. Life Sci J 9(4):3852–3859

Sanders MJ, Morgan GR (1989) Review of the fisheries resources of the Red Sea and Gulf of Aden. FAO, Rome, 138 p

Spaet J (2013) Predictable annual aggregation of longnose parrotfish (*Hipposcarus harid*) in the Red Sea. Mar Biodivers 43(3):179–180

Spaet JLY, Berumen ML (2015) Fish market surveys indicate unsustainable elasmobranch fisheries in the Saudi Arabian Red Sea. Fish Res 161:356–364

Tesfamichael D (2012) Assessment of the Red Sea ecosystem with emphasis on fisheries. PhD thesis, University of British Columbia, Resource Management and Environmental Studies, Vancouver, 241 p

Tesfamichael D, Pitcher TJ (2006) Multidisciplinary evaluation of the sustainability of Red Sea fisheries using rapfish. Fish Res 78(2–3):227–235

Tesfamichael D, Pitcher TJ, Pauly D (2014) Assessing changes in fisheries using Fishers' knowledge to generate long time series of catch rates: a case study from the Red Sea. Ecol Soc 19(1):18

Walczak P (1977) The Yemen Arab Republic. A study of the marine resources of the Yemen ArabRepublic. A report prepared for the Fisheries Development Project, FAO, Rome, 67 p

Zeller D, Rossing P, Harper S, Persson L, Booth S, Pauly D (2011) The Baltic Sea: estimates of total fisheries removals 1950–2007. Fish Res 108(2–3):356–363

Jordan

7

Dawit Tesfamichael, Rhona Govender, and Daniel Pauly

Abstract

Jordan has a short coastline in the Gulf of Aqaba, to the north of the Red Sea. It is the only access the country has to sea, and thus numerous economic activities are concentrated there. The fisheries catch taken within the small Exclusive Economic Zones (EEZ) of Jordan in the inner Gulf of Aqaba is presented, based on a variety of government and non-government sources for the years 1950–2010, and compared with the catch Jordan reports annually to the Food and Agriculture Organization of the United Nations (FAO). The different sectors of the fisheries are treated separately and the composition of the catches estimated. There are strong discrepancies between the reconstructed catch and FAO data, in part due to Jordan's fishing fleets having operated outside of their waters, in Saudi Arabian waters, from 1950 to 1984, and to an overall lack of interest for what are rather small operations, generating very low catches. The catch was around 150 t·year^{-1} from 1950 to the mid-1960s, then declined, due to conflicts in the area, and started to increase again in the mid-1980s, with some fluctuations. The highest catch of 330 t was achieved in 2009. Overall, the estimated reconstructed catch of Jordan from 1950 to 2010 was 1.7 times what is reported in the FAO database. This reconstruction exercise, with its explicitly stated procedures and assumptions and accounting all the sectors comprehensively, can serve as starting point to improve the quality of the data for Jordan, and for better management of its marine resources, which are under increasing pressure from fishing and other developments in the region.

Keywords

Catch time series • Fisheries • Catch composition • Mis-reported catch • Gulf of Aqaba

Introduction

The Hashemite Kingdom of Jordan has access to the north-eastern tip of the Gulf of Aqaba, Red Sea, via a small Exclusive Economic Zones of about 95 km^2 (Al Ouran 2005) (Fig. 7.1). This unique region harbors the world's northern-most coral reef ecosystem (Khalaf and Disi 1997; MoE 2002) and its waters are hypersaline, psu averaging 42 (Sneh and Friedman 1985). This is in part due to its semi-enclosed nature (Hargreaves 1981), which allows for a complete renewal of the water basin only once every 20 years (Lapidoth-Eschelbacher 1982). These factors not only lead to a high degree of endemism in the region, but also increase

D. Tesfamichael (✉)
Fisheries Centre, University of British Columbia, Vancouver, BC, Canada

Department of Marine Sciences, University of Asmara, Asmara, Eritrea
e-mail: dawittes@gmail.com

R. Govender • D. Pauly
Fisheries Centre, University of British Columbia, Vancouver, BC, Canada

D. Tesfamichael, D. Pauly (eds.), *The Red Sea Ecosystem and Fisheries*, Coral Reefs of the World 7,
DOI 10.1007/978-94-017-7435-2_7

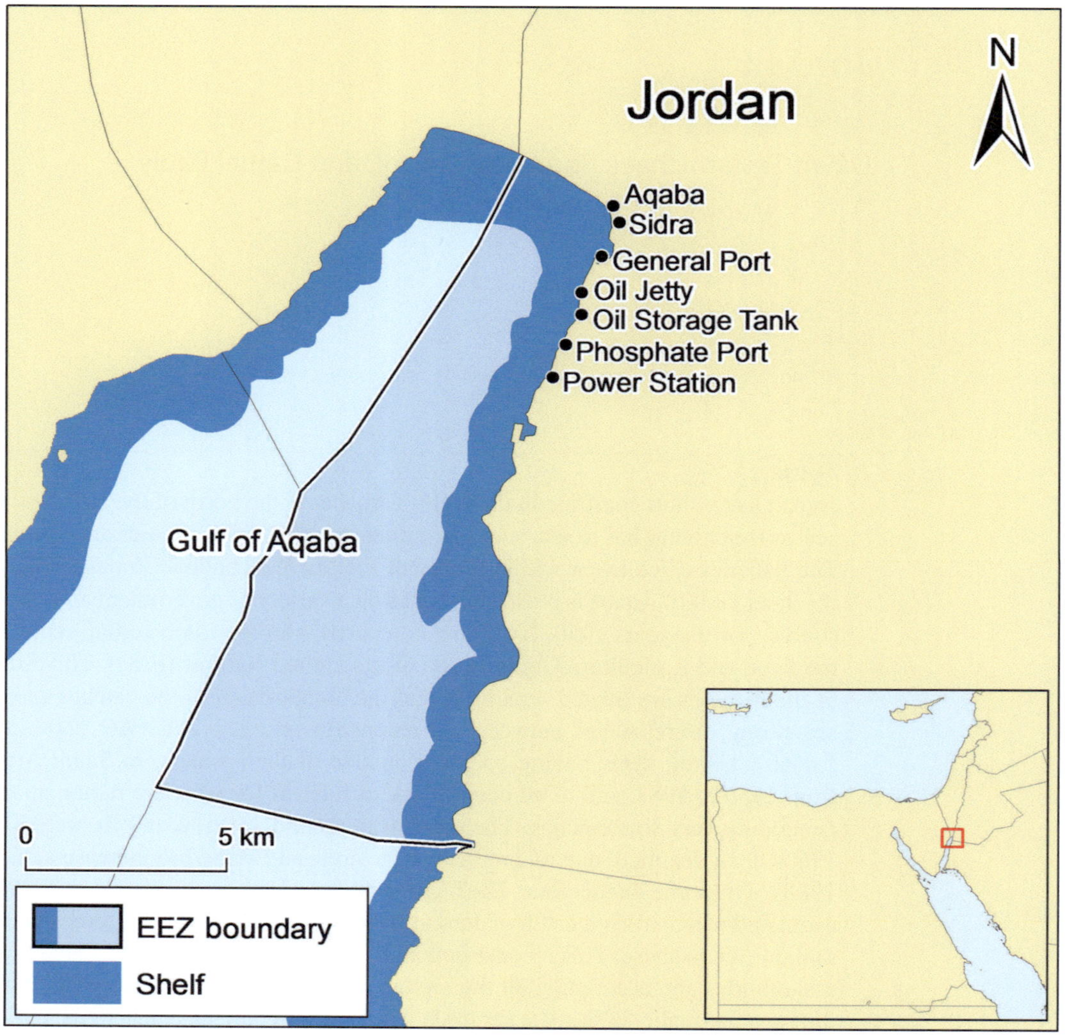

Fig. 7.1 The shelf areas, Exclusive Economic Zone (EEZ) and major sites of Jordan in the Gulf of Aqaba, Red Sea

the susceptibility to pollution, which, jointly with overfishing, affect the diversity and abundance of the fish stocks, as well as the health of the coral reefs (Tellawi 2001). The major settlement is the city of Aqaba (Fig. 7.2), the sole port in Jordan. The fish landing site is at Sidra, near the town of Aqaba, where fish are sold directly to merchants, hotels and restaurants based in Aqaba (PERSGA 2002).

One of the main human activities impacting the Jordanian Red Sea coast is fishing. However, the state of the fisheries in the inner Gulf of Aqaba cannot be assessed using only the catch data Jordan submits to the Food and Agriculture Organization (FAO) of the United Nations (Pauly and Zeller 2003). Rather, as for other countries and territories, it is necessary to 'reconstruct' historic catch trends to acquire an understanding of the evolution of these fisheries (Zeller et al. 2006, 2007). Thus, we present in this chapter an alternative catch estimates by reconstructing the Jordanian catch in the Red Sea from 1950 to 2010.

The fishery is divided into sectors and for each the total catch and its composition are presented. In order to have a good understanding of the ecosystem within which the fisheries are embedded, a brief description of the coral reef ecosystem of Jordan in the Gulf of Aqaba is given, followed by an introduction to the fisheries. Then the results are presented; the sources and methods used to estimate the catches are presented at the end.

Coral Reef Ecosystems

About 50 % of the Jordanian coast in the Gulf of Aqaba is covered by fringing coral reefs, with a high diversity of corals (about 180 species) and fishes (well over about 500 species). The live coral cover can be as high as 90 % and butterflyfishes have the highest abundance followed by parrotfishes (Kotb et al. 2008). The coral reef ecosystems on the Jordanian coast in the Gulf of Aqaba, along with those in the

Gulf of Suez (Riegl et al. 2012), are the northernmost reefs in the western Indo-Pacific (Khalaf and Disi 1997; MoE 2002). The main threats to the coral reefs of Jordan, besides fishing, are industrial development and tourism.

Some of the anthropogenic factors impacting the Jordanian coastal ecosystem over the past few decades include commercial fish farms, sewage outflow, and phosphate emissions from nearby industrial terminals, which led to increasing levels of eutrophication (MoE 2002; Al Ouran 2005). Other stressors are oil spills (Kotb et al. 2008) and littering, especially plastics (Abu-Hilal and Al-Najjar 2009). As in many northern Red Sea countries, the development of tourism (Fig. 7.3) industry in Jordan has been rapid (Hawkins and Roberts 1994). These factors are having an adverse effect on the coral reefs. For example, in 1996, 70 % of the coral were reported to be alive, yet in 2002 the reef was comprised of only 30 % live coral in Jordan (MoE 2002). Khalaf and Kochzius (2002) found out, by comparing areas impacted by industrial development and undisturbed areas, that coastal industrial development disturbed coral reef ecosystems through dreading and pollution, which in turn affected the community structure of fishes; thus, e.g., fish abundance in disturbed areas was 50 % less than undisturbed areas, although species richness remained similar. They also found the disturbance affected different trophic levels of the ecosystem differently. Other impact of the stress on the ecosystem was shown by disease outbreaks (Al-Moghrabi 2001). Natural disturbances also affect the coral reefs of Jordan. The most recent example is the bleaching and mortality on reef flats in extreme low tide in 2007 (Kotb et al. 2008), which was thought to be aggravated by climate change. These unfavorable marine conditions, coupled with fishing (Tellawi 2001) are responsible for the increase of pressure on the ecosystem. The extent of the impact of these pressures needs to be studied and monitored. Some of the impacts are obvious, for example in 2006 half of all the Jordanian fishing boats were permanently anchored (IRIN 2006).

Fig. 7.2 The city of Aqaba, Jordan (Photo: Kht01, Wikipedia Commons)

Jordan has been involved in active management of its marine ecosystem, with emphasis on sustainable development through planning and education, notably through the 'On-ground Projects Programme' implemented by The

Fig. 7.3 Diving in coral reef area is one of the most popular tourist activities along the Jordanian Red Sea coast (Photo: Marina Plaza Hotel)

Regional Organization for the conservation of the Environment of the Red Sea and Gulf of Aden (PERSGA) in 2006. Also, Jordan has declared 30 % of its coast to be marine protected areas (Kotb et al. 2008). The results of these initiatives, however, will be revealed by long-term monitoring.

Fisheries

Due to the limited area available for fishing (Farid 1984), the fisheries of Jordan in the Gulf of Aqaba are generally small. In the past, Jordanian fishers operated beyond the inner Gulf of Aqaba, in the waters of neighboring countries, almost exclusively in Saudi Arabia (PERSGA 2002), as Jordanian fishers are rarely seen in Egyptian and Yemeni waters. The predominant gears utilized by Jordanian fishers are beach seines and gillnets, but trammel nets, hand lines, longlines, various traps, and (illegally) spear guns and explosives are also used. The main target species are snappers (Lutjanidae), groupers (Serranidae) and emperors (Lethrinidae), along with numerous incidental species (Khalaf and Disi 1997). The catch, which is not iced, is landed by boats ranging from 5.5 to 11 m (PERSGA 2002), and sold immediately to merchants in the port of Aqaba. The fisheries of Jordan can be categorized into three sectors; the first is the artisanal sector, with fishers who tend to work full time, and who may occasionally venture into the waters of neighboring countries. The second is the subsistence sector consisting of two components, one being the catch from small boats for direct consumption by family or neighbors, with a small part being sold or bartered, the other being that part of the catch of the artisanal fishery which is consumed by the crew, or given to family members and friends. The third fishery sector is recreational fishing, which is occasionally engaged in the weekends by the people who live on the coast. These three sectors are treated separately in the catch reconstruction.

Fisheries Catches

Our results suggest that the catch in Jordanian waters ranged from slightly less than 150 t in 1950 to a maximum of 300 t in 2009 (Fig. 7.4). The total catch showed an increasing trend from its lowest values in the late 1970s until its peak in 2009. In contrast, the data supplied to FAO were lower than the reconstructed total catch for most of the period, except in the early period, where a total of 200 t was reported for a few years. These years correspond with the time the Jordanian fishers were fishing outside the Jordanian waters, i.e., overwhelmingly in Saudi Arabia. Since the origin of these catch was outside Jordan, we did not include them as Jordanian catch from Jordanian waters in our reconstruction. Instead, they were accounted in their main country of origin, Saudi Arabia (see Chap. 6). In the last few years, Jordanian fisheries hardly fish outside Jordanian waters. Similar to the reconstruction, the highest catch reported in the FAO database was for 2009. The FAO database has zero catches for some years. This is due to a lack of reporting by Jordan to FAO and does not mean that there were no fishery catches. Such data gaps can be very misleading in the assessment and management of fishery (Pitcher et al. 2002). From the late 1990s, the FAO data were equal to the total artisanal fishery, which is due to the subsistence and recreational not being reported at all.

When the reconstructed total catch of Jordan in Jordanian waters is divided into different sectors, the artisanal fishery contributes 62 %, followed by the subsistence fishery at 20 % and the recreational fishery at 18 % (Fig. 7.4). Jordan does not have an industrial fishery (PERSGA 2002). The artisanal fishery has the strongest influence on the pattern of the overall total. The subsistence fishery follows a pattern similar to that of the artisanal fishery because it was calculated mainly as a ratio of the artisanal fishery. The recreational fishery started in 1974 and its contribution became noticeable only in the later years.

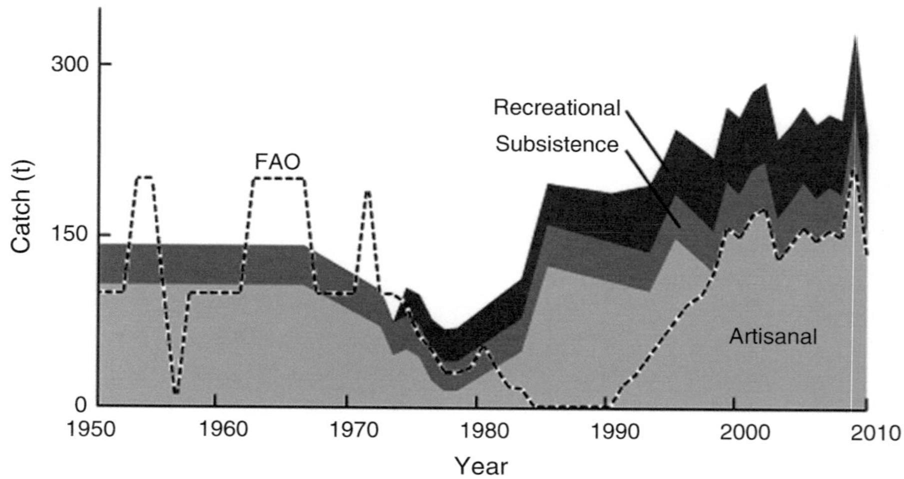

Fig. 7.4 Total catch of the Jordanian fishery in the Gulf of Aqaba by sectors and the total catch Jordan reported to FAO from 1950 to 2010

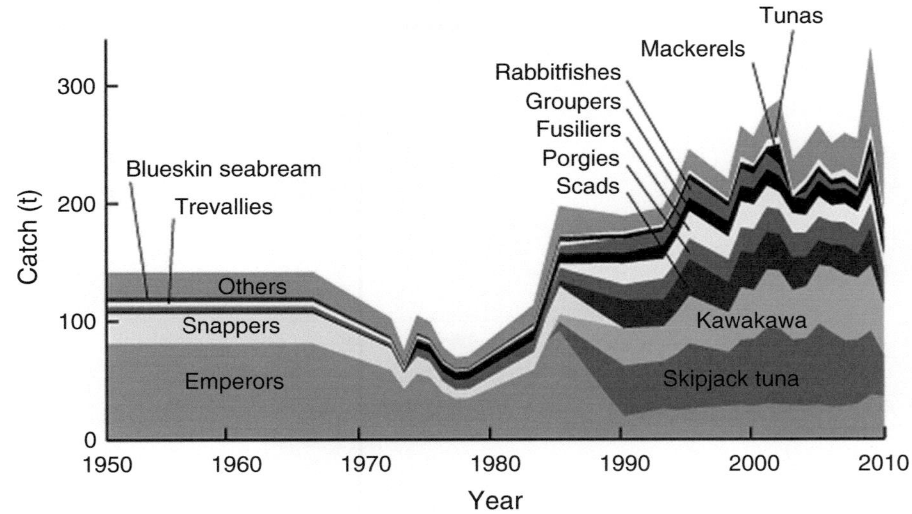

Fig. 7.5 Composition of the total catch of Jordanian fisheries in the Gulf of Aqaba from 1950 to 2010

The total marine catch within Jordanian waters was found to be 1.7 times the catches reported to the FAO from 1950 to 2010. This can be attributed to two main reasons: non-reporting for some years, which translated as zeroes in the FAO database, and the fact that the subsistence and recreational fisheries are not included in the FAO database at all. The unreported catch accounted for 52 % of the reconstructed total catch. The reported catch (part of the reconstructed catch accounted in the FAO data) represented only 48 % of the total catch.

The total catch of Jordan in the Gulf of Aqaba is dominated by few taxa, although the total number of taxa in the catch is very high. The dominant taxon is emperors (Lethrinidae, 32 %), the distant second is skipjack tuna (*Katsuwonus pelamis*, 12 %), followed by kawakawa (*Euthynnus affinis*, 9 %) and snappers (Lutjanidae, 8 %). These four taxa jointly constitute 61 % of the total catch. Of the four taxa, emperors and snappers are coral-reef associated and their catch is visible from 1950, although their contributions decreased later. On the other hand, skipjack tuna and kawakawa appear only after 1980 (Fig. 7.5). This is a clear result of the shift of target species by the fishers from coral-reef associated fishes to pelagic in the mid-1980s. The composition of the total catch of Jordan in the Red Sea strongly resembles that of the artisanal fishery (Figs. 7.5 and 7.6a), because the artisanal fishery has the highest contribution to the total catch.

Looking at the fisheries separately, there are 21 identified taxonomic groups that made up the total artisanal fishery catch of Jordan in the inner Gulf of Aqaba (Fig. 7.6a). The most dominant taxon is emperors (Lethrinidae) at 29 %, followed by skipjack tuna (*Katsuwonus pelamis*) at 15 % and kawakawa (*Euthynnus affinis*) at 12 %. An explanation for the partial shift from coral reef to pelagic fish around 1985 is that in the early period, when Jordanian fishers used to venture out to neighboring countries, their target was coral-reef associated fishes. The gear and technology they used must have influenced their operation in the domestic waters as well, in Jordanian water, or the fishers who used to fish in Saudi waters would fish in Jordanian waters when they were not fishing in Saudi waters. However, later, when they were no longer allowed to fish in the waters of neighboring countries, the Jordanian fishers had to focus on local resources and target the more abundant resource – the pelagic fishes.

The catch composition of the subsistence fishery is similar to the artisanal fishery (Fig. 7.6b), as the same catch composition ratios were used. However, it does not show strong fluctuations like the artisanal fishery; it stays more or less stable. This is quite realistic for a subsistence fishery. As its name indicates, this is catch consumed by locals, is a source of food security, and it is usually less affected by external factors (e.g., market and politics) compared to the other fisheries. The catch composition of the recreational fishery has only three identified taxa (Fig. 7.6c).

Jordan has been plagued by numerous conflicts since 1950. This tumultuous history is mirrored in the state of the fisheries. Some of the fluctuations of its fisheries are attributed to lack of political stability in the region. The small coastal access to the Gulf of Aqaba attributed to one of the lowest total catch of all the countries bordering the Red Sea. Hence, there is little motivation for the country to collect detailed fisheries statistics, and assess and manage its fisheries. Most of the fisheries data appear to be collected by university researchers and sporadically by the fisheries administrations (Tellawi 2001). However, for effective policies to manage the fisheries to be developed and implemented, consistent time series data are essential (Caddy and Gulland 1983; PERSGA 2002; Tesfamichael 2012).

This study showed considerable discrepancies between the reconstructed catch and data supplied to the FAO by

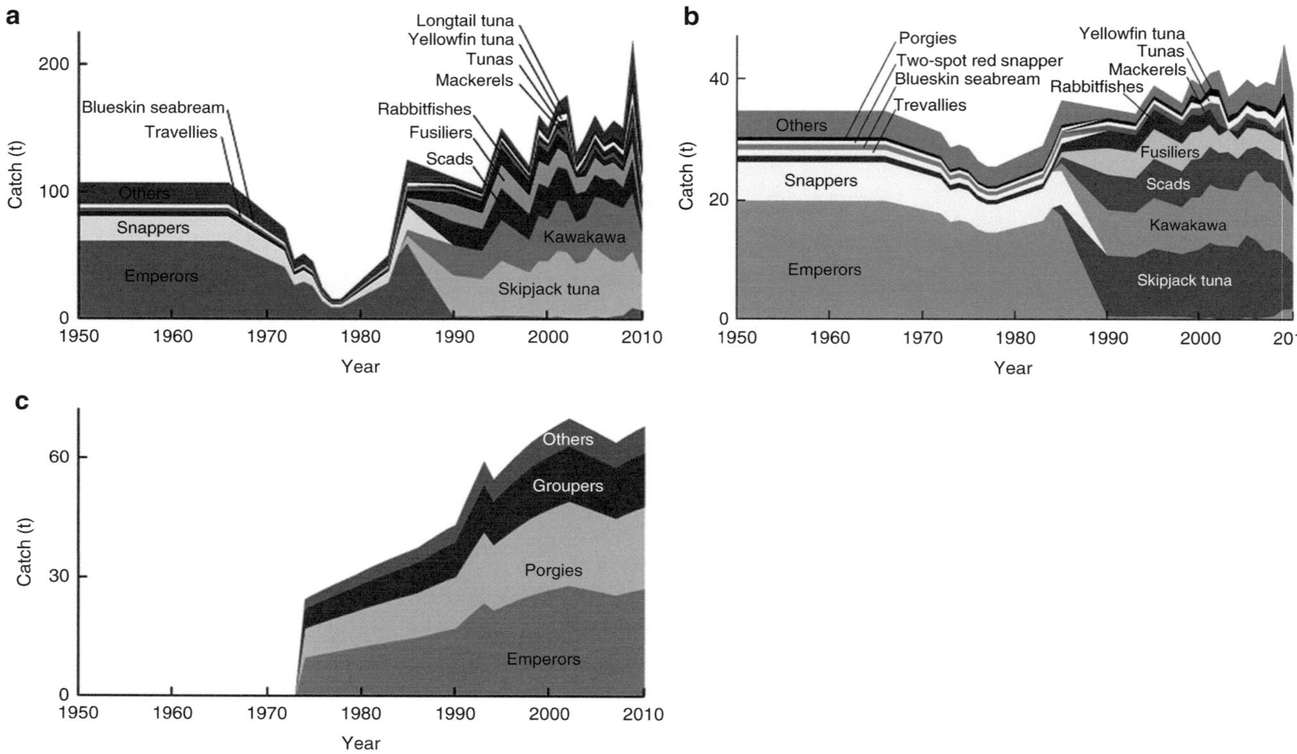

Fig. 7.6 Catch composition of (**a**) artisanal (**b**) subsistence and (**c**) recreational fisheries of Jordan in the Gulf of Aqaba from 1950 to 2010

Jordan. One source of discrepancy is the fact that the official data refer to a larger spatial area (FAO area 51) than is suitable for management. For the period after Jordanian fishing in Saudi Arabia ended, the reconstructed catch is 148 % of that reported to the FAO. We note that such 'official' underestimates are common (see contributions in Zeller and Haprer 2009).

The sharp decline of the Jordanian fisheries catch in the 1970s and 1980s may be due to political instability, but also may be a result of phosphate dust emissions (PERSGA 2006), compounded by harmful fishing practices. Jordan has substantial phosphate mines. Industrial pollution of the marine environment is a recurrent theme on the Jordanian side of the Gulf of Aqaba, and affects fishery catches (IRIN 2006). Tourism is an important contributor to the local economies (Al Ouran 2005), and it will be impacted if pollution and excess fishing continue to harm the reefs which form major fish habitats in the inner Gulf of Aqaba.

It will be beneficial to strike a balance between industrial development, fisheries and tourism for the inner Gulf of Aqaba which will permit the coexistence of these sectors, along with a revival of the natural habitats and their fauna and flora. Examples exist of such beneficial coexistence, one being Monterey Bay, in California, USA, which experienced a tourism-led revival following its near destruction by a succession of out-of-control industries (Palumbi and Sotka 2010). This may serve as a model for the inner Gulf of Aqaba.

Sources and Methods

The total marine catches by Jordan within its EEZ in the Gulf of Aqaba were estimated for the period from 1950 to 2010. The required data were primarily obtained from government reports and the scientific literature. Since the fisheries of Jordan are dominated by artisanal fisheries, which use selective gear (PERSGA 2002), there is no discarded catch or, if it exists, it is negligible. Emphasis was given to 'hard' estimates of the catch of distinct fisheries, which were used as 'anchor points' (Zeller et al. 2006, 2007) between which estimates for missing years could be obtained by interpolation. The reconstructed catch derived here is proposed as an improvement over the data currently available. All our estimates are conservative and our assumptions explicitly stated. A brief account of the catch reconstruction procedure for each sector is presented below. A spatial distribution of the catch is given at www.seaaroundus.org.

Artisanal Fishery

For some years, the Jordanian Red Sea fisheries data were obtained from peer-reviewed articles, government reports and grey literature. The catch estimates for years without sources were derived by interpolation between years with known catch. Also, the 1965 value was carried back to 1950, as the fishing fleet and environmental conditions have experienced few changes from 1950 to 1964 (Barrania 1979). After 1985, the fishery statistical reporting ceased and the catch values available were based on researchers from Jordanian universities (PERSGA 2002). These values match what are available in the FAO database for the later years. Thus, the FAO values were used from 1998 to 2010.

The landings in the Jordanian port of Aqaba prior to 1985 included both catches in Jordanian waters and catches from Saudi Arabian waters up to 300 km south of Aqaba. In 1985, access to these southern fishing grounds ceased, as did the collection of fisheries statistics by the government (PERSGA 2002). Until 1984, approximately half of the catch was obtained in Saudi Arabian waters (Barrania 1979) and thus only half of the reconstructed catch was applied to the Jordanian catch. The other half which was caught in Saudi waters was accounted in Saudi Arabia's catch reconstruction (see Chap. 6).

In addition to fishing in foreign waters, there are mentions of high spoilage rates due to low handling standards and insufficient amount of ice to preserve the catch. Estimated rates of spoilage ranged from 0.4 % to 54 % (Barrania 1979). The values from 1972 to 1978 included spoilage, whereas the data for the previous years did not. A conservative estimate of 19 %, calculated by taking the average spoilage values from 1972 to 1976, was added to the catch in all years prior to 1972.

Finally, the reconstructed catch was disaggregated into major taxa. There were two sources of catch composition data, one for the early period (Barrania 1979) and for the later period (1998–2010) from the FAO database. The Jordanian fishery in the Gulf of Aqaba had two distinct periods: before and after 1985. This coincides with the time the fishers were allowed to operate in the waters of neighboring countries (mainly in Saudi waters), or not. When they fished outside their waters, most of the local fishers were focused on inshore and coral reef fishes (Barrania 1979). Once they were prevented from fishing outside of Jordanian waters, in 1985, most of their catch focused on pelagic fishes (PERSGA 2002). Accordingly, the composition from 1950 to 1984 was calculated based on the data from Barrania (1979). Only local names were available in the report and their corresponding scientific and English common names were obtained from the matching names tables in Chap. 10, FishBase (Froese and Pauly 2012) and Dr. Dori Edelist (University of Haifa, Israel, pers. comm.). The report by Barrania (1979) had catch composition estimates for only eight of the taxa, while the rest were put under 'others'. However, local names for some of the components of the 'others' were available, and we assumed a value of 0.3 % for each of the taxa for which we were able to find the corresponding scientific names in our sources; only the rest were categorized under 'others'. The FAO catch composition was used from 1998 to 2010. From 1990 to 1997, the values of 1998 were used. From 1985 to 1989, we interpolated the catch composition to allow a relatively smoother transition of the fishery from inshore coral reef to pelagic. Although there is evidence that the fishery switched its target fishes from reef-associated to pelagic, it is unrealistic to assume the shift happened within a year, as there are capital investment and technical issues that need to be considered. We allowed 5 years for the shift to occur.

Subsistence Fishery

The catch of the subsistence fishery consists of two components, the first is the catch of small boats mainly for direct consumption by the fishers' families and communities. A small portion of their catch may be sold or bartered. These catches are not reported at all. The second component of the subsistence catch is part of the catch of the larger boats that is consumed by the crew and given to family and friends. For the small boats it was calculated based on a survey (Barrania 1979), where it was estimated that a total of about 12 boats made day trips and caught 6–10 kg (average 8 $kg \cdot day^{-1}$). We assumed the total number of fishing days per year to be 250 and thus could compute the total catch to be 24 $t \cdot year^{-1}$. For the bigger boats, we assumed that 10 % of their catch was consumed by crew and/or given to family members. The percentages used to calculate the artisanal fishery catch composition were used to calculate the composition of subsistence catch composition as well, because they employ similar gears.

Recreational Fishery

A recreational fishery exists in the Jordanian Gulf of Aqaba and it is growing fast (FAO 2003). However, data on its size and catches do not exist. The recreational fishery catch was estimated using the population of Aqaba, the main coastal settlement on the Jordanian coast, taken from www.populstat.info and Wikipedia (2012). Data were not available for the whole time series, and interpolation was used to fill the gaps. The recreational fishery was assumed to start in 1974, after the war between Israel and Arab countries in 1973, referred to as '6th of October War', so zero was assigned from 1950 to 1973. From 1974 onward, it was calculated using a participation ratio of 0.08, based on the regional value of 0.12 (Cisneros-Montemayor and Sumaila 2010). Note the participation ratios in Cisneros-Montemayor and Sumaila (2010) are given for the whole countries in percentages. In our calculations, we used population size of only Aqaba, the main coastal settlement because the coastal people are the ones to be involved more in recreational fishing than people far from the coast. The participation ratio was adjusted for the population size of Aqaba using the ratio of Aqaba population to the total Jordanian population from 1974 to 2010, which was 1.5 % (i.e., the total population is 65 times that of Aqaba). Hence the participation ratio for Jordan in total (0.12 %) was multiplied by 65, making the participation percentage only for Aqaba to be 8 %. In addition, we assumed that recreational fishers go fishing a total of 15 days · year^{-1}. As for the catch rate, we assumed a rate of 1 kg · day^{-1} for 1974 and 0.5 kg · day^{-1} for 2010. The rate was interpolated between those 2 years. Finally, the annual recreational fishery estimate was calculated as a product of population, participation ratio, number of days per year, and catch rate. The catch composition of recreational fishery was adapted from Saudi Arabia's recreational fishery in the Red Sea, where emperors accounted for 40 %, sea breams 30 %, groupers 20 % and 'others' 10 % (see Chap. 6).

Comparing Reconstructed Catches with FAO Statistics

The reconstructed catches, i.e., what were caught in the EEZ's of Jordan and excluding what were caught in other countries' EEZs, were compared to the data for Jordan in the FAO database. Since the subsistence and recreational fisheries were not reported at all, only the reconstructed artisanal catches were compared to the FAO data. In order to be able to compare taxon by taxon, the FAO data, when only totals were given, were divided into the components using the ratios in the reconstructed catches. For any taxon, if the amount in the reconstruction was higher than its value in the FAO database, the difference is assigned as 'unreported catch'; if it is the opposite, then it is 'over-reported catch'. The part of the reconstructed catch that is accounted in the FAO data is referred as 'reported catch' in our result.

Acknowledgements We would like to thank the fisheries administration of Jordan for their collaboration. Our thanks goes to Izzat Feidi for reviewing the chapter. This is a contribution of *Sea Around Us*, a scientific collaboration between the University of British Columbia and The Pew Charitable Trusts.

References

Abu-Hilal A, Al-Najjar T (2009) Marine litter in coral reef areas along the Jordan Gulf of Aqaba, Red Sea. J Environ Manage 90(2):1043–1049

Al Ouran N (2005) Environmental assessment, documentation and spatial modeling of heavy metal pollution along the Jordan Gulf of Aqaba using coral reefs as environmental indicator. PhD thesis, University of Würzburg, Department of Hydrogeology and Environment, Würzburg, 158 p

Al-Moghrabi S (2001) Unusual black band disease (BBD) outbreak in the northern tip of the Gulf of Aqaba (Jordan). Coral Reefs 19(4):330–331

Barrania AA (1979) Report on the socioeconomic survey of the Gulf of Aqaba, the Hashemite Kingdom of Jordan. Development of Fisheries in Areas of the Red Sea and Gulf of Aden, United Nations Development Programme (UNDP) and Food and Agriculture Organization (FAO) of the United Nations, Cairo

Caddy JF, Gulland JA (1983) Historical patterns of fish stocks. Mar Policy 7(4):267–278

Cisneros-Montemayor A, Sumaila UR (2010) A global estimate of benefits from ecosystem-based marine recreation: potential impacts and implications for management. J Bioecon 12(3):245–268

FAO (2003) Hashemite Kingdom of Jordan: fishery country profile. Food and Agriculture Organization of the United Nations (FAO), Rome. Available at: http://www.fao.org/fi/oldsite/FCP/en/JOR/profile.htm

Farid AM (1984) The Red Sea: prospects for stability. St. Martin's Press, New York, 173 p

Froese R, Pauly D (eds) (2012) FishBase. Available at: www.fishbase.org

Hargreaves P (ed) (1981) The Red Sea and Persian Gulf. Wayland Publishers, Hove, 67 p

Hawkins JP, Roberts CM (1994) The growth of coastal tourism in the Red Sea: present and future effects on coral reefs. Ambio 23(8):503–508

IRIN (2006) Jordan: fishermen catching less in polluted Aqaba water. Integrated Regional Information Networks. Available at: http://www.irinnews.org/Report.aspx?ReportId=61933

Khalaf MA, Disi AM (1997) Fishes of the Gulf of Aqaba. Marine Science Station, Aqaba, 252 p

Khalaf MA, Kochzius M (2002) Changes in trophic community structure of shore fishes at an industrial site in the Gulf of Aqaba, Red Sea. Mar Ecol Prog Ser 239:287–299

Kotb MMA, Hanafy MH, Rirache H, Matsumura S, Al-Sofyani AA, Ahmed AG, Bawazir G, Al-Horani FA (2008) Status of coral reefs in the Red Sea and Gulf of Aden region. In: Wilkinson C (ed) Status of coral reefs of the world: 2008. Global Coral Reef Monitoring Network and Reef and Rainforest Research Centre, Townsville, pp 67–78

Lapidoth-Eschelbacher R (1982) The Red Sea and the Gulf of Aden, International straits of the world. Martinus Nijhoff Publishers, The Hague, 265 p

MoE (2002) Protecting the Gulf of Eilat/Aqaba. Ministry of the Environment. Israel Environ. Bullet. 25(3):5

Palumbi S, Sotka C (2010) The death and life of Monterey Bay: a story of revival. Island Press, Washington, 216 p

Pauly D, Zeller D (2003) The global fisheries crisis as a rationale for improving the FAO's database of fisheries statistics. In: Zeller D, Booth S, Mohammed E, Pauly D (eds) From Mexico to Brazil: Central Atlantic fisheries catch trends and ecosystem models. Fisheries Centre Research Reports, vol 11(6), Fisheries Centre, University of British Columbia, Vancouver, pp. 1–9

PERSGA (2002) Status of the living marine resources in the Red Sea and Gulf of Aden and their management. Strategic action programme for the Red Sea and Gulf of Aden, Regional Organization for the Conservation of the Environment of the Red Sea and Gulf of Aden (PERSGA), Jeddah, 134 p

PERSGA (2006) Status of the living marine resources in the Red Sea and Gulf of Aden and their management. Strategic action programme for the Red Sea and Gulf of Aden, Regional Organization for the Conservation of the Environment of the Red Sea and Gulf of Aden (PERSGA), Jeddah, 260 p

Pitcher TJ, Watson R, Forrest R, Valtysson HP, Guénette S (2002) Estimating illegal and unreported catches from marine ecosystems: a basis for change. Fish Fish 3(4):317–339

Riegl BM, Bruckner AW, Rowlands GP, Purkis SJ, Renaud P (2012) Red Sea coral reef trajectories over 2 decades suggest increasing community homogenization and decline in coral size. PLoS One 7(5), e38396

Sneh A, Friedman GM (eds) (1985) Hypersaline ecosystems: the Gavish Sabkha. Springer, New York

Tellawi A-MM (2001) Conservation and sustainable use of biological diversity in Jordan: first national report of The Hashemite Kingdom of Jordan on the implementation of Article 6 of the convention on biological diversity. The General Corporation for the Environment Protection, Amman, 118 p. Available at: www.cbd.int/doc/world/jo/jo-nbsap-01-en.doc

Tesfamichael D (2012) Assessment of the Red Sea ecosystem with emphasis on fisheries. PhD thesis, University of British Columbia, Resource Management and Environmental Studies, Vancouver, 241 p

Wikipedia (2012) Wikipedia: the free encyclopedia. Wikimedia Foundation. Available at: http://en.wikipedia.org

Zeller D, Harper S (eds) (2009) Fisheries catch reconstructions: Islands, Part I. Fisheries Centre Research Reports, vol 17(5), Fisheries Centre, University of British Columbia, Vancouver, 104 p

Zeller D, Booth S, Craig P, Pauly D (2006) Reconstruction of coral reef fisheries catches in American Samoa, 1950–2002. Coral Reefs 25(1):144–152

Zeller D, Booth S, Davis G, Pauly D (2007) Re-estimation of small-scale fisheries catches for U.S. flag island areas in the Western Pacific: the last 50 years. US Fish Bull 105(2):266–277

Israel

8

Dawit Tesfamichael, Rhona Govender, and Daniel Pauly

Abstract

In addition to its main coast on the Mediterranean Sea, Israel has the very short coast on Red Sea, in the northern, inner part of the Gulf of Aqaba. The fisheries catch taken within the small Exclusive Economic Zones (EEZ) of Israel, in the Gulf of Aqaba is presented, based on catch estimates reconstructed from a variety of published papers, government and non-government sources for the years 1950–2010, and compared with the catch it reports annually to the Food and Agriculture Organization of the United Nations (FAO). The different sectors of the fisheries are treated separately and the composition of the catches estimated. Israel's total catch in its Red Sea EEZ was less than 100 t·year^{-1} in the early 1950s, increased until it reached its peak of around 300 t·year^{-1} in the early 1980s, then decreased abruptly (when Israel signed a peace treaty with Egypt and left the Sinai Peninsula) to around 100 t·year^{-1} in the late 2000s. Overall, the estimated reconstructed catch of Israel from 1950 to 2010 was 1.4 times what is reported in the FAO database after accounting for the fact that, from 1957 to the early 1970s, and Israeli vessels operated in Eritrean waters, but pooled their landing with those from the Israeli EEZ. This catch reconstruction with explicitly stated procedures and assumptions, accounting all the sectors comprehensively, should serve as a starting point to improve the quality of the data and for better management of resources.

Keywords

Time series • Fisheries • Catch • Catch composition • Mis-reported catch • Israel • Red Sea • Gulf of Aqaba

Introduction

Israel shares the north tip of the Gulf of Aqaba, Red Sea, with Jordan. It has small Exclusive Economic Zone of 29 km^2 (Fig. 8.1). This unique region harbors one of the world's northernmost coral reef ecosystems (Khalaf and Disi 1997; MoE 2002; Loya 2004) and is characterized as hypersaline, with waters averaging 42 psu (Sneh and Friedman 1985). This is in part due to its semi-enclosed nature (Hargreaves 1981), which allows for a complete renewal of the water basin only once every 20 years (Lapidoth-Eschelbacher 1982). These factors not only lead to a high degree of endemism in the region (Sheppard et al. 1992), but also increase its susceptibility to pollution, which, jointly

D. Tesfamichael (✉)
Fisheries Centre, University of British Columbia, Vancouver, BC, Canada

Department of Marine Sciences, University of Asmara, Asmara, Eritrea
e-mail: dawittes@gmail.com

R. Govender • D Pauly
Fisheries Centre, University of British Columbia, Vancouver, BC, Canada

D. Tesfamichael, D. Pauly (eds.), *The Red Sea Ecosystem and Fisheries*, Coral Reefs of the World 7,
DOI 10.1007/978-94-017-7435-2_8

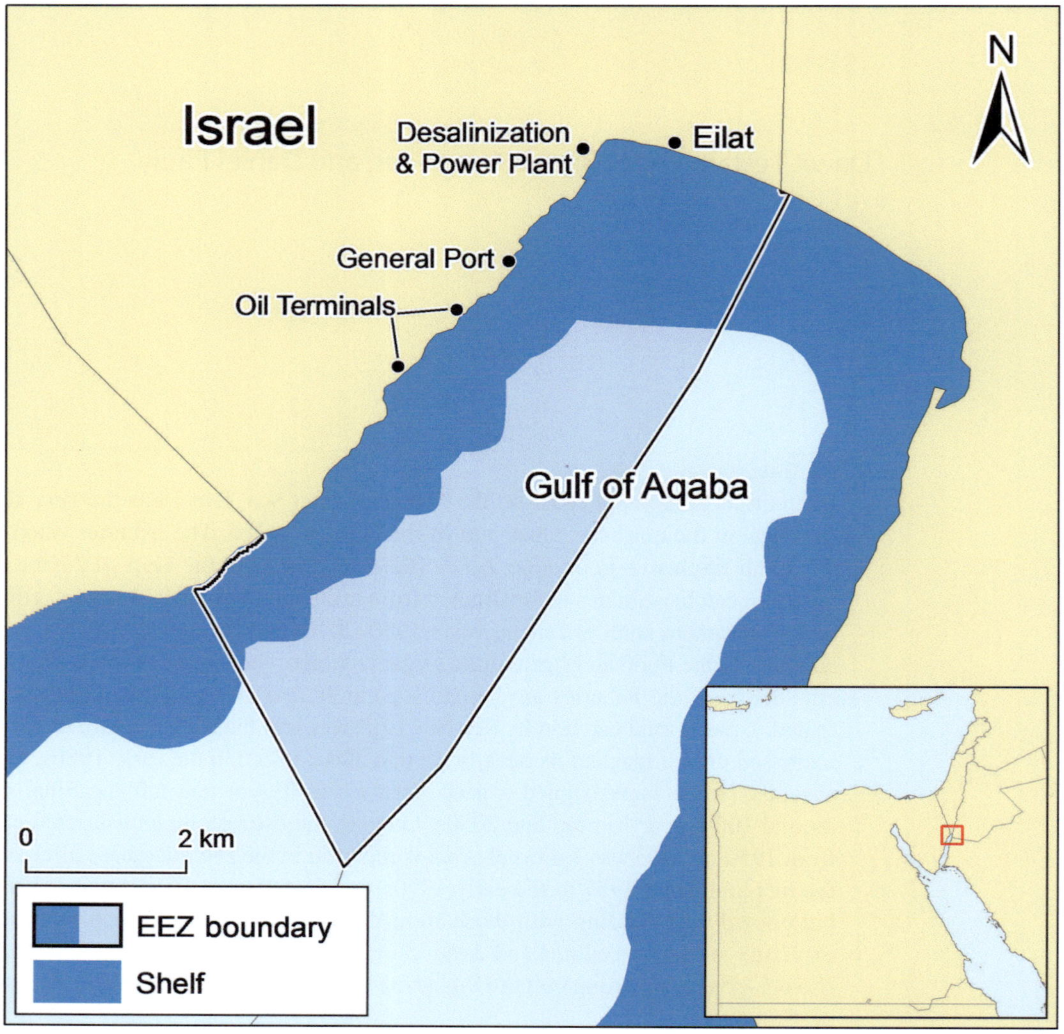

Fig. 8.1 The shelf areas, Exclusive Economic Zone (EEZ) and major sites of Israel in the Gulf of Aqaba, Red Sea

with overfishing, affect the diversity and abundance of the fish stocks, as well as the health of the coral reefs (Tellawi 2001; Loya 2004). Eilat is the major settlement and is the focus of coastal tourism along the Israeli Red Sea coast (Fig. 8.2). As Israel has a relatively high income level, the coast is highly developed for tourism and the population of Eilat has increased dramatically in the last few decades (Loya 2004).

Although the Israeli fishery in the Gulf of Aqaba is small, it should not be ignored. Rather, like all fisheries, it needs to be properly documented and catches reported on regular basis. A reliable assessment of the fisheries in the inner Gulf of Aqaba cannot be done using only the catch data which Israel submit to the Food and Agriculture Organization (FAO) of the United Nations (Pauly and Zeller 2003). As for other countries and territories, it is necessary to 'reconstruct' historic catch trends to understand the evolution of Israel's Red Sea fisheries (Zeller et al. 2006, 2007). Hence, in this chapter, we present an estimate of the Israeli catch in the Gulf of Aqaba by reconstructing the catches from a variety of sources. A comprehensive account is given for each fishery sector. We first present a brief introduction of the ecosystem and the fisheries, then the catches and their composition are given. The sources and methods used for catch reconstruction are presented at the end of the chapter.

Coral Reef Ecosystem

Israel's Red Sea coast is fringed with diverse and spectacular coral reefs (Fig. 8.3). These reefs are among the most northern of western Indo-Pacific coral reefs. The semi-enclosed nature of the Red Sea gives it unique characteristics resulting in high endemism (Ormond and Edwards 1987; Sheppard et al. 1992; Khalaf and Disi 1997; MoE 2002; Loya 2004). The coral reefs of the Gulf of Aqaba are among the best studied reefs in the world in general and in

Fig. 8.2 The Red Sea coast of Israel is a popular tourist destination (Photo: Sasi Horesh/Israel HaYom)

Fig. 8.3 A diverse coral reef in the Gulf of Aqaba, Israel (Photo: Daviddarom, Wikipedia Commons)

the Red Sea in particular, and over 50 % of the coral reef ecology papers published on the Red Sea deal with the Gulf of Aqaba, although it represents less than 2 % of the area of the Red Sea (Bahartan et al. 2010; Berumen et al. 2013). Nevertheless, there are many issues that remain understudied, and which need attention if an understanding of the ecology of the coral reefs of the region is to be attained (Loya et al. 2014); fishing is among the least studied issue in the region.

Notably, these issues concern the many stressors to the health of the coral reefs in the Gulf of Aqaba, due to natural environmental disturbances such as extreme low tides and above average water temperatures, which result in coral bleaching and mortality (Loya 2004). Another set of issues is related to stresses from human activities, either direct (e.g., diving) or indirect, originating from terrestrial activities. These factors have degraded the health of coral reefs of Israel in the Gulf of Aqaba especially in the shallow, easily-accessible areas (Hawkins and Roberts 1994; Loya 2004). The coral reefs along the short Israeli coastline (around Eilat) are among the most frequently 'dived' reefs in the world, involving over 250,000 dives·year^{-1} (Zakai and Chadwick-Furman 2002). The impacts are mainly raising sediments during diving and direct breakage. Such perturbations have resulted in a high percentage of algal cover (up to 72 %) in shallow reefs off the coast of Eilat, while in the nearby area of Aqaba turf cover is only around 6 % (Bahartan et al. 2010). Also, studies in the Gulf of Aqaba showed that coral recruits are negatively correlated to algal biomass (Stambler et al. 2008).

Israel has taken major management steps to improve the health of the coral reef ecosystems in its coastal waters. Some practical actions include the complete cessation of sewage pollution from the city of Eilat since 1995 and the 2008 closure of sea bream farms, which had been a major source of nutrient enrichment (Loya 2004; Stambler et al. 2008). Israel has declared the area of the shallow fringing reefs (ca 1.2 km long) in its Gulf of Aqaba as a nature reserve. In fact, it was the first nature reserve in Israel, declared as early as 1960 (Loya 2004).

Fisheries

In 1949, Israel conducted an experimental fishing expedition from the port of Eilat which confirmed the potential for a commercial fishery in the region, previously exploited only by poorly-documented traditional fisheries. Due to the limited area available for fishing (Farid 1984), Israel expanded its fishery beyond the inner Gulf of Aqaba, south to the coast of Eritrea beginning in 1957 and continued until the beginning of the 1970s when it stopped due to political instability in the region (Ben-Tuvia 1968; Giudicelli 1984). The predominant gears utilized by fishers in Israel are gill-nets, and trammel nets (Ben-Tuvia 1968). In the past, beach seines were used but they are now illegal. The few active fishing boats are 7–11 m long (Shapiro 2007). The fishing vessels used to fish in the Eritrean waters were larger trawlers with higher capacity than the boats used in Israeli Gulf of Aqaba waters (Ben-Yami 1964). The Gulf of Aqaba is not very suitable for trawl fishing as it has narrow shelf directly adjacent to very deep waters. The limited shallow waters are also covered with coral reefs and rocky bottoms. Thus all fishing is done by small scale fisheries using small boats and it occurs in the northernmost coast, where the bottom is covered in seagrass rather than coral reefs.

The fisheries in the Israeli part of the Gulf of Aqaba can be divided into three sectors. The main sector is the artisanal fishery, which accounts for the majority of the catch. The artisanal fishers use mainly gill-nets and trammel nets as well as some traps. They rarely set their nets in or close to the coral reef areas to target reef fishes such as snappers, and groupers. More often, they target pelagic fishes such mackerels and tunas or demersal fish such as goatfishes and rabbitfish (Sarig 1982; Shapiro 2008). The artisanal fishery was reduced from 12 licenced vessels in 2007 to only 5 currently. The second sector is the subsistence fishery, which includes small fishing operations where the catch is mainly for direct consumption, and not for sale. This type of fishery is declining in Israel as its society is becoming more affluent. However, we include in this group the catch by the artisanal fishery that is consumed by the crew or given to family and friends, which can be substantial (Snovsky and Shapiro 2004). The third sector is the recreational fishery, which is relatively recent and is a leisure activity. Recreational fishing has gained prominence as Israeli income increased (Zakai and Chadwick-Furman 2002), and is practiced both by the coastal communities in and around Eilat and by tourists visiting the area, who pay to go on fishing trips in the Gulf of Aqaba and target mostly tuna.

Fisheries Catches

The reconstructed total catch of Israel in the Gulf of Aqaba increased continuously from its low value of a little more than 60 t·year^{-1} at the beginning of the 1950s to a peak of 346 t in 1980 (Fig. 8.4). Thereafter, it exhibited more fluctuations and remained at relatively lower values, except for a smaller peak in 1996. There are large discrepancies between

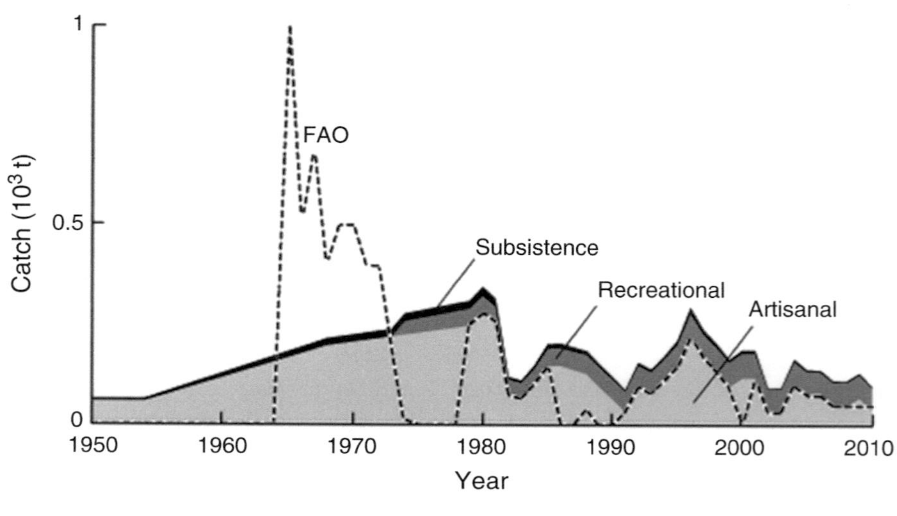

Fig. 8.4 Total catch of the Israeli fishery in the Gulf of Aqaba by sectors and the total catch Israel reported to FAO from 1950 to 2010

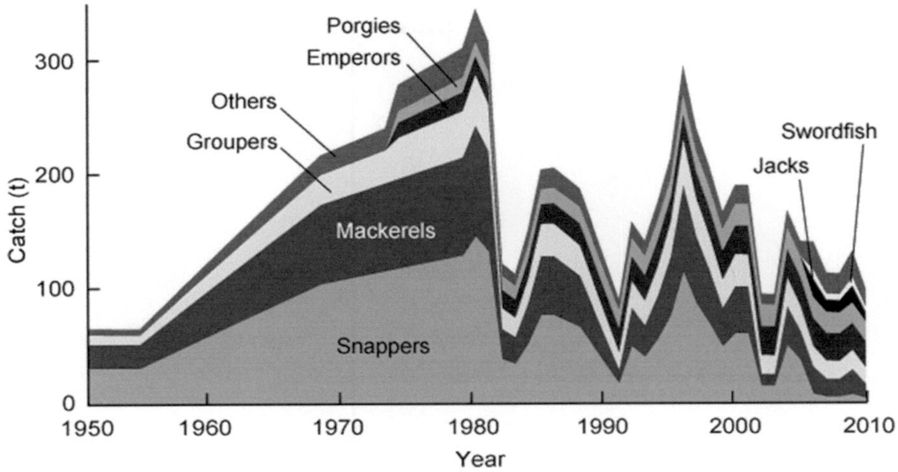

Fig. 8.5 Composition of the total catch of the Israeli fisheries in the Gulf of Aqaba from 1950 to 2010

the reconstructed catch and the data submitted to FAO by Israel, with the latter ranging from zero catches from 1950 to 1964 to a peak of 1,000 t in 1965. While the early catches of zero show the absence of reporting and do not mean there was no fishing, the high values include what Israel caught outside its EEZ, in the Eritrean EEZ, because FAO's catches are categorized by FAO's 'statistical area', and the entire Red Sea falls into one of these statistical areas (the Western Indian Ocean). However, our catch reconstruction focuses on the use catch data for ecosystem management, which makes it imperative to report the EEZ where the catch originated, preferably with a clear indication who caught the fish as well. Thus, the Israeli catches from Eritrean waters are reported in the catch reconstruction of Eritrea (see Chap. 4). Starting 1979, the FAO data matched with the artisanal catches, the only sector reported to FAO, except when the FAO values were zero, which again was due to lack of reporting.

The artisanal sector had the highest contribution to the reconstructed total catch of the Israeli fishery in the Gulf of Aqaba, contributing 76 %. The second was the recreational fishery with 18 %; the subsistence fishery (6 %) was last (Fig. 8.4). Artisanal and subsistence fisheries operated for the entire period, 1950–2010, while the recreational fishery began only in 1974, it shares a good proportion of the total catch at the present.

Overall, the reconstructed catch was 1.4 times the catches reported to FAO. If the years where Israel was fishing in Eritrean waters, from 1958 to 1968, are excluded from the analysis, the reconstructed catch is 1.8 times the catches reported to FAO. The unreported catch accounts for 62 % of the reconstructed catch. The reported catch (part of the reconstructed catch accounted in the FAO data) represented only 38 %.

The catch composition of the total catch reflected more or less that of the artisanal fishery. The contribution of snappers (Lutjanidae) was reduced from 47 % in the artisanal fishery to 38 % in the total catch and that of mackerels (tuna-like) fishes was also reduced from 32 % to 26 %. The contributions of these taxa were reduced by an increased contribution of the recreational fishery, but they still were the dominant taxa in the total catches (Fig. 8.5).

When we look at the sectors separately, the catch composition of the artisanal fishery was dominated by snappers (Lutjanidae) accounting for 47 % of the reconstructed total catch. The second most important group was pelagic species mackerels (Scombridae), with a contribution of 32 % and groupers (Serranidae) was third with 12 % (Fig. 8.6a). These three groups accounted for more than 90 % of the reconstructed total catch. The diversity of the total catch was higher in the later years with some taxa, mainly pelagic, that were not dominant in the early years were represented with a higher percentage in the later years. The catch composition of the subsistence fishery was the same as that of the artisanal fishery (Fig. 8.6b), because the catch composition ratios used in the artisanal fishery were used for the subsistence fishery as well.

The total recreational catch estimate for Israel in the Gulf of Aqaba was generally low, a maximum of 63 $t \cdot year^{-1}$ in the early 2000s. It increased continuously until it reached its peak and then declined slowly (Fig. 8.6c). As compared to the Israeli recreational fishery in Mediterranean, where the catch of recreational fishery was comparable to the that of the artisanal fishery (Edelist et al. 2013), our estimate for the Gulf of Aqaba was very conservative; only 24 % of the artisanal catch. The composition of the recreational fishery was dominated by three taxa.

Some of the fluctuations of the fisheries are attributed to political instability it in the region. The magnitude of the fishery being small, it does not get a lot of attention from the government to assess and manage the resource, especially compared to the resources allocated to coral reef conservation in Eilat. For effective policies to manage the fisheries to be developed and implemented, consistent time series data

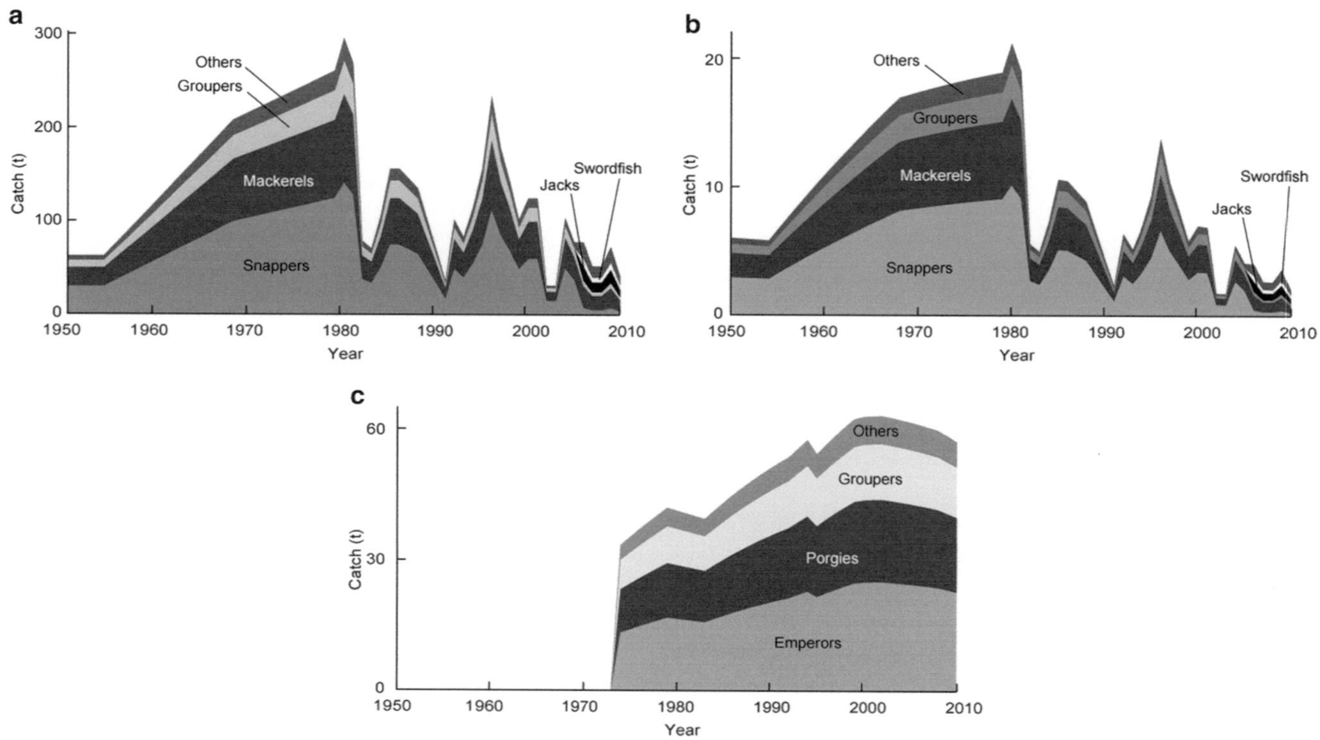

Fig. 8.6 Catch composition of (**a**) artisanal (**b**) subsistence and (**c**) recreational fisheries of Israel in the Gulf of Aqaba from 1950 to 2010

are essential (Caddy and Gulland 1983; Tesfamichael 2012). This study showed differences between the reconstructed catch and data supplied to the FAO by the Israeli Ministry of Agriculture. One source of discrepancy is the fact that the 'official data' is for a larger spatial area (an FAO area) than what we are trying to reconstruct (an EEZ), and therefore do not necessarily refer to Israeli waters, but also to waters further south, in Eritrea. However, even for the period after fishing in the southern Red Sea waters ended, the reconstructed catch is 65 % higher than what Israel reported to the FAO; such 'official' underestimates are common (Zeller and Haprer 2009).

Sources and Methods

The total marine catches of Israel within its EEZ in the Gulf of Aqaba were estimated for the period from 1950 to 2010. The required data were primarily obtained from government reports and the scientific literature. Since the fishery is dominated by artisanal fisheries, which use selective gear, there is no discarded catch or, if it exists, it is negligible. Emphasis was given to 'hard' estimates of the catch of distinct fisheries, which were used as 'anchor points' (Zeller et al. 2006, 2007) between which estimates for missing years could be obtained by interpolation. A spatial distribution of the catch is given at www.seaaroundus.org.

Artisanal Fishery

Fishery catch data were obtained from '*Bamidgeh*', a publication of the Israeli Department of Agriculture for some years between 1954 and 1985, after which we used reports from '*The Fisheries and Aquaculture of Israel*'. The 1954 total catch estimate, from Cohen (1957), was carried back to 1950 as the gear and fishing grounds appear to have remained similar during those years. The following sources were used for 1968 (Ben-Tuvia 1968), 1979 and 1980 (Sarig 1982), 1986 (Sarig 1987), 1988 (Anonymous 1992), 2003 (Snovsky and Shapiro 2004), 2004 (Shapiro 2005), 2005 (Shapiro 2006), 2006 (Shapiro 2007) and 2007 (Shapiro 2008). For periods where data were missing, FAO data were used selectively. The FAO data for Israel have some inconsistency, mainly reports of zero catches for some periods; however, we were able to get non-zero estimates from national sources. On the other hand, for some periods, the catch values reported in the FAO database matched the sources from the Israeli fishery administration. It is safe to conclude that Israel reported its fishery catch in the Red Sea accurately to FAO, for some periods, vis-à-vis its national reports. Thus, we used FAO data from 1981 to 1985 because the FAO data and other sources for neighboring years (1979, 1980 and 1986) were the same. Similarly, FAO data were used from 1991 to 2002 and 2008. FAO data were not used for 1955–1967, 1969–1978, 1987, 1989–1990, where FAO data contained numerous zero catches. Instead the catches were estimated using interpolation. For 2009 and 2010, data from Israeli Department of Fisheries Statistical Yearbooks, DoFSY (Dr. Dori Edelist, pers. comm.) were used; these values differed from the FAO dataset.

The composition of the catch, predominantly comprised of snappers was obtained from Sarig (1982) and was used from 1950 to 2005. However, since 2006, the catch composition changed, as pelagic species started to become abundant in the catch. Thus, for 2006–2010, catch composition data obtained from DoFSY (Dr. Dori Edelist, pers. comm.) were used. Israel fished in Eritrean waters starting in 1958 (Ben-Tuvia 1968; Sarig 1969). Since these catches originated from Eritrean waters, they are reported in the Eritrea catch reconstruction (see Chap. 4 for details on this fishery).

Subsistence Fishery

Israel publishes its annual fishery statistics for its fishery in the Gulf of Aqaba, and that is also what is reported to FAO, at least for most years. The reports clearly state that the data do not include part of the catch that is consumed by the crew and catch given to families and friends (Snovsky and Shapiro 2004). This constitutes the subsistence fishery in our report. The subsistence catch in the Israeli Gulf of Aqaba fishery was estimated as a percentage of the artisanal catches. Based on interviews with fishers (Dr. Dori Edelist, pers. comm.), 5 % of the total catch of artisanal fishers is consumed by the crew or given freely to family and friends, which was used as an anchor point for 2010. As observed in other Red Sea countries, the ratio is generally higher for earlier years (see other chapters in this volume). We assumed 10 % in 1950. The ratio was interpolated from 1950 to 2010. The composition of the subsistence fishery was adapted from that of the artisanal fishery

Recreational Fishery

The recreational fishery of Israel in the Gulf of Aqaba was calculated based on the population of its largest coastal city, Eilat on the Red Sea coast. The recreational fishery was assumed to start in 1974, after the 1973 war. A participation ratio of 0.12 was used in the calculations (Cisneros-Montemayor and Sumaila 2010) and the number of days fished per year was assumed to be 20; these data were applied to the population of Eilat, which had an average of a little less than 1 % of the total Israeli population from 1974 to 2010. We assumed 1 % and multiplied the participation ratio by 100. The catch rate per day was assumed to be 1 $kg \cdot day^{-1}$ for 1974 and 0.5 $kg \cdot day^{-1}$ in 2010. Catch rates were interpolated for the intervening years. The catch composition of the recreational fishery was adapted from Saudi Arabia's recreational fishery in the Red Sea, where emperors accounted for 40 %, sea breams 30 %, groupers 20 % and 'others' 10 % (see Chap. 6).

Comparing Reconstructed Catches with FAO Statistics

The reconstructed catches of Israel in its Red Sea EEZ were compared to the catch data reported for Israel in the FAO database. Since the subsistence and recreational fisheries were not reported at all, only the reconstructed artisanal catches were compared to the FAO data. In order to be able to compare taxon by taxon, the FAO data, when only totals were given, were divided into the components using the ratios in the reconstructed catches. Large quantities of brushtooth lizardfish (*Saurida undosquamis*) and narrow-barred Spanish mackerel (*Scomberomorus commerson*) catches were reported in the FAO data, mainly from 1965 to 1972, which were higher than the reconstructed catch. This is the period when Israel fished in Eritrean waters, and thus they were excluded from the taxon-to-taxon comparison as they were deemed to be caught outside the Israeli EEZ. For other taxa, if the amount in the reconstruction was higher than its value in the FAO database, the difference was assigned as 'unreported catch'; if it was the opposite, then it was considered 'over-reported catch'. The part of the reconstructed catch that is accounted in the FAO data is referred as 'reported catch' in our result.

Acknowledgements We would like to thank Dr. Dori Edelist from the University of Haifa for providing insights and data for the Israeli fishery in the Gulf of Aqaba and reviewing the chapter, which was enriched by his comments. We also thank Dr. James Shapiro for supplying data on Israel's fisheries. This is a contribution of *Sea Around Us*, a scientific collaboration between the University of British Columbia and The Pew Charitable Trusts.

References

Anonymous (1992) Israel fisheries in 1988. Thessaloniki, Greece

Bahartan K, Zibdah M, Ahmed Y, Israel A, Brickner I, Abelson A (2010) Macroalgae in the coral reefs of Eilat (Gulf of Aqaba, Red Sea) as a possible indicator of reef degradation. Mar Pollut Bull 60(5):759–764

Ben-Tuvia A (1968) Report on the fisheries investigation of the Israel South Red Sea expeditions, 1962. Sea Fish Res Stn Haifa Bull 52:21–55

Ben-Yami M (1964) Report on the fisheries in ethiopia. Ministry of Foreign Affairs, Department for International Cooperation, Jerusalem, 106 p

Berumen ML, Hoey AS, Bass WH, Bouwmeester J, Catania D, Cochran JEM, Khalil MT, Miyake S, Mughal MR, Spaet JLY, Saenz-Agudelo P (2013) The status of coral reef ecology research in the Red Sea. Coral Reefs 32(3):737–748

Caddy JF, Gulland JA (1983) Historical patterns of fish stocks. Mar Policy 7(4):267–278

Cisneros-Montemayor A, Sumaila UR (2010) A global estimate of benefits from ecosystem-based marine recreation: potential impacts and implications for management. J Bioecon 12(3):245–268

Cohen SB (1957) Israel's fishing industry. Geogr Rev 47(1):66–85

Edelist D, Scheinin A, Sonin O, Shapiro J, Salameh P, Rilov G, Benayahu Y, Schulz D, Zeller D (2013) Israel: reconstructed estimates of total fisheries removals in the Mediterranean, 1950–2010. Acta Adriat 54(2):253–264

Farid AM (1984) The Red Sea: prospects for stability. St. Martin's Press, New York, 173 p

Giudicelli M (1984) The Ethiopian fisheries: situation, development needs and opportunities. A report prepared for the Fishing Planning and Development Project, FAO, Rome, 157 p

Hargreaves P (ed) (1981) The Red Sea and Persian Gulf. Wayland Publishers, Hove, 67 p

Hawkins JP, Roberts CM (1994) The growth of coastal tourism in the Red Sea: present and future effects on coral reefs. Ambio 23(8):503–508

Khalaf MA, Disi AM (1997) Fishes of the Gulf of Aqaba. Marine Science Station, Aqaba, 252 p

Lapidoth-Eschelbacher R (1982) The Red Sea and the Gulf of Aden, International straits of the world. Martinus Nijhoff Publishers, The Hague, 265 p

Loya Y (2004) The coral reefs of Eilat – past, present and future: three decades of coral community structure studies. In: Rosenberg E, Loya Y (eds) Coral health and disease. Springer, Berlin/Heidelberg, pp 1–34

Loya Y, Genin A, el-Zibdeh M, Naumann MS, Wild C (2014) Reviewing the status of coral reef ecology of the Red Sea: key topics and relevant research. Coral Reefs 33(4): 1179–1180

MoE (2002) Protecting the Gulf of Eilat/Aqaba. Ministry of the Environment. Israel Environ. Bullet. 25(3):5

Ormond R, Edwards A (1987) Red Sea fishes. In: Edwards AJ, Head SM (eds) Red Sea. Pergamon Press, Oxford, pp 251–287

Pauly D, Zeller D (2003) The global fisheries crisis as a rationale for improving the FAO's database of fisheries statistics. In: Zeller D, Booth S, Mohammed E, Pauly D (eds) From Mexico to Brazil: Central Atlantic fisheries catch trends and ecosystem models. Fisheries Centre Research Reports, vol 11(6), Fisheries Centre, University of British Columbia, Vancouver, pp 1–9

Sarig S (1969) Fisheries and fish culture in Israel in 1968. Isr J Aquacult Bamidgeh 1(40):95–113

Sarig S (1982) Fisheries and fish culture in Israel in 1980. Isr J Aquacult Bamidgeh 34(1):1–32

Sarig S (1987) Fisheries and fish culture in Israel in 1986. Isr J Aquacult Bamidgeh 34(1): 1-35

Shapiro J (2005) The fisheries and aquaculture of Israel 2004. Ministry of Agriculture, Department of Fisheries, Tiberias, 33 p

Shapiro J (2006) The fisheries and aquaculture of Israel 2005. Ministry of Agriculture and Rural Development, Department of Fisheries and Aquaculture, Tiberias, 41 p

Shapiro J (2007) The fisheries and aquaculture of Israel 2006. Ministry of Agriculture, Department of Fisheries, Tiberias, 47 p

Shapiro J (2008) The fisheries and aquaculture of Israel 2007. Ministry of Agriculture, Department of Fisheries, Tiberias, 48 p

Sheppard C, Price A, Roberts C (1992) Marine ecology of the Arabian region: patterns and processes in extreme tropical environments. Academic, London

Sneh A, Friedman GM (eds) (1985) Hypersaline ecosystems: the Gavish Sabkha. Springer, New York

Snovsky G, Shapiro J (2004) The fisheries and aquaculture of Israel 2003. Ministry of Agriculture, Department of Fisheries, Tiberias, 37 p

Stambler N, Zakai D, Shaked Y, Genin A (2008) The coral reefs of Eilat, Israel. In: Wilkinson C (ed) Status of coral reefs of the world: 2008. Global Coral Reef Monitoring Network and Reef and Rainforest Research Centre, Townsville, p 158

Tellawi A-MM (2001) Conservation and sustainable use of biological diversity in Jordan: first national report of The Hashemite Kingdom of Jordan on the implementation of Article 6 of the convention on biological diversity. The General Corporation for the Environment Protection, Amman, 118 p. Available at: www.cbd.int/doc/world/jo/jo-nbsap-01-en.doc

Tesfamichael D (2012) Assessment of the Red Sea ecosystem with emphasis on fisheries. PhD thesis, University of British Columbia, Resource Management and Environmental Studies, Vancouver, 241 p

Zakai D, Chadwick-Furman NE (2002) Impacts of intensive recreational diving on reef corals at Eilat, northern Red Sea. Biol Conserv 105(2):179–187

Zeller D, Harper S (eds) (2009) Fisheries catch reconstructions: Islands, Part I. Fisheries Centre Research Reports, vol 17(5), Fisheries Centre, University of British Columbia, Vancouver, 104 p

Zeller D, Booth S, Craig P, Pauly D (2006) Reconstruction of coral reef fisheries catches in American Samoa, 1950–2002. Coral Reefs 25(1):144–152

Zeller D, Booth S, Davis G, Pauly D (2007) Re-estimation of small-scale fisheries catches for U.S. flag island areas in the Western Pacific: the last 50 years. US Fish Bull 105(2):266–277

An Exploration of Ecosystem-Based Approaches for the Management of Red Sea Fisheries

9

Dawit Tesfamichael

Abstract

The Ecopath with Ecosim (EwE) modelling tool was used to simulate trophic interactions in the Red Sea ecosystem, with emphasis on its fisheries. Time-dynamic simulations were run to quantify the impact of fisheries, which represent the main anthropogenic impact on the ecosystem. The model was fitted to a time series of observed catch and effort data to improve its ability to mimic changes in the Red Sea ecosystem. EwE was also used to predict the consequences of different fishing scenarios: maintaining the status quo, banning all fishing, and projecting into the future at the present growth rate of the fisheries. Monte Carlo simulations were used to examine the sensitivity of the predictions to changes in model input parameters and the risk of fish abundance falling below selected thresholds. Equilibrium surplus-yield analyses were carried out on the major groups affected by the fishery. Finally, the model was used to examine the conflict between artisanal and industrial fisheries in the Red Sea by running scenarios where the fishing effort of each of these sectors was doubled.

Keywords

Ecosystem based management • Ecosystem modeling • Ecopath with Ecosim • Trophic interactions • Fisheries management • Time series fitting • Red Sea

Introduction

Quantitative assessment of fisheries has evolved in the last five decades from single-species assessment (Beverton and Holt 1957) to multispecies evaluation and recently into ecosystem-based management (Browman 2000; Pikitch et al. 2004). Each step in this progression addressed certain questions pertinent at the time of their development. This progression is continuing as new knowledge is acquired about ecosystems, including human interactions, and drawbacks of the already existing approaches are identified. The more recent approach, ecosystem-based management (EBM), attempts to put fisheries management into a 'holistic' framework, which attempts to avoid the pitfalls of reductionism. Lots have been written about EBM, with some authors attempting to define and/or frame it (Link 2002; Pikitch et al. 2004), while others have developed conceptual or software tools for its implementation (Brodziak and Link 2002; Smith et al. 2007). Overall, EBM's acceptance has grown over time, and it is now under serious consideration by researchers and practitioners as well, although still poorly implemented (Pitcher et al. 2009). Ecosystem modelling is an important component of EBM that enables us to translate the ideas of EBM into workable quantitative assessment tools (Plagányi 2007). Ecopath with Ecosim (EwE) is one of these tools (Pauly et al. 2000), and it has been used widely for different ecosystem types. Here, the construction and

D. Tesfamichael (✉)
Fisheries Centre, University of British Columbia, Vancouver, BC, Canada

Department of Marine Sciences, University of Asmara, Asmara, Eritrea
e-mail: dawittes@gmail.com

D. Tesfamichael, D. Pauly (eds.), *The Red Sea Ecosystem and Fisheries*, Coral Reefs of the World 7,
DOI 10.1007/978-94-017-7435-2_9

application of an EwE model of the Red Sea is documented, and then used to assess the fisheries in an ecosystem-based framework.

The Red Sea is one of areas identified as Large Marine Ecosystems (LME), large regions of the world oceans delineated based on their physical parameters, ecology, and exploitation history (Sherman and Alexander 1986). Although the management of the fisheries is performed by individual countries in their own Exclusive Economic Zones (EEZs), it would be helpful to obtain a general ecological understanding of the ecosystem as a whole. Thus, the model incorporates the entire Red Sea and the major components of its ecosystems, from primary producers to top predators, as well as humans, impacting the Red Sea through fisheries.

Habitat and trophic parameters of organisms are very important for ecosystem modelling. The following habitat definitions based on FishBase (Froese and Pauly 2014) are used explicitly in building the model and to categorize organisms by the depth of their habitats:

Reef associated: Organisms living and/or feeding on or near coral reefs between 0 and 200 m;

Pelagic: Organisms occurring mainly in the water column between 0 and 200 m, and not feeding on benthic organisms;

Demersal: Organisms living and/or feeding on or near the bottom between 0 and 200 m;

Benthopelagic: Organisms living and/or feeding on or near the bottom, as well as in mid-water between 0 and 200 m;

Bathypelagic: Organism living in the oceanic zone between 1,000 and 4,000 m, between the mesopelagic layer above and the abyssopelagic layer below, and/or living or feeding in open waters at depths between 1,000 and 4,000 m;

Bathydemersal: Organisms living and/or feeding on or near the bottom below 200 m.

These are habitat descriptions in relation to the location of mainly fishes in the ecosystem types listed in FishBase. However, this is not an exhaustive list of habitats; thus, e.g., seagrass, sea weeds and other habitats are explicitly included in the model.

The Ecopath Model

Ecopath is an ecosystem modelling tool which accounts for the energy transfers in an ecosystem (Polovina 1984; Christensen and Pauly 1992). Its basic feature is that energy can be transferred from one ecosystem group to another, but the overall transfers and biomass are in steady-state for a period of arbitrary duration, i.e., that the ecosystem remains self-similar. This is in line with the first law of thermodynamics, which states that energy can be changed from one form to another, but not created or destroyed. The first Ecopath model (Polovina 1984) was developed to study the ecosystem of the French Frigate shoals, an atoll in the Northwestern Hawaiian islands. There, a group of scientists were researching and estimating different aspects of the ecosystem, and Ecopath was developed as a framework to assemble their various field estimates into a coherent quantitative representation of the atoll's ecosystem. Ecopath has since been applied to a wide range of ecosystems (Christensen and Pauly 1993). In the early development of Ecopath, its steady-state or equilibrium assumption was understood to mean that the mean annual biomass for each species group does not change from year to year (Polovina 1984). In the later development of EwE (Christensen and Pauly 1992), this assumption was replaced by an emphasis on 'mass- balance', implying that there could be change in biomass over time (i.e., biomass accumulation or decline), as long as it was accounted for, i.e., the net change over the whole system (including exchanges with adjacent ecosystem) remained zero.

Ecopath has two master equations. The first one states that biological production within a group equals the sum of mortalities by predation and fisheries, net migration, biomass accumulation and other unexplained mortality, i.e.:

$$B_i \cdot (\frac{P}{B})_i Y_i + \sum_{n}^{j=1} B_j \cdot (\frac{Q}{B})_j \cdot DC_{ij} + E_i + BA_i + B_i (\frac{P}{B})_i \cdot (1 - EE_i) \tag{9.1}$$

where B_i and B_j are biomasses of prey (*i*) and predator (*j*), respectively; P/B_i is the production/biomass ratio; Y_i is the total fishery catch rate of group (*i*); Q/B_j is the consumption/biomass ratio; DC_{ij} is the fraction of prey (*i*) in the average diet of predator (*j*); E_i is the net migration rate (emigration – immigration); and BA_i is the biomass accumulation rate for group (*i*). EE_i is the ecotrophic efficiency, i.e., the fraction of group mortality explained by predation (or grazing) within the model.

The second equation states that the consumption within a group equals the sum of its production, respiration and unassimilated food, i.e.:

$$B \cdot (\frac{Q}{B}) = B \cdot (\frac{P}{B}) + (1 - GS) \cdot Q - (1 - TM) \cdot P + B(\frac{Q}{B}) \cdot GS \tag{9.2}$$

where GS is the fraction of the food that is not assimilated; and TM is the trophic mode expressing the degree of heterotrophy (0 and 1 represent autotrophs and heterotrophs, respectively; intermediate values represent facultative consumers).

Predation (or grazing) mortality is the process that connects the different groups in the system. What is predation mortality for the prey is consumption to the predator, and Ecopath uses a set of algorithms to simultaneously solve the above linear equations for all the functional groups under the assumption of mass balance. The basic inputs of Ecopath are biomass, production per unit biomass (P/B) and consumption per unit biomass (Q/B). Because of the mass-balance assumption, Ecopath can estimate one free parameter of the basic input for each group. Diet composition must be entered as input to Ecopath, i.e., the elements of diet compositions are not usually estimated by the model.

Ecosim

Ecopath gives a snapshot of the ecosystem at one time. Ecosim, on the other hand, runs time dynamic simulations (Walters et al. 1997) and can be used in policy exploration. A mass-balanced Ecopath model is used for Ecosim runs, which are driven by changes in fishing mortality. Change in biomass over time and the flux of biomass among the groups is expressed by varying biomasses and harvest rates. Simulations are used to fit the predicted biomass to independent time series data, but changes in ecosystem can also be driven by climate change or fluctuation in nutrient supply. In the policy exploration routine of Ecosim, four policy options are included: maximize the rent (profits) from fisheries; maximize jobs in fishing, maximize the biomass of certain species, and maximize ecosystem 'health', i.e., the biomass of groups with low turnover rates, such as marine mammals or sharks (Christensen et al. 2008). The basic differential equation used in Ecosim is:

$$\frac{dB_i}{dt} = g_i \sum_{n}^{j=1} f(B_j,B_i) - \sum_{n}^{j=1} f(B_j,B_i) + I_i - (M_i + F_i + e_i) \cdot B_i \tag{9.3}$$

where dB_i/dt represents biomass change rate of group (i) during the interval dt; g_i represents the net growth efficiency (production/consumption ratio); I_i is the immigration rate; M_i and F_i are natural and fishing mortality rates of group (*i*), respectively; e_i is emigration rate; and $f(B_j,B_i)$ is a function used to predict consumption rates of predator (*j*) on prey (*i*) according to the assumptions of foraging arena theory (Walters and Martell 2004; Walters and Christensen 2007), as modulated by the predator-prey vulnerability parameter.

Besides providing a snapshot of the ecosystem (Ecopath) and time dynamics (Ecosim), the EwE package also has a routine for dynamic spatial simulations called Ecospace (Walters et al. 1999), which overcomes the assumption of homogenous spatial distribution of organisms that is implicit in Ecopath and Ecosim. The use of Ecospace so far has been mainly in placement and evaluation of marine protected areas (MPA) (Walters 2000; Beattie et al. 2002; Varkey et al. 2012). Ecotracer is another component of EwE dealing with the movement and accumulation of contaminants and tracers in the food web (Christensen and Walters 2004). For further accounts of EwE, particularly for the theoretical and mathematical backgrounds, see Walters et al. (1997) and Christensen et al. (2008). Plagányi and Butterworth (2004) and Plagányi (2007) also present critical reviews of the EwE approach.

So far, an ecosystem model for the entire Red Sea does not exist, although a model exists for the Eritrean coast (Tsehaye and Nagelkerke 2008). The main objective of this chapter is to assess the Red Sea fisheries in ecosystem-based framework. This was accomplished by building an ecosystem model of the Red Sea that does the following:

- Presents a quantitative description of the structure of the ecosystem in terms of the 'players' (groups, or 'state variables'), which include the organisms living in that sea and the fisheries, and their interactions, i.e., the flux of energy from one group to another, and including basic ecosystem parameters for each group in the model;
- Quantifies and evaluates the effect of fisheries on the system;
- Explores the interaction between the different fisheries, and their policy implications. The specific question addressed is whether the industrial and artisanal fisheries have negative impact on each other (the assumption that they do has been a frequent cause of conflict).

Materials and Methods

Ecopath

Defining the boundaries of an ecosystem to be modeled can be difficult, especially in marine systems, where the boundary can be blurred, or vary through time. However, this is not a problem here, as the whole Red Sea is taken into consideration. The fact that the Red Sea is an enclosed sea with little exchange with neighboring ecosystems makes it ideal to be modeled as a unit.

The data needed to build an Ecopath model are extensive. The Red Sea organisms included in the model are divided into two categories, fish and non-fish, for the convenience of data acquisition (e.g., FishBase was used for fish) and approaches for calculating required parameters.

Fish Species

The Red Sea, a subtropical system, has relatively high species biodiversity. There are more than 1,400 fish species reported for the Red Sea (Froese and Pauly 2014). However, it is neither practical nor necessary for each species to be represented as a group by itself in the model; rather, only groups of similar species ('guilds') that function in similar manner need to be identified. Here, grouping was performed using characteristics that define the trophic interaction of the organisms: habitat, trophic level and size of fish. Using these parameters the fish species were grouped into 20 ecologically distinct functional groups (Table 9.A1 in the Appendix). The fish species that are major contributors to the catch of the different major gears in the Red Sea were kept in distinct groups, so that detailed analyses of their behaviours could be carried out.

The two important Ecopath input parameters, consumption and production per unit of fish biomass were calculated using population parameters from FishBase. Priority was given to data from research carried out in the Red Sea (as illustrated in Fig. 9.1); however, when data from the Red Sea could not be found, they were taken from similar ecosystems, i.e., coral reef ecosystems with similar mean annual temperature.

Consumption

Food consumption per unit biomass (Q/B) values for the fish species were taken from FishBase, preferably from the Red Sea. When the Q/B value was not given, the empirical equation developed by Palomares and Pauly (1998) was used:

$$\frac{Q}{B} = 7.964 \cdot 0.204 \log W_{\infty} + 1.965T + 0.083A + 0.532h + 0.398d \quad (9.4)$$

Fig. 9.1 Researcher taking picture of a large jack, Family Carangidae, Saudi Arabia (Photo: Andrew Bruckner)

where W_{∞} is asymptotic weight of the species (in g), T is mean annual temperature of the Red Sea, 27.71 °C, expressed as 1,000/(T °C + 273.1), A is the aspect ratio obtained from FishBase, h and d refer to the types of food consumed (i.e., for herbivores h = 1, d = 0; for carnivores h = 0, d = 0; for detritivores d = 1, h = 0).

When W_{∞} was not directly given it was calculated from length-weight relationship:

$$W_{\infty} = a \cdot L_{\infty}^{\ b} \quad (9.5)$$

where L_{∞} is the asymptotic length, and *a* and *b* are constants from FishBase.

When the aspect ratio was not available, a different empirical equation developed by Pauly (1986) was used to calculate the consumption per unit biomass (Q/B):

$$\frac{Q}{B} = 10^{6.37} \cdot 0.0313(\frac{1,000}{T}) \cdot W_{\infty}^{\ -0.168} \cdot 1.38^{Pf} \cdot 1.89^{Hd} \quad (9.6)$$

where T is the Red Sea mean annual temperature in degree Celsius (27.7 °C), Pf is feeding mode parameter set to 1 for predators and zooplankton feeders, and zero for other fish species, Hd is a diet composition parameter, set to 1 for herbivores and zero for omnivores and carnivores.

Production

Production per unit biomass (P/B) is equal to total mortality, which is the sum of natural mortality and fishing mortality (Z = M + F). For species that are not exploited P/B equals M. For all species, M estimates were taken from FishBase; when these were not available, the empirical formula of Pauly (1980) was used, i.e.,

$$\mathrm{M} = \mathrm{K}^{0.65} \cdot \mathrm{L}_{\infty}^{\ -0.279} \cdot \mathrm{T}^{0.463} \quad (9.7)$$

where K ($year^{-1}$) is a von Bertalanffy growth parameter and L∞ (total length, cm) is the asymptotic length both obtained from FishBase and T is Red Sea mean annual temperature (27.71 °C).

Biomass

Biomass data were not available for most of the fish species included in the model. However, searches resulted in some estimates, and these were used as starting values to parameterize the model. For pelagic fishes, data from an acoustic survey in the southern Red Sea (Massé and Araia 1997) were used. Also, for demersal fish, a trawl survey (Blindheim 1984) was used, as well as visual censuses for coral reef fishes (Roberts and Ormond 1987; Bouchon-Navaro and Bouchon 1989; Zekaria 2003). Abundance values of a wider range of organisms were also available (Antoine et al. 1997; Price et al. 1998; Tsehaye 2007).

Non-fish Groups

The non-fish groups include marine mammals, turtles (Fig. 9.2), birds, invertebrates and primary producers. Shrimp are the most important non-fish groups for fisheries. Hence, they are given their own functional group, as the main focus of the model is ecosystem-based assessment of fisheries in the Red Sea. Similar to the fish group, priority for non-fish group data was given to research from the Red Sea; in some cases, data from similar ecosystems were also used. For invertebrates, SeaLifeBase (Palomares and Pauly 2012) and a benthic invertebrate population dynamics database (Brey 2001) were used as sources. The list of the non-fish groups, jointly with their parameters and sources is given in Appendix (B).

Diet Matrix

Diet data for the fish species, unless specified otherwise, were obtained from FishBase. Priority was given for data from the Red Sea, but when not available, data from similar ecosystems were used. For the non-fish group, diet compositions were compiled based on similar coral reef ecosystem models of the Eritrean Red Sea coast (Tsehaye 2007), the Caribbean (Opitz 1996; Arias-González 1998), Indonesia (Buchary 1999; Ainsworth et al. 2007), and French Frigate Shoals in the Northwestern Hawaiian Islands (Polovina 1984). The complete diet matrix is given in Tesfamichael (2012).

Fishery

Fishery data for the model were taken from the catch reconstruction of Red Sea fisheries (see previous chapters). These fisheries can be divided into two main categories: artisanal and industrial. The major fishing gears from each group are represented in the model. For the artisanal sector the major gears are handlines, gillnets and beach seines, while the major industrial fishing methods are bottom trawling and purse seining. The three major artisanal gears reflect the ecosystem and behaviour of the fish they target, i.e., small pelagic fish are caught by beach seines, large pelagic by gillnet, and coral reef-associated fishes by handlines. These are not all the gears in these fisheries, but representative and the major ones in each habitat; there are other gears employed in the Red Sea, e.g., trammel net for pelagic species. This categorization is helpful for practical ecosystem-based management. As the main objective of the model is to explore the Red Sea fisheries at ecosystem level, the species that contribute the highest proportion to the catch of the various fishing gears were assigned to distinct functional groups in the model (for each gear). These groups are identified in the model by the gear name followed by 'fishes' e.g., fishes targeted by handlining are called 'handlining fishes'. The major taxonomic groups for each gear that are treated as separate functional groups in the model accounted for more than 80 % of the catch by the respective gears. These functional groups are: handline fishes, gillnet fishes, beach seine fishes, trawl fishes, purse seine fishes, shark, and shrimp. Sharks and shrimps are treated separately, because sharks are targeted by two types of gears, i.e., handlines and gillnets; while shrimps are targeted by only one gear (trawl); however, because of their importance for the fisheries they are assigned to separate group. The remaining species were divided among other functional groups by matching the catch compositions to the functional groups. The shark catch was similarly divided between handlining and gillnet, as both gears are used to catch sharks in the Red Sea. Discards from the trawl fishery were included in the model and were made to flow to detritus. The total catch values were expressed per unit area ($t \cdot km^{-2} \cdot year^{-1}$). The five fisheries are

Fig. 9.2 Sea turtle hatchlings heading to the sea, Eritrea (Photo: Yohannes Tecklemariam)

identified by the names of their respective gears: handline, gillnet, beach seine, trawl and purse seine,

Parameterizing/Balancing the Model

Parameterizing a model is the process wherein the mass-balance requirements for each group are simultaneously met. The model was parameterized following the procedure outlined in the Ecopath with Ecosim manual (Christensen et al. 2008), i.e., the input that were less reliable or whose value had been assumed were modified first. The diet matrix, being the most uncertain, was the input that was adjusted the most during balancing of the model, while P/B and Q/B were changed less, if at all. Model balancing was terminated when it fulfilled the requirements of balanced models: all EE less than 1, gross food conversion efficiency (GE, i.e., production/consumption) with in the range of 0.1–0.3 for fish, and all respiration/biomass ratios within a physiologically reasonable range.

Ecosim

Unlike Ecopath, which is static, Ecosim is a time-dynamic simulation, which can be fitted to time series data. This enabled verifying the parameterization of the Ecopath model, and after some adjustments, performing an equilibrium analysis with Ecopath, and an exploration of fishery policy scenarios with Ecosim.

Fitting to Time Series Data

A time series simulation was made to fit model predictions to independently calculated catch time series. This fitting exercise helps to validate the ability of the model to mimic the actual process in the ecosystem, including fisheries. A time series of fishing effort was needed for this exercise (Table 9. C1 in Appendix). The effort data for industrial fisheries were extracted from the database compiled in support of the publication by Anticamara et al. (2011).

In order to test the ability of the model to mimic the functioning of the Red Sea ecosystem, e.g., to predict the catch data from 1950 to 2010, it was made to run from 1930 to 2006, i.e., first to let the model mimic the situation before 1950 (with restored biomass of the predators that have been depleted by the fishery), so that it will be ready for the procedure of fitting to the independent data (Cox et al. 2002; Villy Christensen, pers. comm.). The year 2006 was used as a reference for the simulations because it was the latest year for which detailed data were available, especially for fishing effort (Note that the reference year, 2006 in this model, does not affect the procedure as the simulations take into account the whole period, but it is structurally necessary for the model to perform the calculations).

The procedure involves first scaling the time series of effort between 0 and 1, and taking the effort of 2006 to be 1. Then, the relative effort of 1950 is carried backward for few years (20–30 years, i.e., starting 1920 or 1930), and simulations were run in Ecosim, until biomasses are stabilized in 1950. Because the simulation stabilized when it was run from 1930, the simulation from 1920 was discarded and all simulations were done from 1930 to 2006. However, the time series fitting was done only from 1950 to 2010. Fishing effort levels of 1950 were very small compared to 2006, except for beach seining (Table 9.C1 in the Appendix), which was a vibrant fishery in the 1950s, especially in Eritrea, until it was largely abandoned. Thus, for the effort ratio of beach seine, instead of the high value of 1950, an arbitrary low ratio of 0.02 was used for the period from 1930 to 1950. The small effort values for all the fisheries from 1930 to 1950 allowed the model to assume an equilibrium characterized by high biomasses of top predators by the time the actual simulation started in 1950. That the biomasses of top predators were higher in 1950 before they were fished out in the following decades is a reasonable assumption. More importantly, those values were to be used only as a starting point for the time series fitting, which works by minimizing the sum of squares of the differences between the observed catch and CPUE data and the trends predicted by the model.

During the time series fitting, some of the basic Ecopath input parameters (biomass, P/B, Q/B and diet composition) were modified and the fit rechecked. This procedure was iterated a few times, and the model fine-tuned (primarily diet compositions, and secondarily the P/B ratios) until the best fit was achieved. Catch per unit effort (CPUE) was used as proxy for biomass to guide the time series fitting. Note that the emphasis of the fitting process was not on CPUE, but on the catch time series data, which appeared more reliable, given the detailed catch reconstruction work documented in the previous chapters.

Trophic Flow Parameter

A key parameter to be adjusted during time series fitting is vulnerability, a foraging arena parameter that regulates the flow between different trophic level groups (Walters and Martell 2004; Walters and Christensen 2007). The minimum value used is 1 when an increase in the biomass of predator does not cause noticeable change in predation mortality, a situation known as 'bottom-up control'. The other extreme occurs when an increase in biomass of predator produces noticeable change in predation mortality, i.e., 'top-down control'. Here, the parameterization of the vulnerability values for the Red Sea was done using both the automated vulnerability search routine in EwE and manually. The vulnerability search routine is an iterative procedure to identify predator-prey interactions that are necessary for the model (and presumably the ecosystem) to function. It uses a least-square method to optimize those critical vulnerabilities in order to recreate the observed time series of catch and CPUE. The search begins with all the interactions in the diet matrix, but later it focuses on the few that are highly influential. Another parameter which affects the feeding behavior of the animals and was adjusted during the fitting process was 'feeding time factor'. It is a measure of how fast organisms adjust their feeding behavior (i.e., their feeding times) so as to stabilize consumption per unit biomass. It ranges between 0, causing feeding time and hence time exposed to predation risk to remain constant, to 1, causing fast time response, which reduces vulnerability to predation (Christensen et al. 2008).

Model Stability and Uncertainty Analysis

The stability of the model was tested by subjecting it to three scenarios: (i) maintaining baseline fishing rates, (ii) assuming zero fishing rates for all the gears, and (iii) increasing fishing rates of each gear by 5 % each year, which is an overall increase of recent fishing rates. The stability test showed the model's behavior under varying functional group parameters and fishing pressure. Generally, if the model behaves in a realistic fashion, then it can be used for fishing policy exploration; otherwise, if unstable results are produced by the model, its use for policy development will not be warranted.

Under the three scenarios, the sensitivity of the model to changes in the basic input parameters was explored using Monte Carlo simulations. The biomasses of all the functional groups were allowed to vary within ±20 % of their original Ecopath values, while P/Q, Q/B and EE were varied ±10 %, then 100 Monte Carlo draws were made from a uniform distribution. The mean and the standard deviation (SD) were calculated for each simulation to establish a range of error for predictions. In addition, the depletion risk of the fishery groups in a population was explored through a viability analysis, i.e., an estimation of the probability that the biomass can drop, within a certain time frame, below a certain ratio of the original biomass.

Equilibrium Analysis

Once the model's stability was established and uncertainty analyses were performed, equilibrium analysis was carried out, which provides important results to assess the fisheries. The pertinent routine calculates the biomass and catch of the functional groups at different fishing mortality rates. EwE allows this analysis either by taking one group at a time and keeping all the other groups constant (which is thus similar to traditional single species stock assessment; or allowing interaction between groups (which is similar to multi-species stock assessment). For the Red Sea model, both options were explored.

Fishing Policy Exploration

Besides the three scenarios mentioned above, two additional scenarios were run using Ecosim simulations to explore the interaction between the artisanal and industrial fisheries in the Red Sea. This is very important for the region as conflicts between the two fisheries types are common, which has serious impact on the decision-making process. The two scenarios involved were: one where the fishing effort of the industrial sector was doubled without changing the artisanal effort, the other where the fishing effort of artisanal sector was doubled without changing the industrial effort. The simulations were run to predict the biomasses of all the groups until 2030.

Results

Ecopath

The key result of the Ecopath modelling part is a snapshot of the ecosystem with all the basic parameters satisfying all mass balance criteria as outlined above, i.e., all the ecotrophic efficiencies (EE) are less than 1 and respiration values are positive. This balanced model of the Red Sea (Table 9.1) was used to describe the Red Sea ecosystem using the diagnostic tools provided in EwE.

The food web with all the flows of energy among different groups sorted in the order of the trophic level of the groups is given in Fig. 9.3. The size of the squares is proportional to the biomass of the groups. Of all the living groups, the primary producers (phytoplankton, seagrass and algae) have biomasses that are considerably larger than all other groups. This is summarized in the food web pyramids both in terms of the biomasses and flows (Fig. 9.4; left and right pyramid, respectively). The left pyramid documents the high biomass at trophic level 1, which tapers off at higher trophic levels. Similarly, the pyramid to the right, which quantifies the ecosystem throughput, has a bottom level representing the flows to the first-order consumers (i.e., the herbivores), with the volume encompassed by subsequent level representing the higher trophic level flows. When drawn to the same scale, pyramids are useful to compare different systems, especially when the top angle of the flow pyramid, as is the case here, is inversely proportional to the geometric mean of the transfer efficiencies between trophic levels observed in the system (Christensen et al. 2008). The Red Sea model is compared with some tropical ecosystem models built using EwE and whose files were available in the

Table 9.1 Basic parameters of the balanced Red Sea model

Group No.	Group name	Trophic level	Biomass ($t \cdot km^{-2}$)	P/B ($year^{-1}$)	Q/B ($year^{-1}$)	EE	GE
1	Cetaceans	3.84	0.0610	0.044	5.914	0.025	0.007
2	Dugongs	2.00	0.0029	0.025	11.000	0.000	0.002
3	Birds	4.04	0.0068	0.380	20.000	0.026	0.019
4	Turtles	2.69	0.0555	0.150	3.500	0.137	0.043
5	Trawler fishes	3.38	0.0402	2.680	11.380	0.972	0.236
6	Purse seine fishes	3.53	0.0210	3.085	14.150	0.945	0.218
7	Beach seine fishes	3.09	0.1080	3.250	15.000	0.800	0.217
8	Handlining fishes	3.54	0.0700	1.300	7.887	0.688	0.165
9	Gillnet fishes	4.07	0.0265	2.000	8.000	0.950	0.250
10	Whale shark	3.28	0.0038	0.035	4.000	0.500	0.009
11	Sharks	4.16	0.0076	0.750	4.371	0.950	0.172
12	Rays	2.88	0.0040	0.373	3.000	0.400	0.124
13	Reef top predators	3.76	0.0197	1.052	4.000	0.950	0.263
14	Large reef carnivores	3.51	0.1100	1.240	5.500	0.344	0.225
15	Medium reef carnivores	3.43	0.1380	1.728	7.324	0.576	0.236
16	Small reef carnivores	3.21	0.3800	2.800	10.000	0.636	0.280
17	Reef omnivores	2.88	0.2630	2.700	13.890	0.950	0.194
18	Reef herbivores	2.00	0.2880	3.200	16.000	0.950	0.200
19	Large pelagic carnivores	3.82	0.1050	0.722	6.508	0.960	0.111
20	Small pelagic carnivores	3.44	0.2740	3.162	10.000	0.950	0.316
21	Pelagic omnivores	2.64	0.2660	2.828	10.000	0.950	0.283
22	Demersal top predators	3.58	0.0073	1.300	6.000	0.946	0.217
23	Large demersal carnivores	3.31	0.0160	1.500	7.000	0.439	0.214
24	Medium demersal carnivores	3.04	0.0620	1.990	8.000	0.920	0.249
25	Small demersal carnivores	2.96	0.2230	3.189	12.000	0.960	0.266
26	Demersal omnivores	2.16	0.2960	3.200	14.000	0.940	0.229
27	Demersal herbivores	2.00	0.3600	3.500	16.500	0.975	0.212
28	Benthopelagic fish	2.78	0.2350	1.800	6.000	0.970	0.300
29	Bathypelagic fish	3.11	0.0020	1.749	12.720	0.126	0.138
30	Bathydemersal fish	2.91	0.0040	1.260	6.940	0.831	0.182
31	Shrimp	2.09	0.0100	9.000	25.000	0.609	0.360
32	Cephalopods	2.92	0.3990	3.500	12.000	0.549	0.292
33	Echinoderms	2.10	0.5960	2.500	8.000	0.553	0.313
34	Crustaceans	2.19	0.8160	6.667	20.000	0.451	0.333
35	Molluscs	2.05	0.3680	9.000	30.000	0.556	0.300
36	Meiobenthos	2.07	0.2950	26.000	100.000	0.402	0.260
37	Corals	2.28	0.9280	2.800	9.000	0.527	0.311
38	Other sessile fauna	2.28	0.8500	3.200	12.000	0.368	0.267
39	Zooplankton	2.11	14.0000	52.000	178.000	0.363	0.292
40	Phytoplankton	1.00	21.5000	110.000	–	0.955	–
41	Seagrass	1.00	11.0000	9.000	–	0.015	–
42	Algae	1.00	38.0000	14.000	–	0.027	–
43	Detritus	1.00	80.0000	–	–	0.034	–

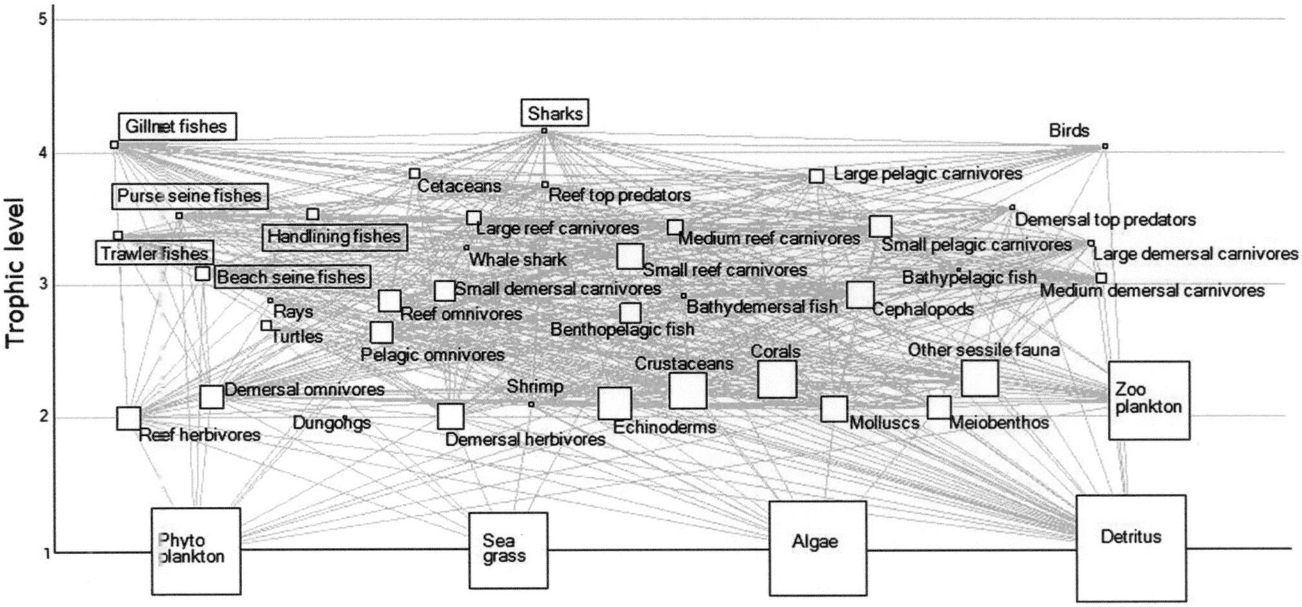

Fig. 9.3 Flow diagram of the food web of the Red Sea ecosystem. The sizes of the *rectangles* are proportional to the biomass of the functional groups. The fish groups names in *rectangle* within *rectangles* represent major fished groups

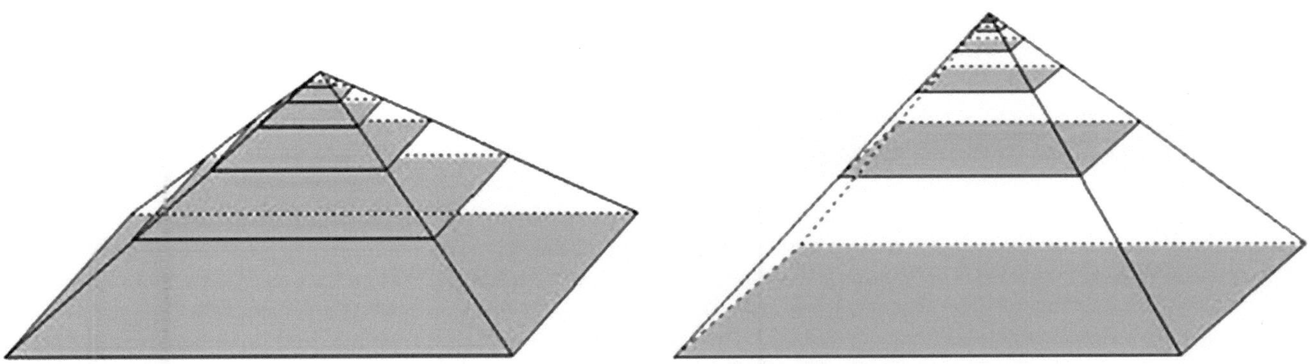

Fig. 9.4 Biomass (*left*, in t · km^{-2}) and flow pyramids (*right*, in t · km^{-2} · year^{-1}) for the Red Sea model

Ecopath website (www.ecopath.org; Table 9.2). In terms of fisheries, it is worth noting that the Red Sea has a very low total catch in relation to total biomass (excluding detritus). However, it should be pointed out here that many of the ecosystem to which the Red Sea may be compared are overfished. Thus, relatively low exploitation rate of the Red Sea should not be viewed as an inducement to massive industrial fisheries developments.

The data requirements for a EwE model are extensive, and thus, models can be categorized by the quality of the data used for constructing them. This is done using pedigree analysis. It is a routine in EwE which allocates the likely uncertainty associated with input parameters based on pre-defined categories according to the sources of the inputs. Parameters from quantitative research in the model area receive a higher pedigree index, implying lower uncertainty. In contrast, parameters estimated by Ecopath receive a lower pedigree index, and are more uncertain. Once the indices are assigned for all input parameters, the routine calculates an overall average ranging between 0 and 1 (inclusive), 1 referring to model built from only local, highly precise data (Christensen et al. 2005) – of which none exists (Morissette 2007). The Red Sea model had an overall pedigree of 0.43; in comparison, the analysis of 50 other models with an average of 27 groups resulted in a mean pedigree of 0.44 (Morissette 2007); thus, the model presented here is typical in its uncertainty, which is surprising, and somehow reassuring, given that the Red Sea is generally viewed as understudied. Table 9.2 gives the pedigrees of four

Table 9.2 Comparison of the Red Sea model with other tropical ecosystem models using system summary statistics

Criteria	Red Sea	Great Barrier Reef	Laguna Bay, Philippines	San Miguel Bay, Philippines	West Florida shelf USA
Total boxes	43.00	32.00	17.00	16.00	59.00
Living groups	42.00	30.00	16.00	15.00	55.00
Pedigree index	0.433	0.139	0.499	0.286	0.630
Sum of all consumption (t/km^2/year)	2,615.82	4,314.13	7,793.81	769.38	18,501.20
Sum of all exports (t/km^2/year)	1,665.10	1,119.89	5,901.51	516.19	903.44
Sum of all respiratory flows (t/km^2/year)	1,330.97	1,732.15	3,137.23	381.56	5,977.33
Sum of all flows into detritus (t/km^2/year)	1,723.53	4,038.89	6,544.32	931.41	17,273.88
Total system throughput (t/km^2/year)	7,335.00	11,205.00	23,377.00	2,599.00	42,656.00
Sum of all production (t/km^2/year)	3,756.00	3,920.00	10,838.00	1,080.00	14,071.00
Mean trophic level of the catch	3.40	2.49	2.08	3.00	3.51
Gross efficiency (catch/net p.p.)	0.000085	0.002971	0.031380	0.016502	0.000051
Total net primary production (t/km^2/year)	2,996.00	2,846.24	8,950.30	897.75	6,986.95
Total primary production/total respiration	2.25	1.64	2.85	2.35	1.17
Net system production (t/km^2/year)	1,665.03	1,114.09	5,813.06	516.19	1,009.62
Total primary production/total biomass	32.49	9.82	49.99	28.65	9.74
Total biomass/total throughput	0.01	0.03	0.01	0.01	0.02
Total biomass, excluding detritus (t/km^2)	92.22	289.87	179.05	31.34	717.61
Total catches (t/km^2/year)	0.25	8.46	280.86	14.82	0.36
Connectance Index	0.31	0.28	0.21	0.34	0.23
System Omnivory Index	0.24	0.23	0.14	0.17	0.26

other tropical and subtropical ecosystem models compared with that of the Red Sea model, where, again, the Red Sea model is intermediate in its uncertainty score.

The mixed trophic impact (MTI) matrix of Ecopath, documenting the effects that a small change in the biomass of one group could have on other groups was used to examine the reciprocal effects of various gears (Fig. 9.5). If we focus only on the fishery groups, it appears that the purse seine fishery has the highest positive impact on the biomass of fish targeted by the beach seine fishery. The handline fishery, on the other hand, has the highest negative impact on fish targeted by the purse seine and gillnet fisheries, and to a smaller extent by the beach seine fisheries. The beach seine fishery has the least impact on other fisheries.

Ecosim

Fitting to Time Series

After fine tuning the basic Ecopath input parameters, estimating vulnerability values and fitting the time series, the best fit between the observed and predicted catch was obtained (Fig. 9.6). The patterns for the observed and predicted data were similar for almost all the fisheries. However, a clear distinction is observed between the artisanal and industrial fisheries. The fit is generally better for the industrial fisheries (purse seine, trawl and shrimp). The best fits are for trawl (fishes) and shrimp. For the groups in the artisanal fishery (gillnet, handlining, shark and beach seine fishes), the fits were poor at the beginning of the fitting run. The model was responding to changes in CPUE, assumed here to be proportional to biomass. The CPUE calculated for the artisanal fishery can be divided into two periods, before and after motorization, which started in the 1960s, but had major impact since the 1970s. The expansion of fishing effort was higher after motorization, and that CPUE calculated after motorization appears to better fit the data for the whole Red Sea. Thus, emphasis was given to the fit after 1970.

For the vulnerability search routine, the most important functional groups were sharks, gillnet fishes and handlining fishes. Changes in these three groups, which are on top part of the food web (Fig. 9.3), had a high impact on the foraging

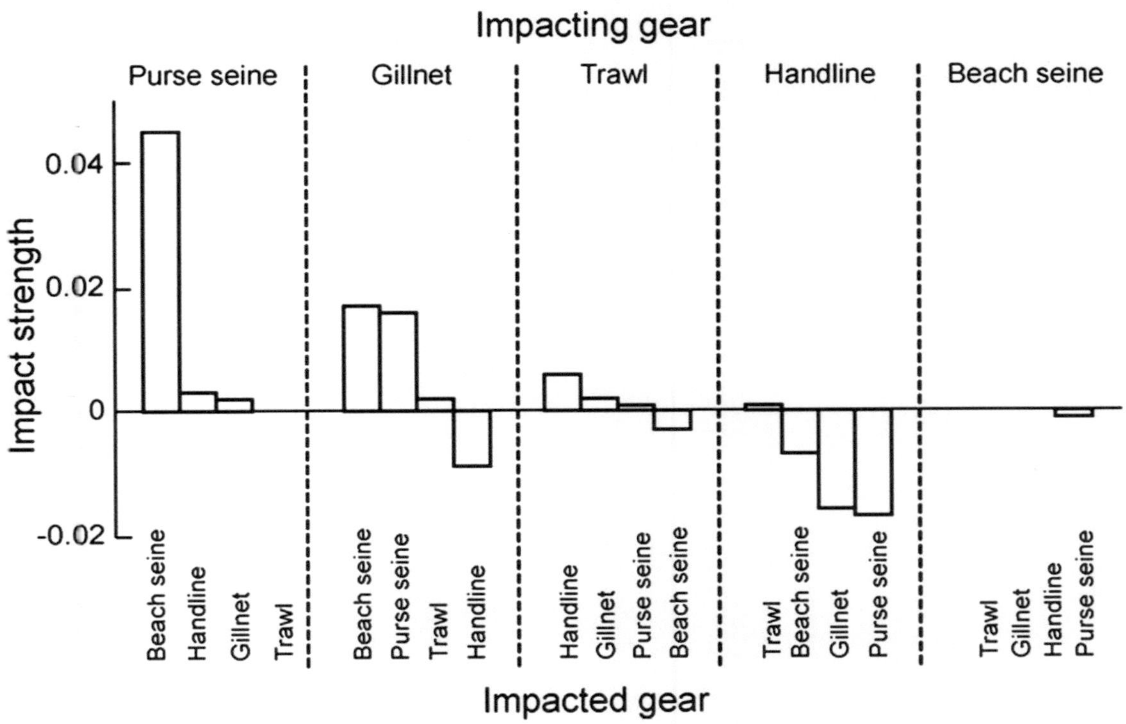

Fig. 9.5 Effect of gear deployment (mixed trophic impact) on the catches of other gears

arena dynamics of the model. Once the vulnerability values for the three groups were adjusted, the minor groups were easily accommodated, along with the feeding time factor. For all groups important to fisheries the latter value was adjusted to zero, which means that the feeding time and hence the time they were exposed to predation risk remained constant. The detail of the final vulnerability settings and feeding time factors are given in Tesfamichael (2012).

Stability and Uncertainty

Three scenarios run to test the stability of the model generated largely predicable results. When the fishing mortality was kept at the baseline, the biomasses of all groups important to fisheries remained more or less constant. When the fishing mortality was set to zero, the biomasses of all the groups increased, except for the fish exploited by beach seines (which decreased slightly at first, then stabilized at a slightly higher level) and the fish exploited by trawlers (which increased drastically at first, then stabilized at a lower level, but still higher than the initial level). This scenario allowed for the rebuilding of the stocks. In the third scenario, when the fishing mortality was increased by 5 % per year, the biomass of all groups decreased, except for those exploited by beach seines, which consist of low trophic-level fishes. Thus, once the biomasses of predators are decreased, the biomasses of the lower trophic level fishes increased, due to reduced predation. The Monte Carlo uncertainty analysis showed that all the estimated values were within ±1 standard deviation. The results of the analysis are given in Fig. 9.7, but only two scenarios are presented. The status quo scenario resulted in flat biomasses, and hence it is omitted from the figure.

The depletion risk of the fishery groups in a population viability analysis, i.e., the probability the biomass falling below a certain fraction of the original biomass within a certain period, was calculated. The zero and baseline fishing scenarios did not cause any depletion beyond 50 % of the baseline biomass within 24 years. On the other hand, in the scenario where fishing was increased 5 % per year, the probability of the biomass in 2030 dropping below 5 % of the baseline was 100 % for purse seine, handlining and sharks. Beach seine fishes would not go below 50 % of the baseline, while trawler, gillnet fishes and shrimps exhibited varying degrees of depletion (Table 9.3).

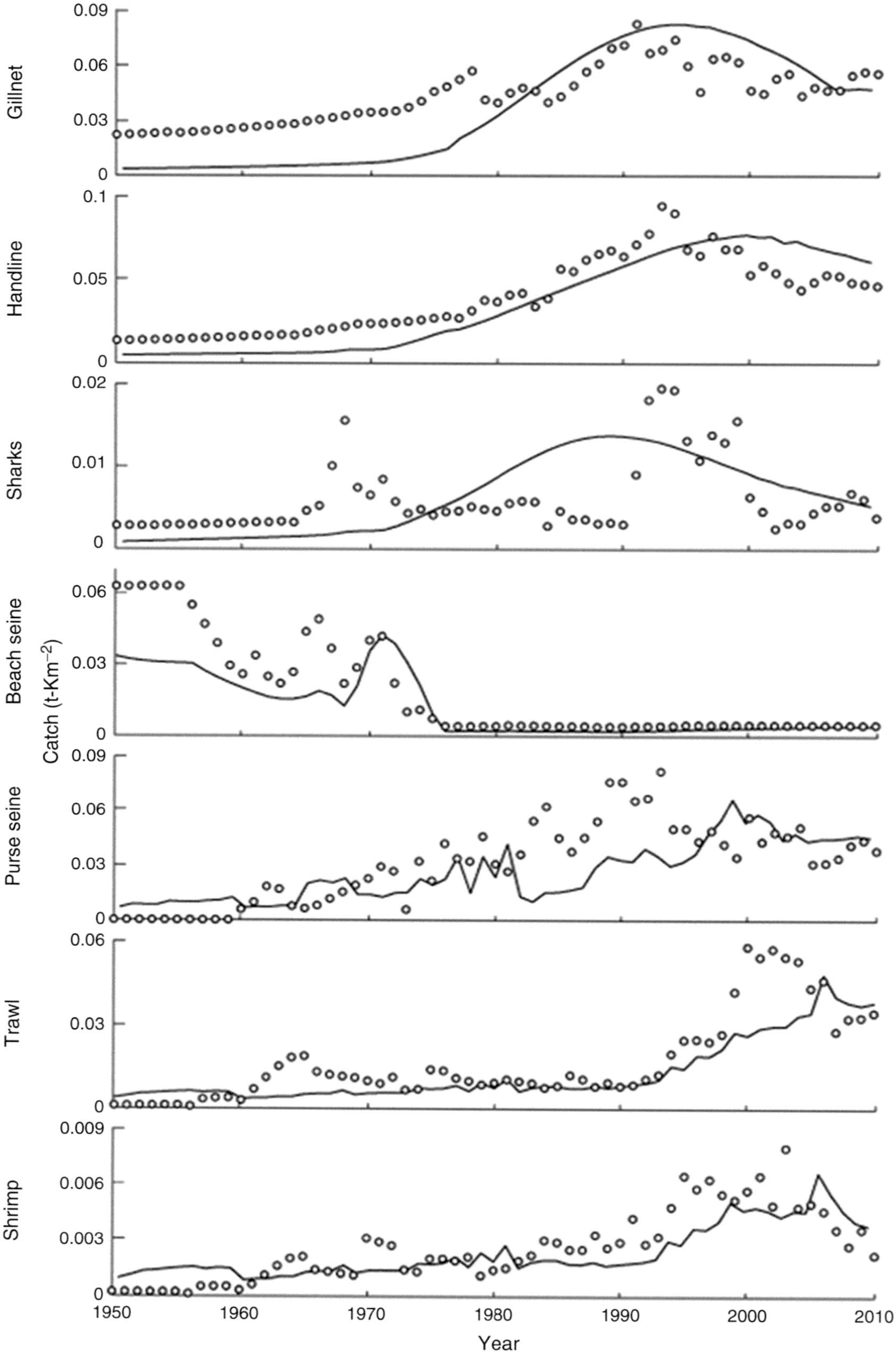

Fig. 9.6 Times series of observed catch data from the Red Sea (*dots*) and catch predicted by the fitted Red Sea Ecosim model (*line*) from 1950 to 2010 for the functional groups important in fisheries. The model was driven by independently estimated fishing effort data

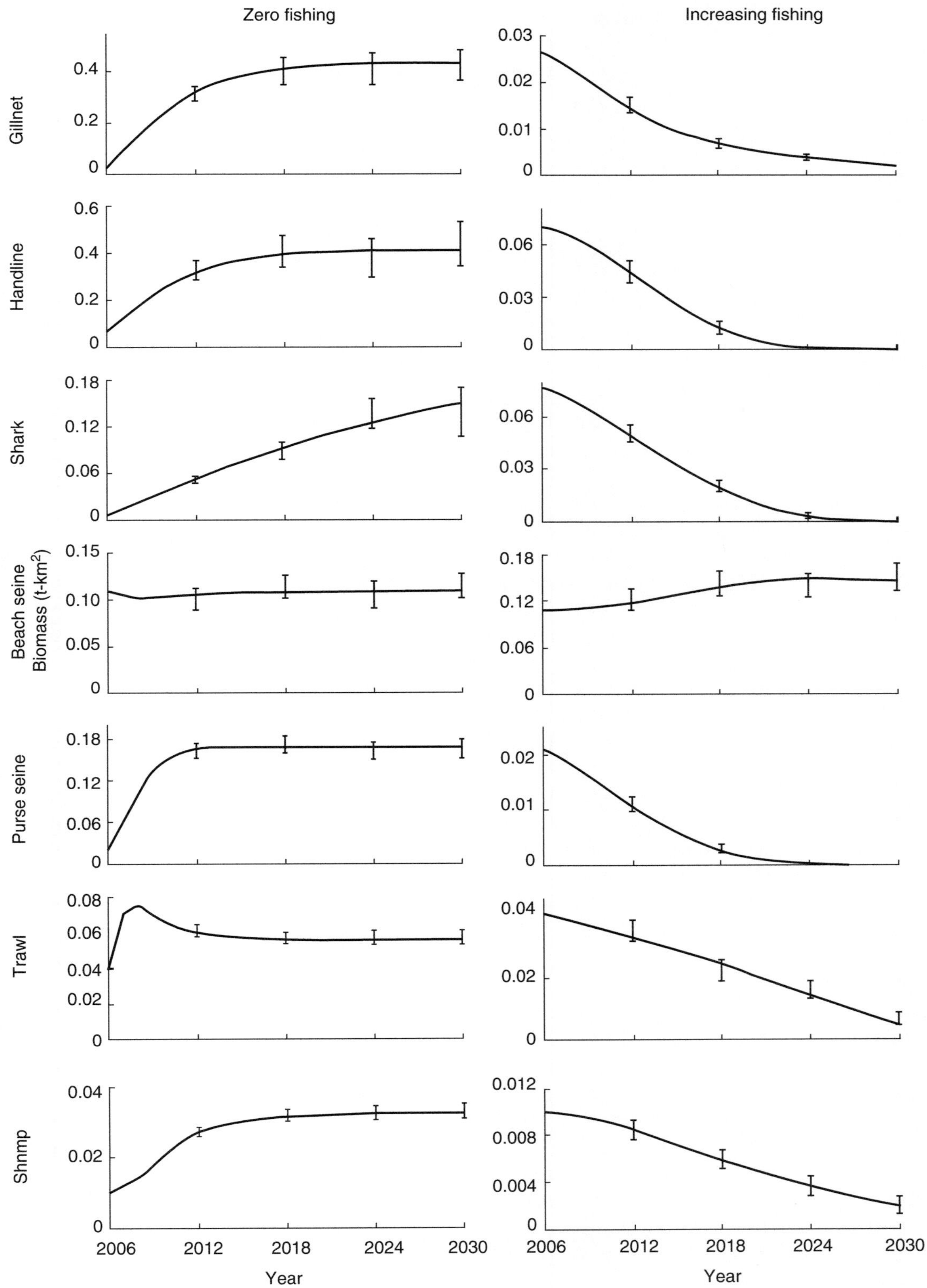

Fig. 9.7 Ecosim simulation test of two scenarios (zero and effort increasing at 5 % per year). The lines are the biomasses of the major fishery groups predicted by the model for 24 year simulations from 2006 to 2030, error bars show 1 SD around the mean

Table 9.3 Biomass depletion risk probabilities for the major fishery groups in the Red Sea below different levels of biomasses, as a ratio of the baseline (2006), at the end of 24 years simulation (2030)

	End state (2030) biomass as a percentage of baseline (2006)						
Groups	5 %	10 %	15 %	20 %	30 %	40 %	50 %
Trawler fishes	0	4	38	78	99	100	100
Purse seine fishes	100	100	100	100	100	100	100
Beach seine fishes	0	0	0	0	0	0	0
Handlining fishes	100	100	100	100	100	100	100
Gillnet fishes	0	74	100	100	100	100	100
Sharks	100	100	100	100	100	100	100
Shrimps	0	5	25	47	90	98	100

Equilibrium Analysis

The equilibrium analysis provided, for all the groups important to fisheries, estimates of equilibrium biomass and catch values at different fishing mortality rates and the value of the current fishing mortality rate in relation to that generating maximum sustainable yield ($F_{msy;}$ Fig. 9.8). Gillnet, handlining, shark and purse seine fisheries are operating at fishing mortality rate beyond $F_{msy,}$ while trawl and shrimp are near F_{msy} level. The beach seine fishery was the only fishery operating at a level much lower than F_{msy} (Fig. 9.9). The baseline fishing mortality rate of the shark fishery is 3.6 times the optimum calculated by the model, the furthest from F_{msy} of all the fisheries, i.e., the sharks are the most depleted resource in the Red Sea. The decline of sharks in the Red Sea was observed by other researches as well (Spaet and Berumen 2015).

The equilibrium analysis considers multispecies interactions, which is more realistic and closer to the actual ecosystem functioning than single species assessment. For this reason, the yields from multispecies are lower than from single species assessments for all the fisheries except for shrimps (Fig. 9.10).

Fishery Policy Exploration

The conflict between artisanal and industrial fisheries was explored by doubling the effort of one sector at a time. This caused, as expected, the biomasses of the groups targeted by the respective sector in question to decrease drastically (Fig. 9.11). What was interesting and contrary to expectations were the effects of one sector on the other. An increase in the effort of one sector did not decrease the biomass of the groups targeted by the other sector; rather, it increased slightly. When industrial fishing effort was doubled, the increase in the biomass of groups targeted by the artisanal fisheries was higher (Fig. 9.11a) than the converse (Fig. 9.11b). Doubling the industrial sector did not have an impact on the shark biomass (Fig. 9.11a), while the beach seine fish biomass benefited from it. The small pelagic beach seine fishes are the main prey for the purse seine fishes, and when the industrial sector effort is doubled, the biomass of the purse seine fishes decreases strongly. This implies that the fishes exploited by beach seines experience a predatory release, resulting in an increased biomass.

Discussion

The Red Sea, as many other coral reef ecosystems, is extremely complex, with a multitude of interactions among the organisms, and with humans within the ecosystem. The EwE model represents the ecosystem quantitatively, even though it did not capture all interactions. However, the model gives a reasonable picture of the dominant interactions, and more specifically, about those that affect the fishery, as intended. The Red Sea ecosystem has a large production at its base (primary production), which tapers off as one ascends the trophic pyramid. All the groups important in the fishery are in the upper part of the food web (upper left corner of Fig. 9.3) and have trophic level >3, except for shrimps. This is reflected in the mean trophic level of fisheries catch, which was 3.4 in 2006 (Table 9.2). This shows that the fishery still focuses on top predators, which is not uncommon for most of the exploited reef ecosystems of the world (Jackson et al. 2001; Worm et al. 2005).

In terms of the impact of change in biomass of one group on another, increase in shark biomass has the most negative impact on certain groups: cetaceans, birds, turtles, whale shark and rays. Shark is the main (for some the only) predator for these groups; hence it has a direct impact on their biomasses through predation. Seagrass has a direct positive impact on the biomass of dugongs, which feed extensively on the seagrass. Most of the other impacts of change in the biomass of one group on others are positive, although at a moderate level. The pedigree value of the Red Sea model is about average when compared to other models in the most extensive pedigree analysis done so far (Morissette 2007), despite the fact that some key parameters (especially biomass) were not available. The most comprehensive source of information on the Red Sea was FishBase, especially for the

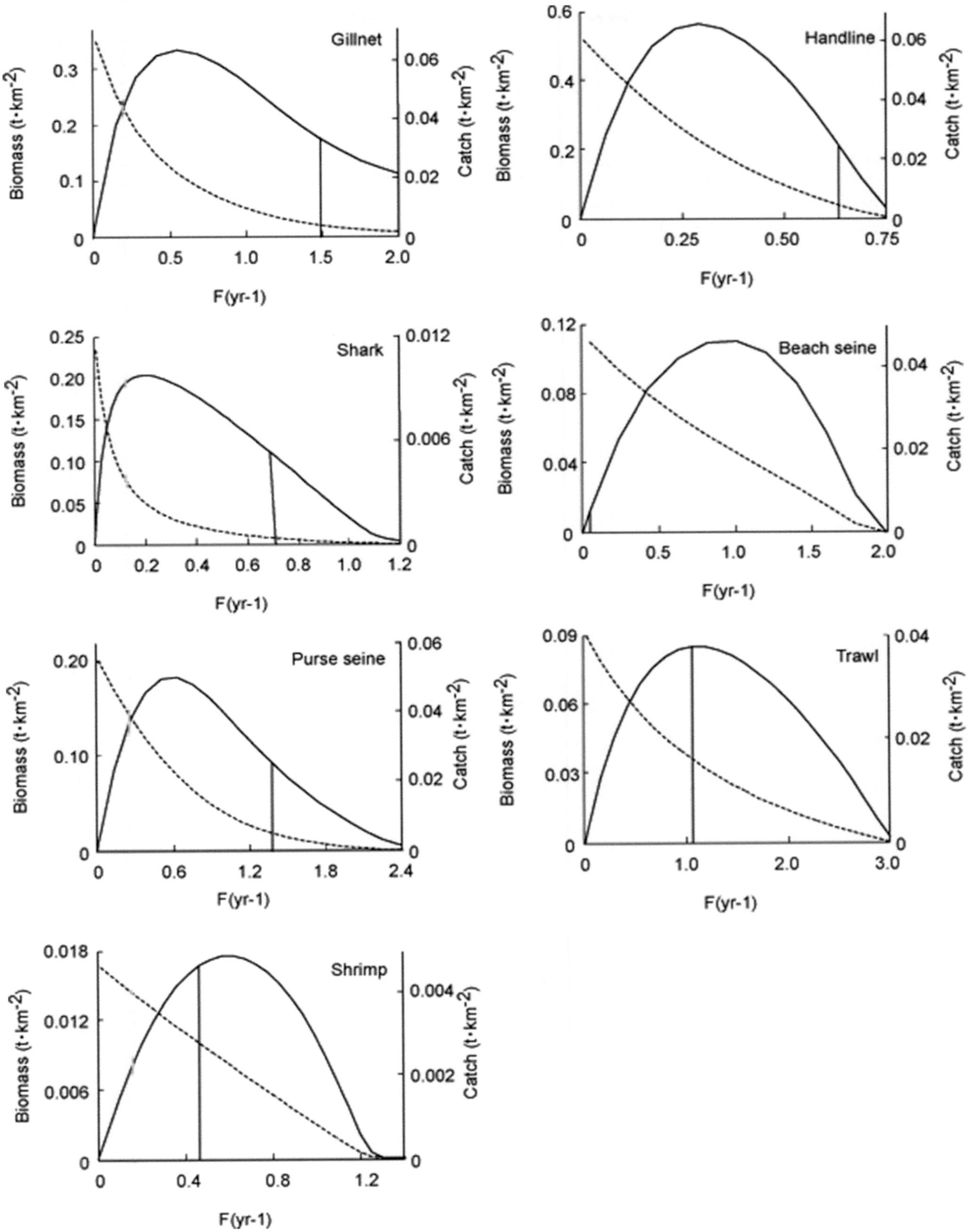

Fig. 9.8 Result of the multispecies equilibrium analysis for major Red Sea fishery groups. Curved *full line* shows surplus yield, *broken line* shows equilibrium biomass and *vertical line* is the baseline fishing mortality rate; read year^{-1} for yr-1 (see text)

three main inputs of the model, i.e., P/B, Q/B and diet composition.

Although a high pedigree value suggests a better quality model, a more useful validation of a model is its ability to predict independent observations, i.e., fit to independently derived time-series data. Indeed, the fitting of the model to time series catch data was the most important part in validating the EwE model of the Red Sea. During the time series fitting, all parameters and possible interactions (diet matrix and trophic flow parameter vulnerabilities) are scrutinized, which led to fine tuning of the model. An interesting observation during the fitting was how difficult it was to fit both

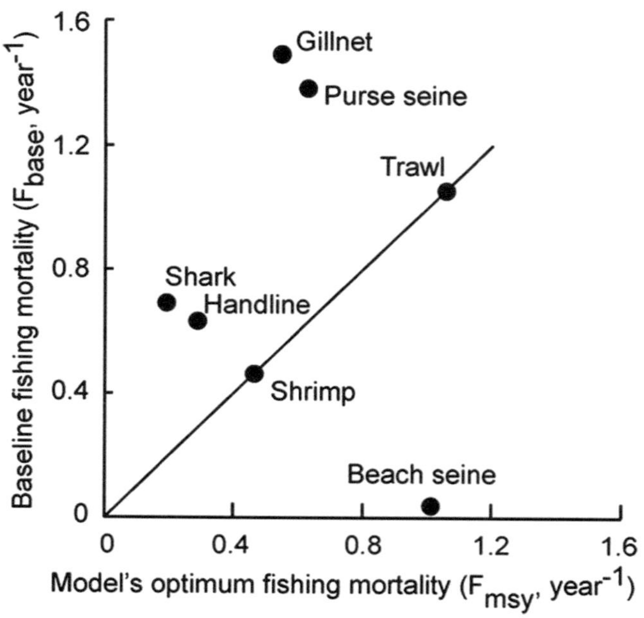

Fig. 9.9 Baseline fishing mortality rate (F_{base}) in relation to the optimum fishing mortality calculated by the model (F_{msy}). The 45° line indicates where F_{base} is equal to F_{msy}

the early years of the time series (1950s and 1960s), and the final decade. When the entire time series was considered, the model at first did not track the independent time series catch at all. Rather, it produced a horizontal line that went through the observed data. This is due to the Red Sea fisheries, prior to 1970s, being non-motorized and having a very small impact on resource species.

With motorization, fishers started to venture out to new fishing grounds further offshore. However, the catch and effort data do not differentiate between inshore and offshore fishing grounds. Hence, the CPUE data do not necessarily reflect trends occurring in entire Red Sea; rather, they reflect changes occurring at the (local) scale at which fisheries operate. Also, Ecosim uses biomass to guide the fitting process. Because a time series of fishery independent data of biomasses of the different groups does not exist for the Red Sea, the temptation was great to use CPUE data as a proxy for biomass. Using CPUE as a proxy for biomass is problematic. A declining CPUE, while locally accurate, may document only a local depletion, leaving the bulk of the biomass of the group in question in its total range in the ecosystem unaffected (Hilborn and Walters 1992) as probably occurred in this case (see below). Thus here, after a few (unsuccessful) attempts to fit the CPUE data, emphasis was given to fitting the catches, as more reliable catch data are available for the Red Sea (see previous chapters). Also, during the fitting process, emphasis was given to the years after 1970, under the assumption that, after motorization, wider areas of the Red Sea were covered, whereas only the inshore waters were fished before 1970.

This brings us back to the issue of localized depletion in the Red Sea, mainly in fishing grounds near major settlements. Even though the Red Sea still has a relatively high predator biomass, some areas which fishers frequent have shown signs of localized depletion (Tesfamichael 2001; Tsehaye 2007). The effect of the spatial distribution of the

Fig. 9.10 Comparison of maximum sustainable yields (MSY) in single species (*open bars*) and multispecies (*black bars*) equilibrium analyses

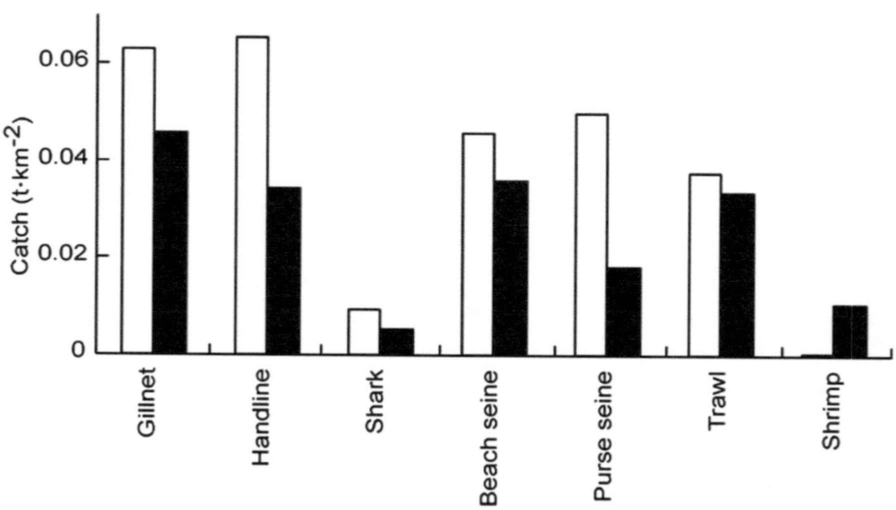

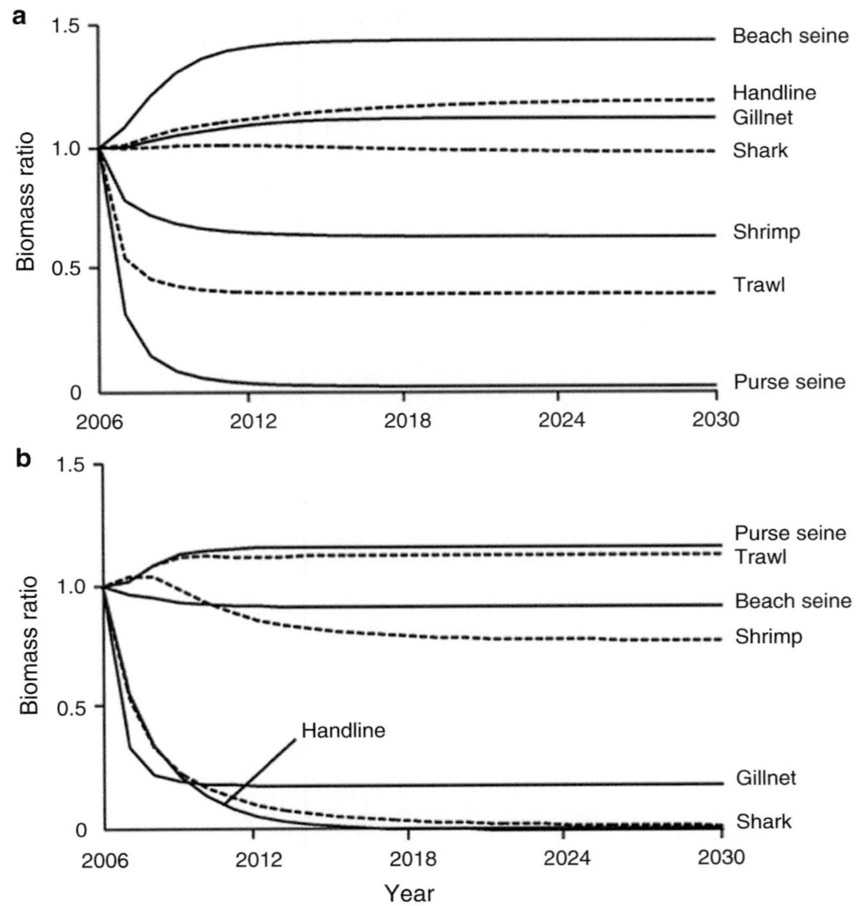

Fig. 9.11 Change in the biomass ratios of the major fishing groups as a result of doubling only the industrial fishery effort (**a**) or the artisanal (**b**)

fishing effort or the fitting procedure can be easily seen by comparing the industrial and artisanal fishery in the Red Sea. Unlike the artisanal fishery, the industrial fishery used motorized vessels from the beginning, giving it a wider coverage. The fits for the industrial fisheries were reasonably good throughout the time series (1950–2010), with no change over time, contrary to the artisanal fisheries, where the fits improve markedly (Fig. 9.6).

Model stability tests, based on three scenarios (zero, baseline and increasing effort fishing) not only showed that the model was behaving well, but also that the result were moderately precise (±1 SD) when the input parameters were allowed to change within certain range in a random fashion. Decreasing fishing effort, for example, is predicted to have a positive impact on the biomasses of the groups that are fished. On the other hand, if the effort is allowed to increase at the rate it has been increasing the last 10 years (about 5 % increase per year), the model predicts that all the groups important to the fisheries (except beach seine fishes) will collapse within the next two decades (Fig. 9.7). The probability that the biomasses of the groups falls below 5 % of the baseline value is very high (100 % for purse seine, handlining and sharks) for all the groups except beach seine fishes. Thus, a continuous increase of effort level, at the same rate as occurred in the last 10 years would have dire consequences.

These predictions were confirmed by analysing the fishing level of each fishery important group using the equilibrium analysis, which showed that most of the fisheries are operating at an effort level higher than that required to generate MSY (F_{msy}; Figs. 9.8 and 9.9), except for beach seines fishes, which is at a very low level, and shrimp and trawl fishes, which operate around F_{msy}. These results are compatible with the general understanding of the situation of the fisheries, and their trends. It seems to be contradicted by the presence of large-sized predatory fishes in the catches of the Red Sea artisanal fisheries. However, this is explained by the fact that the large predators are not common in the catches throughout the Red Sea. While they are still common in Sudan and Eritrea, the countries with the lowest fishing intensity, predominantly from newer fishing grounds; also, there is evidence of localized depletions (Tesfamichael 2001; Tsehaye 2007). Thus, the remaining fishing grounds with still relatively high biomasses are easily overshadowed in an ecosystem modelling exercise analysis covering the entire Red Sea. Still, the occurrence of top predators in the catches of the Red Sea fisheries may be taken as an indicator that the

Red Sea fisheries are doing better than those of otherwise similar ecosystems, which are in worse shape than the Red Sea, for example southeast Asia (Christensen 1998; Pet-Soede et al. 2000) and the Western Indian Ocean (McClanahan 1995).

Using the ecosystem model results in isolation, as if they were the results of stock assessments, may not be advisable. We cannot expect models to generate precise predictions; rather, they provide coherent representations of the system in question, and its dynamics (Christensen et al. 2008). For the Red Sea model caution is needed, particularly in conjunction with the equilibrium analyses of the artisanal fisheries, as they may be still reflecting only the limited area where the fisheries operate, which may not translate to the whole ecosystem. This may hold true even after the motorization of artisanal boats and expansion of their fishing grounds. It will be worth examining this hypothesis with an explicit spatial dynamics of the fishing effort, which is not available at the moment.

One example that stands out clearly is the estimated MSY for the beach seine fishery (Fig. 9.10) is lower than for gillnet, handline and purse seine fisheries (depending on whether one takes the single or multispecies analysis). However, previous stock assessment results indicate that the MSY value of beach seine fishes would be higher than almost all the other fisheries (e.g., Walczak and Gudmundsson 1975; Giudicelli 1984). Indeed, it appears that the representation of the beach seine fishery in the Red Sea EwE model suffered from its limited size, and the absence of reliable data. EwE applications benefit immensely, with regards to the trustworthiness of their prediction, from time series historic fishery data of exploited stocks (Villy Christense, pers. comm.; see also Guénette et al. 2008).

Except for shrimps, all the MSY estimates of the fisheries were lower in the multispecies equilibrium analysis than single species analysis (Fig. 9.10). The former is more realistic representation of the system, and that is why an ecosystem-based fishery assessment and management can produce more reasonable results. One possible explanation for a higher MSY for shrimp in multispecies equilibrium analysis is that shrimp is at the lower trophic level and in multispecies analysis the biomasses of the predators are reduced, which means less mortality by predation, which in turn translates to a higher level of MSY for shrimp.

Perhaps the most important question about the fisheries situation in the Red Sea is whether artisanal and industrial fisheries interact, and if they do, to what extent. The complaints of the artisanal fishers about the industrial fisheries (which are foreign companies in most of the Red Sea countries) are common and sensitive issues. Although their conflict may have many facets, one of the main aspects of the competition between these two fisheries is the effect of the industrial sector on the catch of the artisanal fisheries. In 472 interviews conducted in the Red Sea countries of Sudan, Eritrea and Yemen in 2007 with artisanal fishers, 75 % of them blamed increase in effort, which includes both artisanal and industrial, as the reason for decline in their catch (Tesfamichael et al. 2014). Most of that blame, however, is laid on the industrial sector. The model simulation supported that increase in effort in general is the cause of the decline (Figs. 9.7 and 9.11), but did not support the contention that that one sector is causing the decline of the other (Fig. 9.11). This is contrary to the general perception (e.g., in Pauly 2006); it is also not commonly seen in ecosystem models. Looking at the mixed trophic impact of the fisheries on each other (Fig. 9.5) shows that there is no negative impact on others, except for the slight effect that handlining has on the purse seine and gillnet fisheries, and to a smaller extent on beach seine fishery. Gillnet also showed negative impact on handline fishery.

Another crucial insight comes from the nature of the two sectors, which do not target the same groups, thus avoiding direct competition. These sectors, rather, exploit groups which inhabit different habitats, and even when they target similar habitat (e.g., pelagic by purse seine, gillnet and beach seine), their gears and operations tend to differ. Trawl and handlining fisheries target mainly muddy and reef habitats, respectively, which are not targeted by any of the other fisheries. Possible conflicts would be among the fisheries that target pelagic species. Purse seiners target small and medium pelagic species, but not close to the shore, while gillnet fishery targets large pelagic species using bigger mesh size gillnets than the mesh size used by purse seiners. The main potential conflict would be between beach seine, which also targets small and medium pelagic fishes, and purse seine, which is shown in the mixed trophic impact analysis (Fig. 9.5). This is reflected by the increase in beach seine fish biomass in the simulation when the biomass of purse seine is decreased due to increased industrial fishery effort levels (Fig. 9.11a). However, the beach seine fish biomass increase is not very large, because beach seines operate mainly on shallow beaches as opposed to purse seiners, which operate in relatively deeper water. Thus, there is no overlap of habitats, and hence no large impact of purse seine on beach seine. Second, at the present, the beach seine fishery is almost non-existent, i.e., the group's biomass is almost at its carrying capacity (Fig. 9.8). For the pelagic species, even if beach seine and purse seine fisheries operate in different habitats and use different gears (mesh sizes), one could argue that the very mobile (or migratory) behavior of the target species would cause mixing and possible conflict. Simulation runs where the fishing pressures of the industrial fisheries (trawl and purse seine) were increased tenfolds were run to examine how far the effort can increase before it starts to affect the artisanal fisheries. There was no impact on the biomasses of the groups targeted by the artisanal, except sharks, when the trawl effort was increased ten times.

The lack of major impact among the fisheries is reflected by almost non-existent mixed trophic impact among the groups. This scenario may not be common in many other ecosystems, but the Red Sea still has a wide range of low and high-trophic level fishes appearing in the catch (e.g., the mean trophic level is 3.4; see Table 9.2. A possible hypothesis to explain this singular behaviour of the Red Sea model (and, hopefully, of the Red Sea itself) is that because of the wide range of fish available for the fisheries, the different gears can still target different sections of the ecosystem, with no direct competition. One can conjecture that the fewer top predators, the main target of the artisanal fisheries, are available, the more fishers will start to target lower trophic levels, as they do in many fisheries (Pauly et al. 1998), making them rely on the resources which the industrial sector also exploits. However, it is important to note that these results apply only to the trophic interactions between the industrial and artisanal fisheries. In real life, these two fisheries are not totally separate from each other and there are many non-trophic interactions that are not dealt in with EwE. For example, there are complaints by artisanal fisheries that the industrial fisheries operate close to the shore (forbidden in almost all Red Sea countries) and destroy coastal habitats, and sometimes even the fishing gears of the artisanal fishery. Although, the trophic model of the Red Sea does not deal with such issues, it does deal with an important aspect of the conflict, and thus can be useful, in conjunction with other approaches, for exploring policies for the Red Sea fisheries.

The contribution of fishery catch from the Red Sea to the global catch is not that big. Nevertheless, it is important to the countries in the region. Fish is the main staple food for the coastal communities. It is a cheap source of protein and provides livelihood for the communities. The Red Sea area is very dry and population density is low, which may explain why it is not one of the most exploited seas and why there are still large sized predators in the catch. Since the countries on the Red Sea coast are generally less industrialized, fisheries can be a good source of employment.

Acknowledgements I would like to thank the personnel of the fisheries administrations of the Red Sea countries for allowing me to use their data. I would also like to thank Dr. Villy Christensen for his help with EwE software, Ms. Chira Piroddi and Dr. Divya Varkey for their insights during the building and running of the model, and Dr. Daniel Pauly and Dr. Iyob Tsehaye for reviewing the drafts of this chapter. This research was supported by the *Sea Around Us*, a scientific collaboration between the University of British Columbia and The Pew Charitable Trusts.

Appendix

Table 9.A1 Key data on fish groups of the Red Sea ecosystem model

			Trophic level		L∞ (cm)	
Group No.	Group name	No. of spp.	Min	Max	Min	Max
10	Whale shark	1	3.55	3.55	1,683.0	1,683.0
12	Rays	17	3.1	4.5	68.4	347.4
13	Reef top predators	122	3.98	4.5	9.5	421.1
14	Large reef carnivores	86	3	3.98	51.4	315.8
15	Medium reef carnivores	218	3	3.98	15.0	48.9
16	Small reef carnivores	209	3	3.98	2.1	14.8
17	Reef omnivores	87	2.02	2.99	3.1	115.8
18	Reef herbivores	39	2	2	5.8	94.7
19	Large pelagic carnivores	12	3.47	4.58	105.3	350.5
20	Small pelagic carnivores	35	3	4.5	7.3	87.2
21	Pelagic omnivores	5	2.1	2.95	6.4	26.3
22	Demersal top predators	15	4	4.45	20.6	263.2
23	Large demersal carnivores	24	3.02	3.97	50.1	88.2
24	Medium demersal carnivores	82	3	3.95	15.4	48.4
25	Small demersal carnivores	81	3	3.68	1.8	14.7
26	Demersal omnivores	20	2.45	2.99	3.3	72.3
27	Demersal herbivores	11	2	2.04	6.2	51.5
28	Benthopelagic fish	26	2.13	4.45	3.9	210.5
29	Bathypelagic fish	9	3.03	4.5	4.4	100.0
30	Bathydemersal fish	26	3	4.43	8.5	68.6

Appendix B: Non-fish Taxa Groups Included in the Model

Cetaceans

This group includes the dolphins and whales of the Red Sea, whose list and distributions have been described in the literature (Schmitz and Lavigne 1984; Frazier et al. 1987; Notarbartolo di Sciara 2002). All the reported cetaceans are from the suborder Odontocetea (toothed whales) except *Balaenoptera edeni* (Eden's whale) and *Megaptera novaeangliae* (humpback whale), which are from the suborder Mysticeti. The P/B values for cetaceans were calculated assuming r/2 (Schmitz and Lavigne 1984), where r is the average intrinsic rate of growth (0.088 year^{-1}) for the Red Sea cetaceans species *Stenella attenuata, S. longirostris, S. coeruleoalba* and *Tursiops truncatus* data were available. The estimated P/B for the group equals 0.044 year^{-1}. The r/2 method is commonly used to measure P/B of marine mammals (Guénette 2005; Ainsworth et al. 2007). The Q/B value was estimated based on the body weight of Red Sea cetaceans taken from Schmitz and Lavigne (1984) and Trites and Pauly (1998), from which the ration was determined using the relationship in Trites and Heise (1996). The average Q/B value, 5.91 year^{-1} was used in the model. Biomass data were not available and were estimated by the model.

Dugongs

Dugongs are herbivore marine mammals whose abundance in the Red Sea is estimated to be about 4,000 animals (Gladstone et al. 2003). With an average weight of 320 kg (Frazier et al. 1987), the biomass is calculated to be 0.00292 t·km^{-2}. Similar to the cetaceans, P/B for dugong was calculated using the intrinsic growth rate which is estimated to be 5 % year^{-1} (Marsh et al. 1997), with P/B = 0.025 year^{-1}. The Q/B ratio is taken to be 11 year^{-1} as calculated by Ainsworth et al. (2007), based on body weight.

Birds

The sea birds covering the whole Red Sea are described in Evans (1987) and recent reviews on the status of the Red Sea birds by country are available (PERSGA/GEF 2003; Marchi et al. 2009). However, they are very brief with some list of species sighted and habitat distribution with no estimate of abundance. The P/B value of 0.38 year^{-1} was used based on Russell (1999). Seabird biomass was not available, and was estimated by the model.

Turtles

Five species of sea turtles, hawksbill (*Eretmochelys imbricata*), green (*Chelonia mydas*), loggerhead (*Caretta caretta*), olive ridley (*Lepidochelys olivacea*) and leatherback (*Dermochelys coriacea*), are reported for the Red Sea (Frazier et al. 1987; Tesfamichael 1994). The first two are the most abundant, with known records of nesting on the Red Sea beaches (Frazier and Salas 1984; Frazier et al. 1987; Gladstone et al. 2003). The P/B value for turtles was estimated using the relationship $M = -\ln S$, where M is an estimate of P/B and S is the survival rate, which was 0.948 year^{-1} for green turtle (Mortimer et al. 2000) and 0.867 year^{-1} for loggerhead (Chaloupka and Limpus 2002). This gives an average P/B value of 0.1 year^{-1}. P/B value for all turtles in the Caribbean reef was calculated to be 0.2 year^{-1} (Opitz 1996). Since the P/B estimate using survival rate was only for two species, i.e., it does not include all the five species in the Red Sea, an average of the P/B calculated from survival and the Caribbean value, 0.15 year^{-1}, was used for the model. Q/B value of 3.5 year^{-1} was used based on ecosystem models of the Caribbean reef (Opitz 1996) and west coast of Peninsular Malaysia (Alias 2003). Sea turtle biomass was not available and was estimated by the model.

Invertebrates

The invertebrates most important for the Red Sea fisheries are shrimps. Hence, they was given a separate functional group. The most common species caught are *Penaeus semisulcatus, P. monodon, Marsupenaeus japonicus, Melicertus latisulcatus, Metapenaeus monoceros and Fenneropenaeus indicus*. P/B value of 5 year^{-1} and Q/B of 29 year^{-1} based on Buchary (1999) were used as a starting parameters to balance the model.

The coral reef structure in the Red Sea is important ecologically and is also the main fishing ground for the artisanal fisheries. Thus, the reef forming corals are categorized as a separate functional group. The high and relatively stable temperature of the Red Sea is favourable for the formation of coral reefs. They are home to more than 200 species of corals (Head 1987). The Red Sea coral reef coverage area is estimated to be around 16,030 km^2 (Spalding et al. 2001). Coral reefs are more developed in the northern part starting from the tip of Sinai Peninsula going south parallel to the coast until the central part (Sheppard et al. 1992). The longest continuous fringing reef in the Red Sea extends from Gubal, at the mouth of the Gulf of Suez, to Halaib, at the Egyptian border with Sudan (Pilcher and Alsuhaibany 2000). In the south, more patchy reefs are observed as the turbid water of the shallow shelf does not allow the growth of extensive reefs. Sanganeb Atoll, located in Sudan near the border with Egypt, is the only atoll in the Red Sea. It is unique reef rising from 800 m depth to form an atoll that has been recognized as regionally important conservation area. It was proposed to UNESCO for World Heritage Status in the 1980s (Pilcher and Alsuhaibany 2000). The biomass of corals was calculated based on data from the southern Red Sea (Ateweberhan 2004; Tsehaye 2007) adjusted for the total area of the Red Sea and the north-south abundance gradient giving 2.75 t·km^{-2}. The P/B value of corals was calculated based on daily turnover rate of 0.003 day^{-1} (Crossland et al. 1991), which equals to 1.095 year^{-1}. A Q/B value of 9 year^{-1} was used based on the Caribbean reef model (Opitz 1996).

The other invertebrates included in the model are: non-coral sessile fauna such as sponges, sea anemones, and tunicates; cephalopods: squids, octopuses and cuttlefish; other molluscs; echinoderms: starfish, sea urchins and sea cucumber; crustaceans: representing all crustaceans except shrimps (which have a group of their own); and meiobenthos: polychaetes and nematodes. The P/B and Q/B values of these groups were taken from an ecosystem model of the Eritrean coral reef (Tsehaye 2007) adjusted for the area of the Red Sea fine-tuned during balancing and time series fitting (Table 9.B1).

Primary Producers

There are three functional groups of primary producers in the model: phytoplankton, seagrasses and algae. The phytoplankton biomass of 21.5 t·km^{-2} and a P/B 110 year^{-1} were used based on data in (Weikert 1987; Veldhuis et al. 1997) averaged over all the Red Sea. For seagrass, a biomass of 11 t·km^{-2} and P/B value of 19 year^{-1} were used, based on Wahbeh (1988) and Aleem (1979). The biomass estimate of algae was based on Ateweberhan (2004) and Walker (1987), and was averaged for the whole Red Sea, resulting in 38 t·km^{-2}. The P/B value of 14 year^{-1} was used based on Ateweberhan (2004) and Wolanski (2001), which is similar to the value in other coral reef ecosystems: Caribbean (Opitz 1996) and Indonesia (Ainsworth et al. 2007).

Table 9.B1 Input parameters of some invertebrates groups

	Biomass (t·km^2)	P/B (year^{-1})	Q/B (year^{-1})
Other sessile fauna	0.85	3.2	12
Cephalopods	0.399	3.5	12
Molluscs	0.368	9	30
Echinoderms	0.596	1.6	8
Crustaceans	0.816	3	10
Meiobenthos	0.295	26	100
Zooplankton[a]	14	52	178

[a]Modified after (van Couwelaar 1997)

Table 9.C1 Effort (megawatt·hours) of Red Sea fisheries by gear, 1950–2010

Year	Beach seine	Gillnet	Handlining	Purse seine	Trawl	Others
1950	3,260	2,412	5,515	122	1,685	1,243
1951	3,261	2,482	5,639	153	2,010	1,265
1952	3,262	2,554	5,768	152	2,396	1,287
1953	3,262	2,629	5,900	153	2,487	1,310
1954	3,264	2,707	6,035	190	2,634	1,334
1955	3,265	2,796	6,191	185	2,753	1,361
1956	3,266	2,883	6,345	185	2,843	1,388
1957	2,903	2,975	6,436	201	2,625	1,368
1958	2,582	3,071	6,539	208	2,772	1,355
1959	2,298	3,173	6,656	238	2,677	1,347
1960	2,046	3,282	6,785	132	1,542	1,346
1961	1,824	3,398	6,928	137	1,617	1,350
1962	1,627	3,522	7,084	133	1,635	1,359
1963	1,539	3,656	7,271	146	1,874	1,385
1964	1,532	3,801	7,485	146	1,859	1,424
1965	1,631	3,959	7,733	376	2,276	1,480
1966	1,864	4,132	8,022	429	2,409	1,558
1967	1,689	4,322	8,936	419	2,402	1,661
1968	1,244	4,531	10,357	479	2,977	1,792
1969	2,061	4,762	10,744	291	2,244	1,952
1970	3,623	5,017	10,549	283	2,469	2,132
1971	4,546	5,353	11,175	254	2,525	2,342
1972	4,379	6,045	13,575	300	2,509	2,667

(continued)

Table 9.C1 (continued)

Year	Beach seine	Gillnet	Handlining	Purse seine	Trawl	Others
1973	3,503	6,906	16,820	310	2,511	3,025
1974	2,317	7,952	20,496	476	3,260	3,418
1975	956	9,193	24,494	415	3,195	3,849
1976	213	10,628	28,062	487	3,339	4,323
1977	215	14,944	29,929	791	3,902	5,010
1978	217	18,046	34,068	333	2,922	5,815
1979	220	21,641	38,594	782	4,367	6,717
1980	217	25,769	43,575	534	3,498	7,729
1981	215	30,463	49,010	993	5,165	8,854
1982	213	35,759	54,902	305	2,875	10,097
1983	211	41,692	61,258	213	3,462	11,462
1984	210	48,299	68,088	297	3,723	12,955
1985	209	55,613	75,404	295	3,740	14,581
1986	209	63,666	83,222	313	3,340	16,344
1987	208	72,485	91,557	341	3,320	18,250
1988	208	82,095	100,421	564	3,635	20,303
1989	209	92,517	109,827	729	3,305	22,506
1990	211	103,773	119,787	727	3,516	24,863
1991	223	115,878	130,309	723	3,712	27,379
1992	237	129,588	142,223	910	3,930	30,227
1993	251	143,888	155,021	839	4,382	33,240
1994	265	159,100	168,425	696	6,669	36,421
1995	278	175,251	182,452	724	6,345	39,776
1996	290	191,554	197,936	820	8,640	43,309
1997	301	212,045	216,096	1,110	8,731	47,605
1998	312	230,874	232,641	1,426	10,096	51,536
1999	323	252,358	253,234	2,103	13,446	56,213
2000	334	274,425	273,137	1,993	12,964	60,877
2001	346	292,899	287,072	2,494	14,137	64,480
2002	358	316,250	308,547	2,816	14,396	69,462
2003	371	340,333	312,259	2,526	14,079	72,551
2004	384	366,471	340,405	2,964	15,941	78,584
2005	396	396,431	350,622	3,032	16,653	83,050
2006	409	423,869	374,973	3,727	26,875	88,806
2007	422	455,669	397,698	4,016	27,612	94,865
2008	435	494,756	416,418	4,305	28,350	101,290
2009	448	530,136	441,684	4,594	29,088	108,030
2010	461	564,994	466,442	4,883	29,826	114,655

References

Ainsworth CH, Varkey D, Pitcher TJ (2007) Ecosystem simulation models for the Bird's head seascape, Papua, fitted to field data. In: Pitcher TJ, Ainsworth CH, Bailey M (eds) Ecological and economic analysis of marine ecosystems in the Bird's head seascape, Papua, Indonesia: I. Fisheries Centre Research Reports, vol 15(5), Fisheries Centre, University of British Columbia, Vancouver, pp. 6–174

Aleem AA (1979) A contribution to the study of seagrasses along the Red Sea Coast of Saudi Arabia. Aquat Bot 7:71–78

Alias M (2003) Trophic model of the coastal fisheries ecosystem of the West Coast of Penisular Malaysia. In: Silvestre G, Garces L, Stobutzki I, Ahmed M, Valmonte-Santos RA, Luna C, Lachica-Alino L, Munro P, Christensen V, Pauly D (eds) Assessment, management and future directions for coastal fisheries in Asian countries. WorldFish Centre Conference Proceedings 67, Penang, Malaysia, pp 313–332

Anticamara JA, Watson R, Gelchu A, Pauly D (2011) Global fishing effort (1950–2010): trends, gaps, and implications. Fish Res 107(1–3):131–136

Antoine L, Blanchard F, Boucher J, Massé J, Morizur Y, Araia T, Mahmoud S, Wolday S (1997) Assessment and follow-up of the potential of fisheries resources in Eritrea, Winter campaign report (1996–1997). MOF-IFREMER-CFD, Massawa

Arias-González JE (1998) Trophic models of protected and unprotected coral reef ecosystems in the South of the Mexican Caribbean. J Fish Biol 53:236–255

Ateweberhan M (2004) Seasonal dynamics of coral reef algae in the southern Red Sea: functional group and population ecology. PhD thesis, University of Groningen, Groningen, 187 p

Beattie A, Sumaila U, Christensen V, Pauly D (2002) A model for the bioeconomic evaluation of marine protected area size and placement in the North Sea. Nat Resour Model 15(4):413–437

Beverton RJH, Holt SJ (1993) On the dynamics of exploited fish population. Chapman & Hall, London

Blindheim J (1984) Fishery resources surveys in P.D.R. Yemen, Somalia and Ethiopia, 11 Feb–21 Mar 1984. Institute of Marine Research, Bergen, p 44

Bouchon-Navaro Y, Bouchon C (1989) Correlations between chaetodontid fishes and coral communities of the Gulf of Aqaba (Red Sea). Environ Biol Fishes 25(1):47–60

Brey T (ed) (2001) Population dynamics in benthic invertebrates. A virtual handbook Version 01.2. http://www.thomas-brey.de/science/virtualhandbook

Brodziak J, Link J (2002) Ecosystem-based fishery management: what is it and how can we do it? Bull Mar Sci 70(2):589–611

Browman HI (2000) 'Evolution' of fisheries science. Mar Ecol Prog Ser 208:299–313

Buchary E (1999) Evaluating the effect of the 1980 trawl ban in the Java Sea, Indonesia: an ecosystem based approach. MSc thesis, University of British Columbia, Resource Management and Environmental Studies, Vancouver, 134 p

Chaloupka MC, Limpus CL (2002) Survival probability estimates for the endangered loggerhead sea turtle resident in southern Great Barrier Reef waters. Mar Biol 140(2):267–277

Christensen V (1998) Fishery-induced changes in a marine ecosystem: insight from models of the Gulf of Thailand. J Fish Biol 53(A):128–142

Christensen V, Pauly D (1992) Ecopath II – a software for balancing steady-state ecosystem models and calculating network characteristics. Ecol Model 61:169–185

Christensen V, Pauly D (eds) (1993) Trophic models of aquatic ecosystems. ICLARM Conference Proceedings, No. 26, Manila, Phillipines, 390 p

Christensen V, Walters CJ (2004) Ecopath with Ecosim: methods, capabilities and limitations. Ecol Model 172(2–4):109–139

Christensen V, Walters CJ, Pauly D (2005) Ecopath with Ecosim a user's guide. Fisheries Centre, University of British Columbia, Vancouver

Christensen V, Walters CJ, Pauly D, Forrest R (2008) Ecopath with Ecosim version 6: user guide. Fisheries Centre, University of British Columbia, Vancouver. Available at www.ecopath.org

Cox SP, Essington TE, Kitchell JF, Martell SJD, Walters CJ, Boggs C, Kaplan I (2002) Reconstructing ecosystem dynamics in the central Pacific Ocean, 1952–1998. II. A preliminary assessment of the trophic impacts of fishing and effects on tuna dynamics. Can J Fish Aquat Sci 59(11):1736–1747

Crossland CJ, Hatcher BG, Smith SV (1991) Role of coral reefs in global ocean production. Coral Reefs 10(2):55–64

Evans PGH (1987) Sea birds of the Red Sea. In: Edwards AJ, Head SM (eds) Red Sea. Pergamon Press, Oxford, pp 315–338

Frazier J, Salas S (1984) The status of marine turtles in the Egyptian Red Sea. Biol Conserv 30(1):41–67

Frazier J, Bertram GC, Evans PGH (1987) Turtles and marine mammals. In: Edwards AJ, Head SM (eds) Red Sea. Pergamon Press, Oxford, pp 288–314

Froese R, Pauly D (2014) FishBase. Available at: www.fishbase.org

Giudicelli M (1984) The Ethiopian fisheries: situation, development needs and opportunities. A report prepared for the Fishing Planning and Development Project, FAO, Rome, 157 p

Gladstone W, Krupp F, Younis M (2003) Development and management of a network of marine protected areas in the Red Sea and Gulf of Aden region. Ocean Coast Manag 46(8):741–761

Guénette S (2005) Models of Southeast Alaska. In: Guénette S, Christensen V (eds) Food web models and data for studying fisheries and environmental impacts on Eastern Pacific ecosystem. Fisheries Centre Research Reports, vol 13(1), Fisheries Centre, University of British Columbia, Vancouver, pp 106–178.

Guénette S, Christensen V, Pauly D (2008) Trophic modelling of the Peruvian upwelling ecosystem: towards reconciliation of multiple datasets. Prog Oceanogr 79(2–4):326–335

Head SM (1987) Corals and coral reefs of the Red Sea. In: Edwards AJ, Head SM (eds) Red Sea. Pergamon Press, Oxford, pp 128–151

Hilborn R, Walters CJ (1992) Quantitative fisheries stock assessment. Choice, dynamics and uncertainty. Chapman and Hall, New York

Jackson JBC, Kirby MX, Berger WH, Bjorndal KA, Botsford LW, Bourque BJ, Bradbury RH, Cooke R, Erlandson J, Estes JA, Hughes TP, Kidwell S, Lange CB, Lenihan HS, Pandolfi JM, Peterson CH, Steneck RS, Tegner MJ, Warner RR (2001) Historical overfishing and the recent collapse of coastal ecosystems. Science 293(5530):629–637

Link JS (2002) What does ecosystem-based fisheries management mean. Fisheries 27(4):18–21

Marchi GD, Chiozzi G, Semere D (2009) Wings over the Red Sea: the birds of the Eritrean islands. Società italiana di scienze naturali, Milano, 128 p

Marsh H, Harris ANM, Lawler IR (1997) The sustainability of the Indigenous dugong fishery in Torres Strait, Australia/Papua New Guinea. Conserv Biol 11(6):1375–1386

Massé J, Araia T (1997) An acoustic survey on board research vessel "Europe". MOF-IFREMER-CFD, Massawa

McClanahan TR (1995) A coral reef ecosystem-fisheries model: impacts of fishing intensity and catch selection on reef structure and processes. Ecol Model 80(1):1–19

Morissette L (2007) Complex, cost and quality of ecosystem models and their impact on resilience: a comparative analysis, with emphasis on marine mammals and the Gulf of St. Lawrence. PhD thesis, University of British Columbia, Zoology, Vancouver, 260 p

Mortimer JA, Donnelly M, Plotkin PT (2000) Sea turtles. In: Sheppard CRC (ed) Seas at the millennium: an environmental evaluation: 3. Global issues and processes. Pergamon, Amsterdam, pp 59–71

Notarbartolo di Sciara G (2002) Cetacean species occurring in the Mediterranean and Black Seas. In: Notarbartolo di Sciara G (ed) Cetaceans of the Mediterranean and Black Seas: state of knowledge and conservation strategies. ACCOBAMS, Monaco, pp 6–26

Opitz S (1996) Trophic interactions in Caribbean coral reefs. ICLARM Tech Rep No. 43, Makati City, 341 p

Palomares MLD, Pauly D (1998) Predicting food consumption of fish populations as functions of mortality, food type, morphometrics, temperature and salinity. Mar Freshw Res 49:447–453

Palomares MLD, Pauly D (eds) (2012) SeaLifeBase. http://www.sealifebase.org/ version (02/2012)

Pauly D (1980) On the interrelationships between natural mortality, growth parameters, and mean environmental temperature in 175 fish stocks. ICES J Mar Sci 39(2):175

Pauly D (1986) A simple method for estimating food consumption of fish populations from growth data and food conversion experiments. Fish Bull 4(4):827–840

Pauly D (2006) Major trends in small-scale marine fisheries, with emphasis on developing countries, and some implications for the social sciences. MAST 4(2):7–22

Pauly D, Christensen V, Dalsgaard J, Froese R, Torres F Jr (1998) Fishing down marine food webs. Science (Washington) 279(5352):860–863

Pauly D, Christensen V, Walters C (2000) Ecopath, Ecosim, and Ecospace as tools for evaluating ecosystem impact of fisheries. ICES J Mar Sci 57(3):697–706

PERSGA/GEF (2003) Status of breeding seabirds in the Red Sea and Gulf of Aden. PERSGA technical series no. 8. Regional Organization for the Conservation of the Environment of the Red Sea and Gulf of Aden (PERSGA), Jeddah

Pet-Soede C, van Densen WLT, Pet JS, Machiels MAM (2000) Impact of Indonesian coral reef fisheries on the fish community structure and the resultant catch composition. PhD thesis, Wageningen University, Wageningen, pp 81–100

Pikitch EK, Santora C, Babcock EA, Bakun A, Bonfil R, Conover DO, Dayton P, Doukakis P, Fluharty D, Heneman B, Houde ED, Link J, Livingston PA, Mangel M, McAllister MK, Pope J, Sainsbury KJ (2004) Ecosystem-based fishery management. Science 305(5682):346–347

Pilcher N, Alsuhaibany A (2000) Regional status of coral reefs in the Red Sea and the Gulf of Aden. In: Wilkinson C (ed) Status of coral reefs of the world: 2000. Australian Institute of Marine Science, Dampier, pp 35–64

Pitcher TJ, Kalikoski D, Short K, Varkey D, Pramod G (2009) An evaluation of progress in implementing ecosystem-based management of fisheries in 33 countries. Mar Policy 33(2):223–232

Plagányi ÉE (2007) Models for an ecosystem approach to fisheries FAO fisheries technical paper, no 477. Food and Agriculture Organization of the United Nations (FAO), Rome

Plagányi ÉE, Butterworth DS (2004) A critical look at the potential of Ecopath with Ecosim to assist in practical fisheries management. Afr J Mar Sci 26(1):261–287

Polovina J (1984) Model of a coral reef ecosystem. I. The ECOPATH model and its application to French Frigate Shoals. Coral Reefs 3(1):1–11

Price ARG, Jobbins G, Shepherd ARD, Ormond RFG (1998) An integrated environmental assessment of the Red Sea coast of Saudi Arabia. Environ Conserv 25(01):65–76

Roberts C, Ormond R (1987) Habitat complexity and coral reef fish diversity and abundance on Red Sea fringing reefs. Marine Ecology Progress Series. Oldendorf 41(1):1–8

Russell R (1999) Comparative demography and life history tactics of seabirds: implications for conservation and marine monitoring. In: Musick J.A. (ed) Life in the slow lane: ecology and conservation of long-lived marine animals. American Fisheries Society symposium, Monterey, California, pp 51–76

Schmitz OJ, Lavigne DM (1984) Intrinsic rate of increase, body size, and specific metabolic rate in marine mammals. Oecologia 62(3):305–309

Sheppard C, Price A, Roberts C (1992) Marine ecology of the Arabian region: patterns and processes in extreme tropical environments. Academic, London

Sherman K, Alexander LM (1986) Variability and management of large marine ecosystems. Westview Press, Boulder, 325 p

Smith ADM, Fulton EJ, Hobday AJ, Smith DC, Shoulder P (2007) Scientific tools to support the practical implementation of ecosystem-based fisheries management. ICES J Mar Sci 64(4):633–639

Spaet JLY, Berumen ML (2015) Fish market surveys indicate unsustainable elasmobranch fisheries in the Saudi Arabian Red Sea. Fish Res 161:356–364

Spalding M, Ravilious C, Green E (2001) World atlas of coral reefs. University of California Press, Berkeley

Tesfamichael D (1994) Sea turtles of the Red Sea coast of Eritrea. Ministry of Fisheries and University of Asmara, Massawa

Tesfamichael D (2001) Characterising the Eritrean artisanal fisheries using catch and effort data analysis. MSc thesis, Wageningen University, Fish Culture and Fisheries Group, Wageningen, 42 p

Tesfamichael D (2012) Assessment of the Red Sea ecosystem with emphasis on fisheries. PhD thesis, University of British Columbia, Resource Management and Environmental Studies, Vancouver, 241 p

Tesfamichael D, Pitcher TJ, Pauly D (2014) Assessing changes in fisheries using fishers' knowledge to generate long time series of catch rates: a case study from the Red Sea. Ecol Soc 19(1):18

Trites AW, Heise K (1996) Marine mammals. In: Pauly D, Christensen V, Haggan N (eds) Mass-balance models of North-Eastern Pacific ecosystems. Fisheries Centre Research Reports, vol 4(1), Fisheries Centre, University of British Columbia, Vancouver, pp 21–30.

Trites AW, Pauly D (1998) Estimating mean body masses of marine mammals from maximum body lengths. Can J Zool 76(5):886–896

Tsehaye I (2007) Monitoring fisheries in data-limited situations: a case study of the artisanal reef fisheries of Eritrea. PhD thesis, Wageningen University, Wageningen, 182 p

Tsehaye I, Nagelkerke LAJ (2008) Exploring optimal fishing scenarios for the multispecies artisanal fisheries of Eritrea using a trophic model. Ecol Model 212(3–4):319–333

van Couwelaar M (1997) Zooplankton and micronekton biomass off Somalia and in the southern Red Sea during the SW monsoon of 1992 and the NE monsoon of 1993. Deep-Sea Res II Top Stud Oceanogr 44(6–7):1213–1234

Varkey D, Ainsworth CH, Pitcher TJ (2012) Modelling reef fish population responses to fisheries restrictions in marine protected areas in the coral triangle. J Mar Biol 2012:1–18

Veldhuis MJW, Kraay GW, Van Bleijswijk JDL, Baars MA (1997) Seasonal and spatial variability in phytoplankton biomass, productivity and growth in the northwestern Indian Ocean: the southwest and northeast monsoon, 1992–1993. Deep-Sea Res I Oceanogr Res Pap 44(3):425–449

Wahbeh MI (1988) Seasonal distribution and variation in the nutritional quality of different fractions of two seagrass species from Aqaba (Red Sea), Jordan. Aquat Bot 32(4):383–392

Walczak P, Gudmundsson J (1975) Yemen Arab Republic – summary of fish stock assessment work. FAO, Rome, 14 p

Walker DI (1987) Benthic algae. In: Edwards AJ, Head SM (eds) Red Sea. Pergamon Press, Oxford, pp 152–168

Walters C (2000) Impacts of dispersal, ecological interactions, and fishing effort dynamics on efficacy of marine protected areas: how large should protected areas be? Bull Mar Sci 66(3):745–757

Walters C, Christensen V (2007) Adding realism to foraging arena predictions of trophic flow rates in Ecosim ecosystem models: shared foraging arenas and bout feeding. Ecol Model 209(2–4):342–350

Walters CJ, Martell SJD (2004) Fisheries ecology and management. Princeton University Press, Princeton

Walters C, Christensen V, Pauly D (1997) Structuring dynamic models of exploited ecosystems from trophic mass-balance assessments. Rev Fish Biol Fish 7(2):139–172

Walters C, Pauly D, Christensen V (1999) Ecospace: prediction of mesoscale spatial patterns in trophic relationships of exploited ecosystems, with emphasis on the impacts of marine protected areas. Ecosystems 2(6):539–554

Weikert H (1987) Plankton and the pelagic environment. In: Edwards AJ, Head SM (eds) Red Sea. Pergamon Press, Oxford, pp 90–111

Wolanski E (ed) (2001) Oceanographic process of coral reefs: physical and biological links in the great barrier reef. CRC Press, Boca Raton

Worm B, Sandow M, Oschlies A, Lotze HK, Myers RA (2005) Global patterns of predator diversity in the open oceans. Science (Washington) 309(5739):1365–1369

Zekaria ZA (2003) Butterflyfishes of the southern Red Sea: ecology and population dynamics. PhD thesis, University of Groningen, Groningen, 120 p

Common Names of Exploited Fish and Invertebrates of the Red Sea

10

Dawit Tesfamichael and Hesham Saeed

Abstract

We present 465 local names for 500 distinct fish and invertebrate taxa of the Red Sea, as used in Egypt, Sudan, Eritrea, Yemen Saudi Arabia and Jordan. The local names used in each country are clearly indicated, and the total of scientific-local name combinations is 870. Most of the names are in Arabic, a common language in the region in general and the coastal communities in particular. The names were acquired mainly from published reports and papers. Interviews were made to clarify some names and also to add new names. The process of acquisition, verification and standardization of these names, presented in both Arabic and Roman script along with the corresponding scientific names, is described. A brief discussion of some cultural aspects of these names is presented.

Keywords

Local names • Arabic names • Valid names • Common names • Commercially important fish

Introduction

Traditionally, fisheries science has been focused on assessment of stocks, i.e., estimating changes in catch per effort, estimating potential harvest levels, and generally providing advice to the fishery sector on the exploitation of the resources and fishing gears. Moreover, it has become clear that, once the assessment work is done, there is a need for both incorporating local (fisher) knowledge derived from fishers and communicating the results of analyses to policy makers and ultimately to the users (fishers) in a way they can understand. Such communication will be heavily influenced by the cultural context where the fisheries assessment and management is being done. Most of the fish and fisheries knowledge is documented in English for the simple reason that English has become the undisputed global language of science; see contributions in Ammon (2001).

However, in countries where English is not the main language, it cannot be used for communication with fishers. This is especially true in the case of communication about fish, whose common names is the only key to the knowledge that fishers may have on any given species. However, the common names of fish are, within languages and countries highly variable (Freire and Pauly 2005), Only few countries have standardized official list of fish common names e.g., the US, Canada and Mexico in English and Spanish.

FishBase, the online global encyclopedia of over 33,000 described fish species (www.fishbase.org), whose main contents is in English, has attempted to deal with the language issue by providing an interface that can be toggled to other languages and their associated non-Roman scripts (e.g., Arabic), and especially by providing about 305,000 local common names (228,000 in languages other than English) for over 25,900 species in 338 languages spoken throughout the world. SeaLifeBase, as a database similar to FishBase (see www.sealifebase.org), covering 70,000 marine animals

D. Tesfamichael (✉)
Fisheries Centre, University of British Columbia, Vancouver, BC, Canada

Department of Marine Sciences, University of Asmara, Asmara, Eritrea
e-mail: dawittes@gmail.com

H. Saeed
Marine Research and Resource Center, Hodeidah, Yemen

D. Tesfamichael, D. Pauly (eds.), *The Red Sea Ecosystem and Fisheries*, Coral Reefs of the World 7,
DOI 10.1007/978-94-017-7435-2_10

other than fishes (marine mammals and reptiles, seabirds, and the great invertebrate host), also includes common names in various languages, but its coverage is more spotty, with overall 28,600 common names for 3,500 species, of which 7,400 are in 87 languages other than English.

There is a high level of illiteracy in fishing communities, especially small scale fisheries in tropical and subtropical areas. Fishing communities in those areas, such as the Red Sea, are usually at the bottom of the socio-economic ladder and use oral story telling tradition as a way of communicating and transferring their knowledge from generation to generation (Tonkin 1986). Paying attention to the oral traditions can give a good insight into the society and their interactions with the environment (Johannes 1981; Becker and Ghimire 2003). Language (words and expressions) and stories are expressions of the collective experiences and consciousness of a society (Burns and Engdahl 1998; Freire and Pauly 2003). Words and expressions flourish out of common and collective experience of a society and acceptance of words to represent materials and ideas. When systematically collected and analyzed, the names and stories (knowledge) of communities can be an asset in understanding and managing resources (Palomares et al. 1999; Johannes et al. 2000; Freire and Pauly 2005; Tesfamichael et al. 2014).

To assist in these efforts, we have compiled the local names of fish and invertebrates from the Red Sea, i.e., of the species that are commonly targeted by the fishers. The Red Sea is bordered by seven countries. Although some countries (e.g., Sudan and Eritrea) have diverse languages, Arabic is a common language along most of the coastal areas of the Red Sea and there are strong cultural similarities among the coastal communities of the different countries. Thus, the local names are given in Arabic (written both in Arabic and Roman characters). Although most of the species commonly exploited by the fishers are included, this list is not exhaustive. Notably, the Hebrew names of fishes, used in Israel are not included (but see http://www.dafni.com/fish/list.htm)

Method

The collection and compilation of local names of the Red Sea fishes was carried out based on published reports and interviews. We searched published papers, local fishery reports, technical reports and books that contain corresponding species names and local names. Most of the fishery survey and assessment reports have information on the local names of the most common species of the fishery of the respective country. The earliest report used here were from the 1960s, when systematic exploration of the fishery resources the Red Sea countries started. However, we are aware of earlier expeditions in the Red Sea (see Chap. 1), which introduced Arabic fish names to Europe, e.g., those of Forsskål (1775), Rüppell (1826), and the Mabahiss Expedition (Tesfamichael 2005), where in most of the cases, the scientific name is exactly or loosely based on the local name. Most of the recent works we used for our collection were done by fisheries foreign experts in collaboration with their local counterparts, and most of them added an appendix of the local names of the fishes they encounter during their visits to the region, a practice holding into the 2000s. There are few publications by local researchers, e.g., local names for the fishes of Eritrea (Tesfamichael and Sebahtu 2006), Jordan (Khalaf and Disi 1997) and Sudan (Abu-Gideiri 1984).

We looked for reports that have the local names and corresponding valid scientific names (the latter as provided by FishBase or SeaLifeBase; see below), preferably at species level. When names were not available at species level, the name of the genus or family was used. There were few reports that gave only the common names; they were not included. Another type of sources were national fish catalogues, available for Sudan (Reed 1964; Abu-Gideiri 1984), Eritrea (Tesfamichael and Sebahtu 2006) and Jordan (Khalaf and Disi 1997). Sources that were comprehensive in their coverage were given priority for compiling names.

Fish names were also obtained from fishers at landing sites in Egypt, Sudan, Eritrea and Yemen by using either freshly caught specimens or color photos of fishes. Most of the fishes were already covered by the sources we used and the interviews were mainly used for verify confusing names. There are very few names that were acquired through interview.

Once the local and scientific names were compiled they were checked for the valid scientific names. The main catalogues used for checking valid names were FishBase (Froese and Pauly 2012) for fish and SeaLifeBase (Palomares and Pauly 2012) for invertebrates. Some names were not available in the above two databases and we used other online databases, i.e., Catalogue of life http://www.catalogueoflife.org/, Integrated Taxonomic Information System (ITTS) http://www.itis.gov/, World Register of Marine Species (WoRMS) http://www.marinespecies.org/, and FishWise professional http://www.fishwisepro.com/.

The published sources usually had the scientific name and the local names written in Roman characters; very few had names written in Arabic as well. The local names transcribed in Roman characters by different authors were different based on their transcription process. We standardized those names, and changed a few other names that were misspelled, or consisted of variants considered not different enough to be represented separately, e.g., '*Abu gurz*' and '*Abu gurs*' are represented in our table only by '*Abu gurz*'. However, when similar names were given differently, and we were not sure how to standardize them, we kept the variants, because even if Arabic is spoken all over the Red Sea, local dialects are different; thus, by retaining those differences, we can cover this diversity. The common names are presented on two tables, one sorted by the scientific names, the other by the local names in Arabic script. The names will be included in FishBase (www.fishbase.org) and SeaLifeBase (www.sealifebase.org).

Result and Discussion

We identified 500 distinct taxa with 465 local names, and 870 scientific name – common name combinations, i.e., many taxa have more than one local names and some local names are also used for different taxa (Tables 10.1 and 10.2), as also observed, e.g., by Freire and Pauly (2003). While scientific systematic nomenclature uses detailed characters for identification (e.g., fin ray counts in fishes, genetics) to categorize organisms, local names are usually based on characteristics that are easy to observe (Berlin 1992; Palomares et al. 1999). Hence, local names do not have as high resolution as scientific names.

One major difference of scientific and common names is that the latter are usually not unique. Thus, for example, the generic *Caranx* has the highest number of local names, i.e., 10 (Table 10.1). On the other hand, 'Beyad', the most frequent local name is used for 17 different taxa (Table 10.2) and all the local names with high occurrence frequency in Table 10.2 have higher number of occurrences than the high frequency scientific names with different local names in Table 10.1, except for *Caranx*. Taxa that have more than one local names either have one name per country, but occur in several countries (e.g. *Atule mate* has three different names depending on country) or they are so common and important to local communities that they have more than one local name within the same country (e.g., *Acanthopagrus bifasciatus* has four different names in Eritrea; see Appendix Table 10.3).

Sometimes the different names given to one species may refer to the different developmental stages. In similar study in Brazil, fishes that are commercially important and fishes that inhabit habitats frequented by fishers have on average six local names per species, while fishes that inhabit deeper waters and not fished by locals do not have local names. These discrepancies are believed to affect the data collection system (Freire and Pauly 2005). Similar results were also found for the Philippines (Palomares et al. 1999).

Table 10.1 Taxa with high number of local names

Taxa	No. of local names
Caranx spp.	10
Euthynnus affinis	7
Rachycentron canadum	7
Lethrinus spp.	7
Mugil spp.	7
Lutjanidae	7
Epinephelus tauvina	7
Epinephelus areolatus	6
Haemulidae	6
Cephalopholis miniata	6
Acanthopagrus bifasciatus	6
Diagramma pictum	6

Table 10.2 Number of fish species with the same local names

Local name	No. of spp.
Beyad	17
Muesy	13
Fanas	13
Moscht	13
Arabi	11
Kushar	10
Bagha	10
Abu sheneb	9
Ghabban	9
Shu'ur	8
Um qaren	8

Most of the local names compiled here are based on Arabic language, which is the common language in the coastal communities of the Red Sea. However, few names are based on European (mainly Italian) language. Those names are for taxa (often invertebrates) that were not readily consumed by local residents, at least in the past, but by Europeans when they frequented the Red Sea area. Thus, the locals named these taxa based on their European common names, e.g., in some communities shrimp are called '*Gamberi*', which is based on Italian. Local names are, however, given to some invertebrates not commonly consumed locally, e.g., crabs are generally called '*Abu mekass*' or '*Abu mokas*' meaning scissors. The first author has observed local fishers discarding big size crabs, which they see as nuisance for their fishing operation, and occasionally shrimps used as bait to catch fish.

Only few taxa are reported in many of the countries compared to an ecological check list of fishes in the ecosystem. These are taxa that are common and important to fishers throughout the Red Sea. For example, the two-spot snapper *Lutjanus bohar* is reported from almost all Red Sea countries. Based on interviews with fishers, it is clear that it is an important species to them, a priority target for handlining fishery. The local name is '*Bohar*', which is also the species epithet in the scientific name. Indeed, it is so important to the region that sometimes all *Lutjanus* species are referred to as '*Bohar*'. It is common practice for the fishers to use the same local name to a specific species or as generic name to a group either at genus level or even family level. *Scomberomorus commerson 'Dirak'* is the equivalent of *Lutjanus bohar*, in terms of its importance, for the pelagic fishery, which in the Red Sea is mainly gillnetting. Other names reported in most of the countries include '*Beyad*' (*Caranx* spp.), *Agam* (*Sphyraena barracuda*), *Bagha* (*Rastrelliger kanagurta*), and *Salmani* (*Chanos chanos*). These names would be the names usually known by anyone, even children. The list of all the names sorted in ascending order of the scientific names is given in Appendix Table 10.3 and sorted by the local names in Appendix Table 10.4. Photos of fishes with high frequency of local names, which are also fish important to fisheries, and some charismatic species are added in Appendix 2. These visuals will aid researchers to familiarize themselves to the common fishes and also communicate with fishers, and vice versa.

The usefulness of such local names table is many folds. As language is the registry of the collective consciousness of societies (Burns and Engdahl 1998), which depends on their experiences (their interactions with the environment and each other) and livelihood; looking at local names of fishes, one can learn the fishing habits and behaviours of a fishing community. Although most members of Red Sea coastal communities cannot read and write, their knowledge of the sea is extensive. Tapping into that wealth of experiential knowledge can help in understanding the socio-ecological system for informed planning and management (Johannes et al. 2000). Knowledge gained from fishers can be as informative as knowledge gained through other analytical means (Tesfamichael et al. 2014). In order to tap to fishers' knowledge, knowing the names of the items of discussion (i.e., fish) is critical.

Acknowledgement We would like to thank the fishers who cooperated with us during interviews for their time and willingness to share their knowledge. Thanks also go to Ken for helping entering names into the database, Myriam Khalfallah for double checking the local names written in Arabic and reviewing the chapter, and Daniel Pauly for helping us contextualizing our research, which was supported by the *Sea Around Us*, a scientific collaboration between the University of British Columbia and The Pew Charitable Trust.

Appendix 1

Table 10.3 Common names of Red Sea fish and invertebrates sorted by scientific names

No.	Scientific name	Common name	Local name	Arabic name	Country[a]	Source[b]
1	*Abalistes stellatus*	Starry triggerfish	Hijma	هجمه	Aq	14
2	*Abudefduf sexfasciatus*	Scissortail sergeant	Shabbar	شبار	Aq	14
3	*Abudefduf sordidus*	Blackspot sergeant	Shabbar	شبار	Aq	14
4	*Abudefduf vaigiensis*	Indo-Pacific sergeant	Shabbar	شبار	Aq	14
5	*Acanthopagrus berda*	Picnic seabream	Abu kuhul	ابو كحل	Su	6
6			Abu berite	ابو بريت	Er	11
7	*Acanthopagrus bifasciatus*	Twobar seabream	Rabaag	رباج	Aq,SA	1,14
8			Abu kuhul	ابو كحل	Er,Su	5,7,10
9			Fogil	فوجيل	Su	5
10			Butel hammed	بطل حماد	Er	11
11			Abu berite	ابو بريت	Er	11
12			Abyad	ابيض	Er	10
13	*Acanthurus blochii*	Ringtail surgeonfish	Gahm	جهم	Er	11
14			Zizan	زيزان	Er	11
15	*Acanthurus gahhm*	Black surgeonfish	Gahm	جهم	SA	1
16			Kohom	كوهم	Su	5,7
17	*Acanthurus nigricauda*	Epaulette surgeonfish	Gahm	جهم	Er	11
18			Zizan	زيزان	Er	11
19	*Acanthurus nigrofuscus*	Brown surgeonfish	Juneh	جنه	Aq	14
20	*Acanthurus sohal*	Sohal surgeonfish	Suhal	سوهل	SA	1
21			Sahla	سهله	Aq	14
22			Zizan	زيزان	Er	11
23			Gahm	جهم	Er	11
24	*Acanthurus xanthopterus*	Yellowfin surgeonfish	Gahm	جهم	Er	11
25			Zizan	زيزان	Er	11
26	*Aethaloperca rogaa*	Redmouth grouper	Kushar	كشر	Er	11
27			Katarban	كتربان	Su	5,7
28			Karban	كربان	Er	10
29			Ruga	روجا	SA	1

(continued)

Table 10.3 (continued)

No.	Scientific name	Common name	Local name	Arabic name	Country[a]	Source[b]
30	*Albula vulpes*	Bonefish	Bounouk	بو نوك	SA,Su	1,5,6,7
31	*Alectis ciliaris*	African pompano	Pompano	بمبانو	Er	11
32			Beyad	بياد	Er	10
33	*Alectis indicus*	Indian threadfish	Beyad abu tabag	بياد ابو تابج	SA	1
34			Shawish	شاويش	Su	5,6,7
35	*Alepes djedaba*	Shrimp scad	Beyad	بياد	Er	11
36			Djebbada	دجبدة	Ye	3
37	*Alopias vulpinus*	Thintail thresher	Gursh husseni	قرش حصينى	SA	1
38	*Aluterus monoceros*	Unicorn leatherjacket filefish	Um qaren	ام قرن	Aq	14
39	*Aluterus scriptus*	Scribbled leatherjacket filefish	Kotub	كوتب	Su	0
40	*Amanses scopas*	Broom filefish	Um qaren	ام قرن	Aq	14
41	*Amblygaster leiogaster*	Smooth-belly sardinella	Moza	موزة	Eg	13
42	*Amblygaster sirm*	Spotted sardinella	Sardina marboum	سردينه مبروم	Eg	13
43			Aida	عيدة	Er	11
44			Belem	بليم	Er	11
45	*Amphiprion bicinctus*	Twoband anemonefish	Om Al dukhan	ام الدخان	Aq	14
46	*Anampses caeruleopunctatus*	Bluespotted wrasse	Muesy	ميسي	Aq	14
47	*Anampses lineatus*	Lined wrasse	Muesy	ميسي	Aq	14
48	*Anampses meleagrides*	Spotted wrasse	Muesy	ميسي	Aq	14
49	*Anampses twistii*	Yellowbreasted wrasse	Muesy	ميسي	Aq	14
50	*Anchoviella* spp.	Anchovies	Fagima	فقيمة	Su	5,7
51	*Anthias* spp.	Fairy basslets	Zargh	زرغ	SA	9
52	*Aphareus furca*	Small toothed jobfish	Anteg	عنتق	Er	10
53	*Apogon aureus*	Ring-tailed cardinalfish	Fanas	فانس	Aq	14
54	*Apogon bifasciatus*	Twobelt cardinal	Fanas	فانس	Aq	14
55	*Apogon cyanosoma*	Yellowstriped cardinalfish	Fanas	فانس	Aq	14
56	*Apogon exostigma*	Narrowstripe cardinalfish	Fanas	فانس	Aq	14
57	*Apogon kallopterus*	Iridescent cardinalfish	Fanas	فانس	Aq	14
58	*Apogon nigrofasciatus*	Blackstripe cardinalfish	Fanas	فانس	Aq	14
59	*Apolemichthys xanthotis*	Yellow-ear angelfish	Moscht	مشط	Aq	14
60	*Aprion* spp.	Snappers	Farsi	فارسي	Su	5,7
61	*Aprion virescens*	Green jobfish	Farsi	فارسي	Su	6
62	*Argyrops filamentosus*	Soldierbream	Jarbeeden	جربيدن	Aq	14
63	*Argyrops spinifer*	King soldierbream	Morjan	مرجان	Eg,Su	7,12
64			Fofal	فوفل	Er,Su	5,7,10
65			Najar	نجار	Aq	14
66			Abyad	ابيض	Er	10
67			Najar	نجار	SA	1,9
68	*Ariomma brevimanus*	Pomfret	Maslimani	مسلمانى	Aq	14
69	*Arius* spp.	Sea catfishes	Abu sheneb	ابو شنب	SA	9
70	*Arothron diadematus*	Masked puffer	Hadhroom	حضروم	Aq	14
71	*Arothron hispidus*	White-spotted puffer	Drimma	دريما	SA	1
72			Hadhroom	حضروم	Aq	14
73	*Arothron immaculatus*	Immaculate puffer	Drimma	دريما	SA	1
74	*Arothron stellatus*	Stellate puffer	Hadhroom	حضروم	Aq	14
75			Drimma	دريما	SA	1
76	*Atherinomorus lacunosus*	Hardyhead silverside	Balem	بليم	Aq	14
77			Gurgush	جرجوش	Su	5,7
78	*Atule mate*	Yellowtail scad	Haboot	هابوت	Su	5,7

(continued)

Table 10.3 (continued)

No.	Scientific name	Common name	Local name	Arabic name	Country[a]	Source[b]
79			Saibariya	صعباريا	Ye	15
80			Beyad	بياد	Er,SA	1,11
81	*Auxis thazard thazard*	Frigate tuna	Sherwi	شروي	SA	1
82			Tuna	تونة	Su	5
83	*Balistapus undulatus*	Orange-lined triggerfish	Um qaren	ام قرن	Aq	14
84			Khanzeer	خنزير	SA	1
85			Ajame	عجامة	Er	11
86	*Balistes* spp.	Triggerfishes	Schiyram shiram	شيرام شيرام	SA	9
87			Canzir	كنزير	Er	10
88	*Balistoides viridescens*	Titan triggerfish	Ajame	عجامة	Er	11
89			Faki sharam	فاكي شرام	Su	5,7
90	*Bodianus anthioides*	Lyretail hogfish	Muesy	ميسي	Aq	14
91	*Bodianus axillaris*	Axilspot hogfish	Deek	ديك	SA	1
92	*Bolbometopon muricatum*	Green humphead parrotfish	Harida	حريدا	Er	11
93	*Bothus pantherinus*	Leopard flounder	Hisan al-Bahar	حصان البحر	SA	1
94			Shebet al bahir	شيبت البحر	Er	11
95			Tabaq	طبق	Aq	14
96	Brachyura	Crab	Abu mokas	ابو مقص	Ye	2,3
97	*Branchiostegus sawakinensis*	Freckled tilefish	Theena	تحينا	Su	6
98	*Caesio lunaris*	Lunar fusilier	Bagha	باغة	Aq	14
99	*Caesio striata*	Striated fusilier	Kourab el bahr	كراب البحر	Er	11
100	*Caesio suevica*	Suez fusilier	Bagha	باغة	Aq	14
101			Kourab el bahr	كراب البحر	Er	11
102	*Caesio varilineata*	Variable-lined fusilier	Bagha hamra	باغة حمراء	Aq	14
103	*Callyodon* spp.	Parrotfishes	Harid	حريد	Su	5,7
104	*Calotomus viridescens*	Viridescent parrotfish	Ghabban	غبان	Aq	14
105	*Cantherhines pardalis*	Honeycomb filefish	Um qaren	ام قرن	Aq	14
106	*Canthigaster coronata*	Crowned puffer	Hadhroom	حضروم	Aq	14
107	*Canthigaster margaritata*	Pufferfish	Hadhroom	حضروم	Aq	14
108	*Carangoides armatus*	Longfin trevally	Beyad	بياد	Er	11
109	*Carangoides bajad*	Orangespotted trevally	Beyad	بياد	Er,Su	5,10
110			Reema safra	ريما صفره	Aq	14
111			Beyad gazza	بياد جازة	SA	9
112	*Carangoides coeruleopinnatus*	Coastal trevally	Beyad goutar	بياد جوتر	Er,Su	6,10
113	*Carangoides equula*	Whitefin trevally	Subaria	صعباريا	Ye	3
114	*Carangoides ferdau*	Blue trevally	Beyad	بياد	Er	11
115	*Carangoides fulvoguttatus*	Yellowspotted trevally	Gutur	جوتر	Su	5
116			Reem	ريم	Aq	14
117			Seleikh	سيليخ	Su	5,7
118			Beyad gaz	بياد جاز	SA	1
119			Beyad goutar	بياد جوتر	Er,Su	5,11
120	*Carangoides gymnostethus*	Bludger	Beyad	بياد	Er	10
121	*Carangoides malabaricus*	Malabar trevally	Beyad	بياد	Er	11
122	*Carangoides oblongus*	Coachwhip trevally	Beyad girm	بياد جريم	SA	9
123	*Caranx ignobilis*	Giant trevally	Girim	جيريم	Su	5
124			Beyad	بياد	Er,SA,Su	1,5,10
125	*Caranx melampygus*	Bluefin trevally	Beyad girm	بياد جريم	Er,SA	1,10
126	*Caranx sexfasciatus*	Bigeye trevally	Beyad	بياد	Er,SA,Su	1,5,10
127	*Caranx* spp.	Jacks and pompanos	Tak'oi	تكاوي	Su	7
128			Safloh	سفلوح	Su	7

(continued)

Table 10.3 (continued)

No.	Scientific name	Common name	Local name	Arabic name	Country[a]	Source[b]
129			Haboot	هابوت	Su	7
130			Lamenab	لمناب	Su	7
131			Goutar	قوتر	Su	5,7
132			Seleikh	سيليخ	Su	7
133			Girim	جيريم	Su	7
134			Beyad	بياد	Er,SA,Su,Ye	3,7,9,10
135			Goareit	جواريت	Su	7
136			Karb	كرب	Su	5,7
137	Carcharhinidae	Requiem sharks	Lokhem	لخام	Ye	2
138			Dohoosh	دهوش	Ye	2
139			Girish	قريش	SA,Su	5,8,9
140			Autat	عوتات	Er	10
141	*Carcharhinus amblyrhynchos*	Grey reef shark	Gursh al baba	قرش البابا	SA	1
142	*Carcharhinus limbatus*	Blacktip shark	Autat	عوتات	Er	11
143			Gursh al sahl	قرش السهل	SA	1
144	*Carcharhinus melanopterus*	Blacktip reef shark	Autat	عوتات	Er	11
145			Zingi	زنجى	Er	11
146	*Carcharhinus plumbeus*	Sandbar shark	Autat	عوتات	Er	11
147			Gursh	قرش	Aq	14
148	*Carcharodon carcharias*	Great white shark	Gursh	قرش	SA	1
149	*Centropyge multispinus*	Dusky angelfish	Moscht	مشط	Aq	14
150	*Cephalopholis argus*	Peacock hind	Ghohlab	جو هلاب	Su	6
151			Kushar abu blaha	كشر ابو بلحة	Er,SA	1,11
152	*Cephalopholis hemistiktos*	Yellowfin hind	Mumen	مومن	Aq	14
153	*Cephalopholis miniata*	Coral hind	Ferek	فريق	Er	11
154			Ghohlab	جو هلاب	Su	6
155			Ahmer	احمر	Er	11
156			Kushar	كشر	SA	1
157			Kushar abu adas	كشر ابو عد س	SA	9
158			Shirni	شيرنى	Aq	14
159	*Cephalopholis sexmaculata*	Sixblotch hind	Abu shirni	ابو شرني	Aq	14
160	*Cephalopholis* spp.	Groupers	Gahlab	جهلب	Su	5
161	*Cetoscarus bicolor*	Bicolour parrotfish	Ghabban	غبان	Aq	14
162			Harida	حريدا	Er	11
163	*Chaetodon auriga*	Threadfin butterflyfish	Gringish	جرينجش	SA	1
164			Moscht	مشط	Aq	14
165	*Chaetodon austriacus*	Blacktail butterflyfish	Moscht	مشط	Aq	14
166	*Chaetodon fasciatus*	Diagonal butterflyfish	Moscht	مشط	Aq	14
167	*Chaetodon melannotus*	Blackback butterflyfish	Moscht	مشط	Aq	14
168	*Chaetodon paucifasciatus*	Eritrean butterflyfish	Moscht	مشط	Aq	14
169	*Chaetodon semilarvatus*	Bluecheek butterflyfish	Moscht	مشط	Aq	14
170	*Chaetodon trifascialis*	Chevron butterflyfish	Moscht	مشط	Aq	14
171	*Chanos chanos*	Milkfish	Salmani	سلماني	Er,SA,Su,Ye	1,3,5,6,7,11
172			Bunji	بونجي	Su	5,6,7
173	*Cheilinus abudjubbe*	Abudjubbe wrass	Rabadi	ربادي	Aq	14
174			Arousset el baher	عروسة البحر	Er	11
175			Esha mer'e	عش ميري	Er	11
176	*Cheilinus lunulatus*	Broomtail wrasse	Rabadia	ربادي	Aq	14
177	*Cheilinus mentalis*	Mental wrasse	Muesy	ميسي	Aq	14
178	*Cheilinus trilobatus*	Tripletail wrasse	Abu mulees	ابو مليس	SA	1

(continued)

Table 10.3 (continued)

No.	Scientific name	Common name	Local name	Arabic name	Country[a]	Source[b]
179	*Cheilinus undulatus*	Humphead wrasse	Esha mer'e	عش ميري	Er	11
180			Arousset el baher	عروسة البحر	Er	11
181			Terbaany	تيرباني	SA	1
182			Limalima	ليما ليما	Su	7
183			Abu jibba	ابو جبة	Su	5,6,7
184	*Cheilodipterus lachneri*		Fanas	فانس	Aq	14
185	*Cheilodipterus macrodon*	Large toothed cardinalfish	Fanas	فانس	Aq	14
186	*Cheilodipterus novemstriatus*	Indian Ocean twospot cardinalfish	Fanas	فانس	Aq	14
187	Chelonioidea	Marine turtles	Sulhafa	سلحافه	Ye	3
188	*Chilomycterus spilostylus*	Spotbase burrfish	Hadhroom abu shouka	حضروم ابو شوكة	Aq	14
189	*Chirocentrus dorab*	Dorab wolf-herring	Nakanaf	نكاناف	Ye	3
190			Mekhlef	مخلف	Er	11
191			Abu seif	ابو سيف	SA,Su	1,5,6
192	*Chlorurus gibbus*	Heavybeak parrot fish	Abu greeyan	ابو جريان	SA	1
193	*Chlorurus sordidus*	Daisy parrotfish	Ghabban	غبان	Aq	14
194	*Choerodon robustus*	Robust tuskfish	Far al-Bahar	فار البحر	Aq	14
195	*Cholorurus sordidus*	Daisy parrotfish	Hareeth	حريث	SA	1
196	*Chrysoblephus* spp.	Porgies	Haffar	حفار	SA	9
197	Clupeidae	Herrings, shads, sardines etc.	Aida	عيدة	Ye	3
198			Sardin	سردين	Ye	3
199			Wasif	وزف	Ye	2
200	*Cociella crocodilus*	Crocodile flathead	Rugud	روجود	SA	1
201	*Conger cinereus*	Longfin African conger	Hanish silab	حنيش سيلاب	SA	1
202	*Coris aygula*	Clown coris	Heqab	حيقب	Aq	14
203	*Coris caudimacula*	Spottail coris	Muesy	ميسي	Aq	14
204	*Coris variegata*	Dapple coris	Muesy	ميسي	Aq	14
205	*Coryphaena hippurus*	Common dolphinfish	Saif	سيف	Aq	14
206			Um falloos	ام فلوس	SA,Su	1,5,7
207	*Corythoichthys flavofasciatus*	Network pipefish	Masas ramli	مساس رملي	Aq	14
208	*Corythoichthys schultzi*	Schultz's pipefish	Masas ramli	مساس رملي	Aq	14
209	*Crenidens crenidens*	Karenteen seabream	Haffar	حفار	SA	1
210			Hindook	هيندوك	Su	5,7
211	*Crenimugil crenilabis*	Fringelip mullet	Arabi	عربي	Aq	14
212	*Crenimugil* spp.	Mullet	Arabi	عربي	Er,SA	1,10
213	*Cristacirrhitus punctatus*	Blackspotted hawkfish	Jarbua	جربوعه	SA	1
214	*Ctenochaetus striatus*	Striated surgeonfish	Juneh	جنه	Aq	14
215	*Cynoglossus bilineatus*	Fourlined tonguesole	Hisan al-Bahar	حصان البحر	SA	1
216	*Dactyloptena peterseni*	Starry flying gurnard	Boomet al-Bahar	بومة البحر	Aq	14
217	*Dascyllus trimaculatus*	Threespot dascyllus	Kharayeh	خاراية	Aq	14
218	Dasyatidae	Stingrays	Fahodoo	فهودو	Ye	2
219			Abu remis	ابو رميس	Ye	15
220			Taira	طيرة	Ye	3
221			Bakhat	باخات	Ye	2
222	*Dasyatis* spp.	Stingrays	Abu soot	ابو سوط	Su	5
223			Rugtia	روجتيه	SA	1
224	*Decapterus macarellus*	Mackerel scad	Amia	اميه	Aq	14
225	*Decapterus macrosoma*	Shortfin scad	Sardina	سردينه	Aq	14
226	*Decapterus maruadsi*	Japanese scad	Bagha	باغة	Su	5
227	*Decapterus punctatus*	Round scad	Bagha	باغة	Eg	13

(continued)

Table 10.3 (continued)

No.	Scientific name	Common name	Local name	Arabic name	Country[a]	Source[b]
228	*Decapterus russelli*	Indian scad	Bagha	باغة	Eg	13
229			Shadba	شدبا	Su	5,6,7
230			Sardina aredha	سردينه عريضة	Aq	14
231	*Decapterus* spp.	Scads	Shaduba	شدوبا	SA	1
232	*Dendrochirus brachypterus*	Shortfin turkeyfish	Rani, Abu al-Laban	رنى- ابو اللبن	Aq	14
233	*Diagramma pictum*	Painted sweetlips	Shutaf	شوطاف	Er, Ye	2,11
234			Shakfa	شكفا	Su	5,6,7
235			Caterin	قطرين	Ye	2
236			Sobaity	سوبيتى	Er	11
237			Fataleeta	فتاليتا	Su	5,6,7
238			Istaf	استاف	Aq	14
239	*Diagramma* spp.	Haemulidae	Shutaf	شوطاف	SA	9
240	*Diodon hystrix*	Spot-fin porcupinefish	Drimma	دريما	SA	1
241			Hadhroom abu shouka	حضروم ابو شوكة	Aq	14
242	*Diplodus noct*	Red Sea seabream	Noct	نوكت	Aq	14
243	*Dussumieria acuta*	Rainbow sardine	Belem	بليم	Er	11
244			Aida	عيدة	Er	11
245	*Echeneis naucrates*	Live sharksucker	Qamlet alqersh	قملة القرش	Aq	14
246			Gamla	جمله	SA	1
247	*Echeneis* spp.	Remoras	Kamlet al darfil	قملة الدرفيل	SA	9
248	*Echidna nebulosa*	Snowflake moray	Qmum muraqata	قموم مرقطة	Aq	14
249	*Elagatis bipinnulata*	Rainbow runner	Beyad	بياد	Er	10
250			Adad	اداد	Su	5
251			Muslabah	مسلابه	Er,SA	1,11
252	*Elops affinis*	Pacific ladyfish	Salmani	سلماني	SA	9
253	*Elops machnata*	Tenpounder	Khanny	خاني	SA	1
254			Shagool	شجول	Su	5,6,7
255	*Encrasicholina heteroloba*	Shorthead anchovy	Aida	عيدة	Er	11
256			Wesif	وزف	Er	11
257			Belem	بليم	Er	11
258	*Epinephelus* spp.	Groupers	Kushar	كشر	Su	8
259	*Epinephelus areolatus*	Areolate grouper	Kushar nagel	كشر ناجل	Er,SA	9,11
260			Abu ades	ابو عدس	Er	11
261			Shelwa	شيلوة	Aq	14
262			Kodad	كداد	Su	7
263			Seetiati	سيتياتي	Su	7
264			Gishir shooni	جيشر شونى	Su	5,6,7
265	*Epinephelus chlorostigma*	Brownspotted grouper	Samman	سمن	Er	11
266			Gishir	جيشر	Su	5
267	*Epinephelus fasciatus*	Blacktip grouper	Kushar	كشر	Er	11
268			Daghma	دغمة	Aq	14
269	*Epinephelus fuscoguttatus*	Brown-marbled grouper	Kushar	كشر	Er	11
270	*Epinephelus lanceolatus*	Giant grouper	Kushar twini	كشر طويني	SA	9
271	*Epinephelus malabaricus*	Malabar grouper	Kushar twini	كشر طويني	Er,SA	9,10
272	*Epinephelus merra*	Honeycomb grouper	Ghoshar	قشار	Su	6
273	*Epinephelus morrhua*	Comet grouper	Daghma	دغمة	Aq	14
274			Kushar abu lulu	كشر ابو لولو	SA	1
275	*Epinephelus polylepis*	Smallscaled grouper	Angar	عنقر	Ye	15
276	*Epinephelus polyphekadion*	Camouflage grouper	Gishir	جيشر	Su	5
277			Kushar	كشر	Er	11
278	*Epinephelus radiatus*	Oblique-banded grouper	Daghma	دغمة	Aq	14

(continued)

Table 10.3 (continued)

No.	Scientific name	Common name	Local name	Arabic name	Country[a]	Source[b]
279	*Epinephelus* spp.	Sea basses	Kushar	كشر	Er,SA	9,10
280	*Epinephelus summana*	Summan grouper	Kushar mubal'at	كشر موبالات	Er,SA	1,11
281			Gishir	جيشر	Su	5
282			Aqshar	اقشر	Aq	14
283	*Epinephelus tauvina*	Greasy grouper	Hubog	حوبج	SA	9
284			Tauwina	طوينا	Er,Su	5,6,7,11
285			Toona	تونا	Su	7
286			Gishir tauwina	جيشر توينه	Su	5,6,7
287			Kushar tauwina	كشر توينه	Eg,SA	1,12
288			Aqshar	اقشر	Aq	14
289			Kushar	كشر	Er,Ye	2,11
290	*Etelis carbunculus*	Ruby snapper	Hamaroon	حمرون	Su	5,7
291	*Etrumeus teres*	Red-eye round herring	Sardin	سردين	Su	5
292			Sardina masreya	سردينه مصرية	Aq	14
293	*Euthynnus affinis*	Kawakawa	Ma'agab	ماعجب	Er,SA,Su	1,5,6,7,10
294			Abu dam	ابو دم	Su	5,6,7
295			Fatla	فتلة	Aq	14
296			Zainub	زينب	Ye	2
297			Tun	تن	Ye	2
298			Sherwa	شروة	Er,Ye	2,11
299			Tonno	تونو	Er	11
300	*Exallias brevis*	Leopard blenny	Arfaj	عرفج	Aq	14
301	*Fenneropenaeus indicus*	Indian white prawn	Gamberi	جمبري	Er	11
302	*Fistularia commersonii*	Bluespotted cornetfish	Qasaba	قصبة	Aq	14
303	*Fistularia* spp.	Cornetfishes	Khurm al baaha	خرم الباحه	SA	1
304	*Galeocerdo cuvier*	Tiger shark	Abu nebir	ابو نبير	Er	10
305			Nebrawi	نبراوي	Er	11
306			Gursh nimrany	قرش نمراني	SA	1
307			Numrani	نمراني	Ye	3
308			Gursh	قرش	Aq	14
309	*Genicanthus caudovittatus*	Zebra angelfish	Moscht	مشط	Aq	14
310	*Gerres longirostris*	Strongspine silver-biddy	Mukeresh	مكارش	Er	11
311	*Gerres oyena*	Common silver-biddy	Kass	كس	Er,SA,Su	1,6,7,11
312			Rishan	ريشان	Aq	14
313			Mehara	مهارا	Ye	15
314	*Gerres* spp.	Mojarras	Abu gurz	ابو جرز	SA	9
315			Gash	قش	SA	9
316			Afs	عفس	SA	9
317	*Glaucostegus halavi*	Halavi ray	O'ud	عود	Er	11
318	*Gnathanodon speciosus*	Golden trevally	Bagesh	بجيش	Ye	2
319			Beyad gaz	بياد جاز	SA	1
320			Beyad	بياد	Er,Su	5,11
321	*Gomphosus caeruleus*	Green birdmouth wrasse	Muesy	ميسي	Aq	14
322	*Grammistes* spp.	Fairy basslets	Kushar abu adas	كشر ابو عد س	SA	1
323	*Gymnocranius grandoculis*	Blue-lined large-eye bream	Qamar	قمر	Aq	14
324	*Gymnosarda unicolor*	Dogtooth tuna	Shak abu ein	شك ابو عين	Aq	14
325			Tomad	تمد	Su	5
326	*Gymnothorax johnsoni*	Whitespotted moray	Qmum	قموم	Aq	14
327	*Gymnothorax nudivomer*	Starry moray	Qmum	قموم	Aq	14
328	*Gymnothorax* spp.	Moray eels	Shaaga	شعاجه	SA	1
329	*Gymnothorax undulatus*	Undulated moray	Shaaga	شعاجه	SA	1

(continued)

Table 10.3 (continued)

No.	Scientific name	Common name	Local name	Arabic name	Country[a]	Source[b]
330	Haemulidae	Grunts	Nakem	ناكم	Ye	2
331			Getran	جيتران	Er	10
332			Sobaity	سوبيتى	Er	10
333			Shatef	شاتف	Er	10
334			Shefsh	شيفش	Er	10
335			Koko	كوكو	Er	10
336	*Halichoeres hortulanus*	Checkerboard wrasse	Muesy	ميسي	Aq	14
337			Deek	ديك	SA	1
338	*Halichoeres scapularis*	Zigzag wrasse	Muesy	ميسي	Aq	14
339	*Hemiramphus* spp.	Halfbeaks	Selenti	سيلينتى	Su	5
340	*Heniochus acuminatus*	Pennant coralfish	Gringish	جرينجش	SA	1
341	*Heniochus diphreutes*	False moorish idol	Um shiraa	ام شراع	Aq	14
342	*Heniochus intermedius*	Red Sea bannerfish	Um shiraa	ام شراع	Aq	14
343	*Herklotsichthys punctatus*	Spotback herring	Shagool	شجول	Su	5
344	*Herklotsichthys quadrimaculatus*	Bluestripe herring	Sardina	سردينه	Aq	14
345			Belem	بليم	Er	11
346			Aida	عيدة	Er	11
347			Abu ras	ابو راس	Er	11
348	*Himantura uarnak*	Honeycomb stingray	Abromis	ابروميس	Er	11
349			Halali	هلالى	Er	11
350	*Hippocampus histrix*	Thorny seahorse	Hisan al-Bahar	حصان البحر	Aq	14
351	*Hippocampus kuda*	Spotted seahorse	Hisan al-Bahar	حصان البحر	Aq	14
352	*Hippocampus* spp.	Seahorses	Fara al bahr	فاره البحر	SA	1
353	*Hipposcarus harid*	Candelamoa parrotfish	Harid	حريد	Aq,Su	5,14
354	Holothuriidae	Sea cucumber	Kheiar albahr	خيار البحر	Ye	4
355			Hidra	حيدره	Er	15
356	*Hypoatherina temminckii*	Samoan silverside	Gashgoosha	جشجوشه	SA	1
357	*Hyporhamphus dussumieri*	Dussimier's halfbeak	Zirgaan	زيرجان	SA	1
358			Silinti	سيلنتي	Su	6
359	*Hyporhamphus gamberur*	Red Sea halfbeak	Korom	قورم	Er	11
360			Far	فار	Er	11
361	*Iniistius pentadactylus*	Fivefinger wrasse	Far al-Bahar	فار البحر	Aq	14
362	*Istiblennius edentulus*	Rippled rockskipper	Arfaj	عرفج	Aq	14
363	*Istiompax indica*	Black marlin	Faras al Bahr	فارس البحر	SA	1
364	Istiophoridae	Billfishes	Feraz	فيرز	Ye	2
365	*Istiophorus platypterus*	Indo-Pacific sailfish	Abu feres	أبو فرس	Er	11
366			Abu shiraa	ابو شراع	Su	5,6
367			Faras al Bahr	فارس البحر	SA,Su	1,5,6
368			Faras	فارس	Aq	14
369	*Isurus* spp.	Mackerel sharks or white shark	Gursh deeba	قرش ديبا	SA	1
370	*Katsuwonus pelamis*	Skipjack tuna	Ma'agab	ماعجب	SA	1
371			Fatleh	فتله	Aq	14
372			Tuna	تونة	Su	5
373			Zeinub	زينوب	Er	11
374			Sherwi	شروي	Er	11
375	*Kuhlia mugil*	Barred flagtail	Ghlaimeh	جليمه	Aq	14
376	*Kyphosus cinerascens*	Blue seachub	Tahmal	تحمل	SA	1
377	*Lagocephalus sceleratus*	Silver-cheeked toadfish	Alnaguem	الناقم	Aq	14
378	Leiognathidae	Slimys, slipmouths, or ponyfishes	Afsh	عفش	Er	10
379			Botatos	بوطاطوس	Er	10

(continued)

Table 10.3 (continued)

No.	Scientific name	Common name	Local name	Arabic name	Country[a]	Source[b]
380			Kash	كش	Ye	2
381			Gutat	جوتات	Er	10
382	*Leiognathus* spp.	Slimys, slipmouths, or ponyfishes	Arian	عريان	Su	5
383			Abu gurz	ابو جرز	SA	1
384	*Leptoscarus vaigiensis*	Marbled parrotfish	Hareeth	حريث	SA	1
385	Lethrinidae	Emperors or scavengers	Afsh	عفش	Ye	2
386			Shu'ur	شعور	Er,Su	8,10
387			Terhani	ترهاني	Er	10
388	*Lethrinus borbonicus*	Snubnose emperor	Qeda	قدة	Aq	14
389	*Lethrinus harak*	Thumbprint emperor	Shu'ur	شعور	Er,Su	5,10
390			Abu nugta	ابو نقطة	Er,Su	5,10
391	*Lethrinus lentjan*	Pink ear emperor	Suli	صولي	Er	11
392			Shu'ur	شعور	Er,Su	5,10
393	*Lethrinus mahsena*	Sky emperor	Qeda	قدة	Aq	14
394			Shu'ur	شعور	Er,Su	5,10
395			Mahsena	محسنه	Er	11
396	*Lethrinus microdon*	Smalltooth emperor	Shu'ur	شعور	Er	10
397			Suli	صولي	Er	10
398	*Lethrinus miniatus*	Trumpet emperor	Shu'ur dibi	شعور دبي	SA	9
399	*Lethrinus nebulosus*	Spangled emperor	Shu'ur ramaka	شعور رامكا	SA	9
400			Shu'ur mehseny	شعور محيسنى	SA	1
401			Shu'ur	شعور	Aq,Er,Su	5,10,14
402			Afsh	عفش	Ye	2
403			Suli	صولي	Er	11
404	*Lethrinus* spp.	Emperors or scavengers	Shu'ur abu zahwa	شعور ابو زهوه	SA	1
405			Gash	قش	Ye	3
406			Shu'ur ramaka	شعور رامكا	SA	1
407			Shu'ur khirmiya	شعور خرمية	SA	1
408			Afsh	عفش	Ye	3
409			Shu'ur deeb	شعور ديب	SA	1
410			Shu'ur	شعور	SA,Su	5,7,9
411	*Lethrinus variegatus*	Slender emperor	Bunqus khermawi	بنقوس خرماوي	Aq	14
412	*Liza macrolepis*	Largescale mullet	Arabi	عربي	Er,SA	1,11
413	*Liza vaigiensis*	Squaretail mullet	Arabi	عربي	Er	10
414	*Lobotes surinamensis*	Atlantic tripletail	Abu hajar	أبو حجر	Er	11
415			Rougaah	روجاه	SA	9
416	Loliginidae	Squids	Kalamari	كلاماري	Su	5
417	*Loligo forbesii*	Veined squid	Abu midad	ابو مداد	Er	11
418	Lutjanidae	Snappers	Shaefen	شعفن	Er	10
419			Anteg	عنتق	Er	10
420			Sheik ali	شيخ على	Er	10
421			Farsi	فارسي	Su	8
422			Huberi	حوبيري	Er	10
423			Naisarah	نيصاره	Er	10
424			Bohar	بوهر	Er	10
425	*Lutjanus argentimaculatus*	Mangrove red snapper	Safin	سفين	Su	5,6,7
426			Shaefen	شعفن	Er,SA	1,11
427	*Lutjanus bohar*	Two-spot red snapper	Garabganat	جرابجانت	Su	7
428			Lolab	لو لاب	Su	7
429			Bohar	بوهر	Aq,Er,SA,Su,Ye	1,2,5,7,9,10

(continued)

Table 10.3 (continued)

No.	Scientific name	Common name	Local name	Arabic name	Country[a]	Source[b]
430	*Lutjanus ehrenbergii*	Blackspot snapper	Hebra	حبرا	Aq	14
431			Shaefen	شعفن	Er	11
432	*Lutjanus fulviflamma*	Dory snapper	Hubr	(حوبر(حبر)	SA	1
433			Shaefen	شعفن	Er	11
434			Habair	هبير	Su	5,6,7
435			Hobara	هوبارا	SA	9
436	*Lutjanus gibbus*	Humpback red snapper	Himbuk	هيمبوك	Er	11
437			Asmoodi	عصمودي	SA	1
438			Bohar	بوهر	Er	10
439			Asmoot	عصموت	Er,Su	5,6,11
440			Huberi	حوبيري	Er	10
441	*Lutjanus johnii*	John's snapper	Hubr	(حوبر(حبر)	SA	1
442	*Lutjanus kasmira*	Common bluestripe snapper	Huberi	حوبيري	Er,SA	1,10
443			Ableen asfar	ابلين أصفر	Su	5,6,7
444			Bohar	بوهر	Er	10
445	*Lutjanus lutjanus*	Bigeye snapper	Shukrum	شخروم	Su	5
446	*Lutjanus malabaricus*	Malabar blood snapper	Hamari	حماري	Er	11
447			Gehab	جيهاب	Er	10
448			Bohar	بوهر	Er	10
449	*Lutjanus rivulatus*	Blubberlip snapper	Shaefen	شعفن	Er	11
450	*Lutjanus russellii*	Russell's snapper	Hubr	(حوبر(حبر)	SA	1
451	*Lutjanus sanguineus*	Humphead snapper	Nirjan	نيرجان	Ye	2
452			Morjan	مرجان	Ye	2
453	*Lutjanus sebae*	Emperor red snapper	Himbuk	هيمبوك	Er	11
454	*Lutjanus* spp.	Snappers	Hobara	هوبارا	SA	9
455			Asmoot	عصموت	Su	8
456	*Macolor niger*	Black and white snapper	Kust	كست	Su	5
457			Shaefen	شعفن	Er	11
458			Kut	كت	Su	7
459	*Makaira nigricans*	Blue marlin	Faras al Bahr	فارس البحر	SA	1
460	*Manta* spp.	Eagle and manta rays	Rugia milla	روجي ميله	SA	1
461	*Marsupenaeus japonicus*	Kuruma prawn	Gamberi	جمبري	Er	11
462	*Megalaspis cordyla*	Torpdeo scad	Beyad turfa	بياد طرفه	SA	1
463			Sherwi	شروي	Er	11
464			Khurtum	خرطوم	Su	5,6,7
465	*Melicertus latisulcatus*	Western king prawn	Gamberi	جمبري	Er	11
466	*Metapenaeus monoceros*	Speckled shrimp	Gamberi	جمبري	Er	11
467	*Mobula* spp.	Eagle and manta rays	Rugtia	روجتيه	SA	1
468	*Mola* spp.	Molas or ocean sunfishes	Milla	ميله	SA	1
469	*Monacanthus* spp.	Filefishes	Abu shaukah	ابو شوكة	SA	9
470	*Monodactylus argenteus*	Silver moony	Haymaan	حيمان	SA	1
471	*Monotaxis grandoculis*	Humpnose big-eye bream	Qamar Abu ein	قمر ابو عين	Aq	14
472			Shu'ur abu'ayn	شعور ابو عين	SA	1
473	*Mugil cephalus*	Flathead grey mullet	Arabi	عربي	Er,SA	1,10
474	*Mugil* spp.	Mullets	Ka'oi	كاوي	Su	7
475			Shole	شول	Su	7
476			Fasekh	فاسخ	Su	7
477			Tadab	تدب	Su	7
478			Arabi	عربي	Su	5,7,8
479			Jilan	جيلان	Su	7
480			Sha'aboi	شعبوي	Su	8,7

(continued)

Table 10.3 (continued)

No.	Scientific name	Common name	Local name	Arabic name	Country[a]	Source[b]
481	Mugilidae	Mullets	Arabi	عربي	Er,Ye	10,15
482	Mullidae	Goatfishes	Barbuni	بربونى	Er	10
483			Abu digin	ابو دجن	Er	10
484			Abu sheneb	ابو شنب	Er	10
485			Ambir	امبير	Ye	2
486	*Mulloidichthys flavolineatus*	Yellowstripe goatfish	Abu digin	ابو دجن	Su	5
487			Sabalan	سبلان	Aq	14
488	*Mulloidichthys vanicolensis*	Yellowfin goatfish	Sabalan	سبلان	Aq	14
489	*Mustelus mosis*	Arabian smooth-hound	Gursh	قرش	Aq	14
490	*Myripristis murdjan*	Pinecone soldierfish	Fanas	فانس	Aq	14
491			Iyya sagheera	ايبا صغيرة	SA	1
492	*Narcine bentuviai*	Elat electric ray	Khadala	خداله	Aq	14
493	*Naso brevirostris*	Spotted unicornfish	Abu karn	ابو قرن	Er	11
494			Kurnjal	قرنجل	Er	11
495	*Naso lituratus*	Orangespine unicornfish	Akra abu garn	عكرا ايبو جرم	SA	1
496			Rahu	رحو	Aq	14
497	*Naso* spp.	Unicornfishes	Abu karn	ابو قرن	Su	8
498	*Naso unicornis*	Bluespine unicornfish	Abu garin	ابو جرين	SA,Su	1,5,6
499			Rahu	رحو	Aq	14
500			Kurnjal	قرنجل	Er	11
501	*Neamia octospina*	Eightspine cardinalfish	Fanas	فانس	Aq	14
502	*Nebrius ferrugineus*	Tawny nurse shark	Gursh massassa	قرش مصاصا	SA	1
503	*Negaprion acutidens*	Sicklefin lemon shark	Autat	عوتات	Er	11
504			Zingi	زنجي	Er	11
505	Nemipteridae	Threadfin breams	Sare	سار	Er	10
506			Ser'a	سيره	Er	10
507	*Nemipterus japonicus*	Japanese threadfin bream	Sare	سار	Eg	12
508			Homiara	حوميره	Ye	2
509			Morjan	مرجان	SA	9
510	*Nemipterus marginatus*	Red filament threadfin bream	Fares	فارس	Eg	12
511	*Nemipterus* spp.	Threadfin breams	Morjan	مرجان	SA	9
512			Nofrah	نوفره	SA	9
513	*Neoniphon sammara*	Sammara squirrelfish	Kheha	خيها	Aq	14
514			Iyya sagheera	ايبا صغيرة	SA	1
515	*Netuma thalassina*	Giant seacatfish	Garmout	قرموط	Su	5
516			Kumal	كومل	Er	11
517			Shilan	شيلان	Er	10
518	*Octopus aegina*	Sandbird octopus	Amfesis	امفيسيس	Er	11
519			Akhtebut	اخطبوط	Er	11
520	*Octopus cyanea*	Big blue octopus	Akhtebut	اخطبوط	Er	11
521			Amfesis	امفيسيس	Er	11
522	*Octopus vulgaris*	Common octopus	Akhtebut	اخطبوط	Er	11
523			Amfesis	امفيسيس	Er	11
524	*Odonus niger*	Red-toothed triggerfish	Shuroma	شرومة	Aq	14
525	*Oedalechilus labiosus*	Hornlip mullet	Arabi	عربي	Er	10
526	*Ostichthys hypsipterygion*		Kheha	خيها	Aq	14
527	*Ostracion cubicus*	Yellow boxfish	Abu sandoug	ابو صندوق	SA,Su	1,6
528			Sanduk al-bahar	صندوق البحر	Aq	14
529	*Ostracion cyanurus*	Bluetail trunkfish	Sanduk al-bahar	صندوق البحر	Aq	14
530	*Ostracion* spp.	Boxfishes	Abu sandoug	ابو صندوق	SA	9
531	*Otolithes* spp.	Drums or croakers	Lut	لت	Su	5

(continued)

Table 10.3 (continued)

No.	Scientific name	Common name	Local name	Arabic name	Country[a]	Source[b]
532	*Oxycheilinus diagramma*	Cheeklined wrasse	Abu mulees	ابو مليس	SA	1
533	Palinuridae	Spiny lobsters	Um ruban	ام ربان	Ye	2
534			Sharkha	شركا	Er	10
535	*Panulirus homarus*	Scalloped spiny lobster	Sharkha	شركا	Er	11
536			Langus	لانجس	Er	11
537	*Panulirus ornatus*	Ornate spiny lobster	Langus	لانجس	Er	11
538			Sharkha	شركا	Er	10
539	*Panulirus penicillatus*	Pronghorn spiny lobster	Sharkha	شركا	Er	10
540	*Panulirus* spp.	Spiny lobsters	Estakoza	استاكوزا	Su	5
541	*Panulirus versicolor*	Painted spiny lobster	Sharkha	شركا	Er	10
542			Langus	لانجس	Er	11
543	*Paracaesio xanthura*	Yellowtail blue snapper	Sarra'	ساره	SA	1
544	*Paraexocoetus* spp.	Flyingfishes	Jiraad al bahr	جرده البحر	SA	1
545	*Paramonacanthus pusillus*	Faintstripe filefish	Um qaren	ام قرن	Aq	14
546	*Parapercis hexophtalma*	Speckled sandperch	Dab ramly	دب رملي	Aq	14
547	*Parapercis somaliensis*	Somali sandperch	Dab	دب	Aq	14
548	*Parastromateus niger*	Black pomfret	Alsa	السا	Er	11
549	*Pardachirus marmoratus*	Finless sole	Tabaq	طبق	Aq	14
550	*Pardachirus* spp.	Soles	Hisan al-Bahar	حصان البحر	SA	1
551	*Parexocoetus brachypterus*	Sailfin flyingfish	Farash	فراش	Aq	14
552	*Parupeneus cyclostomus*	Gold-saddle goatfish	Sabalan asfar	سبلان اصفر	Aq	14
553	*Parupeneus forsskali*	Red Sea goatfish	Sabalan abu nocta	سبلان ابو نقطة	Aq	14
554			Abu digin	ابو دجن	SA	1
555			Inber baladi	عنبر بلدي	Eg	12
556	*Parupeneus heptacanthus*	Cinnabar goatfish	Sabalan	سبلان	Aq	14
557			Abu sheneb	ابو شنب	Er	11
558	*Parupeneus macronemus*	Long-barbel goatfish	Sabalan ahmar	سبلان احمر	Aq	14
559	*Parupeneus rubescens*	Rosy goatfish	Barbuni	بربونى	Er	10
560			Abu sheneb	ابو شنب	Er	10
561			Sabalan	سبلان	Aq	14
562			Abu digin	ابو دجن	Er	10
563	*Parupeneus* spp.	Goatfishes	Abu sheneb	ابو شنب	Su	5
564	*Pempheris vanicolensis*	Vanikoro sweeper	Fanas	فانس	Aq	14
565	*Penaeus monodon*	Giant tiger prawn	Gamberi	جمبري	Er	11
566	*Penaeus semisulcatus*	Green tiger prawn	Gamberi	جمبري	Er	11
567	*Penaeus* spp.	Shrimps	Rubean	روبين	Ye	2
568			Gamberi	جمبري	Su,Ye	2,5
569			Zinga	زينجا	Ye	2
570			Abu gobgab	ابو قبقاب	SA	9
571	*Pervagor randalli*		Um qaren	ام قرن	Aq	14
572	*Platax orbicularis*	Orbicular batfish	Kanaf	كناف	Su	5,7
573	*Platax pinnatus*	Dusky batfish	Kanaf	كناف	SA	1
574	Platycephalidae	Flathead	Ro'ed	رويد	Er	10
575			Ruad	رواد	Er	10
576			Sumar	صومر	Er	10
577	*Plectorhinchus gaterinus*	Blackspotted rubberlip	Qatran, staff	قطرين, ستاف	Aq	14
578			Gadreneb	جدرينيب	Su	7
579			Sobaity	سوبيتى	Er	11
580			Gaterin	جاترين	Er,SA,Su	1,7,11
581	*Plectorhinchus schotaf*	Minstrel sweetlip	Sobaity	سوبيتى	Er	11
582			Shutaf	شوطاف	SA	1
583			Istaf	استاف	Aq	14

(continued)

Table 10.3 (continued)

No.	Scientific name	Common name	Local name	Arabic name	Country[a]	Source[b]
584			Telham	تلهام	Su	5,6,7
585	*Plectorhinchus* spp.	Sweetlips	Gaterin	جاترين	Ye	3
586	*Plectropomus maculatus*	Spotted coralgrouper	Kushar	كشر	Er	11
587			Shu'ur	شعور	Er	10
588			Najil	ناجل	Er,SA,Su	1,5,7,10
589	*Plectropomus pessuliferus*	Roving coralgrouper	Najil	ناجل	Aq	14
590	*Plectropomus* spp.	Groupers	Najil	ناجل	Su	8
591	*Plicofollis dussumieri*	Blacktip catfish	Kumal	كومل	Ye	2
592	*Polysteganus coeruleopunctatus*	Blueskin seabream	Fareedin	فريدين	Aq	14
593	*Pomacanthus imperator*	Emperor angelfish	Moscht	مشط	Aq	14
594	*Pomacanthus maculosus*	Yellowbar angelfish	Moscht	مشط	Aq	14
595	*Pomadasys argenteus*	Silver grunt	Koko	كوكو	Er,Su	5,10
596			Shefsh	شيفش	Er	10
597			Getran	جيتران	Er	10
598			Sobaity	سوبيتى	Er	10
599			Shatef	شاتف	Er	10
600	*Pomadasys commersonnii*	Smallspotted grunter	Koko	كوكو	Su	5
601			Nakem	ناكم	SA	1
602	*Pomadasys kaakan*	Javelin grunter	Sobaity	سوبيتى	Er	11
603			Getran	جيتران	Er	11
604	*Pomadasys* spp.	Grunters	Nakem	ناكم	Ye	3
605			Abu hajar	أبو حجر	Ye	3
606	*Pomadasys striatus*	Striped grunter	Ibin skab	ابن سكاب	Su	5
607	*Pomadasys stridens*	Striped piggy	Khushrum	خشروم	Aq	14
608			Shukrum	شخروم	Eg	12
609	*Pomocentrus trichourus*	Paletail damsel	Kharayeh	خارابة	Aq	14
610	*Portunus pelagicus*	Flower crab	Abu mekass	ابو مقص	Er	11
611			Kaboriah	كبوريا	Su	5
612			Hinkakre	هنكاكار	Er	11
613			Abu galambo	ابو جلمبو	Su	5
614	Priacanthidae	Bigeyes	Sahla	سهله	Er	10
615			Sahr	سحر	Er	10
616			Sahr el Leil	سحر الليل	Er	10
617	*Priacanthus hamrur*	Moontail bullseye	Hamaroon	حمرون	SA	1
618			Fanas abu-ein	فناس ابو عين	Aq	14
619			Batel	بطل	Su	6,7
620	*Prionace glauca*	Blue shark	Gursh	قرش	SA	1
621	*Pristipomoides filamentosus*	Crimson jobfish	Koreib	كوريب	Su	5,7
622	*Pristipomoides multidens*	Goldband jobfish	Anteg	عنتق	Er	10
623	*Pristipomoides* spp.	Snappers	Fars	فرس	SA	1
624			Sarra'	ساره	SA	1
625	*Pristis pectinata*	Sawfish	Abu minshar	ابو منشار	SA	9
626	*Pristis* spp.	Sawfishes	Gursh abu minshaar	قرش ابو منشار	SA	1
627	*Psettodes erumei*	Indian spiny turbot	Kelb	كلب	Ye	2
628			Muse	موس	Er	10
629			Mousa	موسى	Er	10
630			Biedhikabiela	بيضكبيلة	Er	11
631			Kutian	كتيان	Er	10
632	*Pseudobalistes flavimarginatus*	Yellowmargin triggerfish	Ajame	عجامة	Er	11

(continued)

Table 10.3 (continued)

No.	Scientific name	Common name	Local name	Arabic name	Country[a]	Source[b]
633	*Pseudobalistes fuscus*	Yellow-spotted triggerfish	Hijma	هجمه	Aq	14
634			Ajame	عجامة	Er	11
635	*Pseudorhombus arsius*	Largetooth flounder	Shebet al bahir	شيبت البحر	Er	11
636	*Pterocaesio chrysozona*	Goldband fusilier	Bagha	باغة	Aq	14
637	*Pterois miles*	Devil firefish	Rana	رنا	Aq	14
638	*Pterois radiata*	Radial firefish	Dujaaja al bahar	دجاج البحر	SA	1
639			Rana	رنا	Aq	14
640	*Pterois volitans*	Red lionfish	Dujaaja al bahar	دجاج البحر	SA	1
641	*Pygoplites diacanthus*	Regal angelfish	Moscht	مشط	Aq	14
642	*Rachycentron canadum*	Cobia	Shakan	شكان	Aq	14
643			Sikin	سيكين	Er	10
644			Fitle	فيتل	Er	10
645			Sikla	سيكلا	Er	10
646			Kuml nu'aakhr	كمل نواخر	SA	1
647			Fatla	فتلة	Er	10
648			Hutian	حوتين	Er	10
649	*Rachycentron* spp.	Cobia	Sakhala	سخاله	Ye	3
650	*Rastrelliger kanagurta*	Indian mackerel	Bagha	باغة	Er,SA,Su,Ye	1,2,5,10
651			Scombry	سقمبري	Eg	13
652			Sardina	سردينه	Aq	14
653	*Remora remora*	Shark sucker	Qamlet alqersh	قملة القرش	Aq	14
654	*Rhabdosargus sarba*	Goldlined seabream	Areedh	عريض	SA	1
655			Eibad	عيبد	Su	5,6,7
656			Haffar	حفار	Aq	14
657	*Rhincodon typus*	Whale shark	Battan	بطان	Aq	14
658			Gursh bitaan	قرش بيتان	SA	1
659	*Rhinecanthus assasi*	Picasso triggerfish	Ajame	عجامة	Er	11
660			Khanzeer	خنزير	SA	1
661	*Rhinobatos punctifer*	Spotted guitarfish	Salfooh	صلفوه	Aq	14
662	*Rhinobatos* spp.	Guitarfishes	Orab	عراب	SA	9
663			Gursh abu halawa	قرش ابو حلاوة	SA	1
664	*Rhizoprionodon acutus*	Milk shark	Autat	عوتات	Er	11
665	*Rhynchobatus djiddensis*	Giant guitarfish	Bera	بيرة	Ye	2
666			Bakhat	باخات	Ye	2
667			Fakhadoo	فكهدو	Ye	2
668	*Sarda orientalis*	Striped bonito	Dirak	ديرك	Aq	14
669	*Sardinella gibbosa*	Goldstripe sardinella	Sardin	سردين	Su	5
670			Belem	بليم	Er	11
671			Aida	عيدة	Er	11
672	*Sardinella jussieu*	Mauritian sardinella	Sardina mofatar	سردينه مفطر	Eg	13
673	*Sardinella longiceps*	Indian oil sardine	Sardin	سردين	Su	5
674	*Sardinella melanura*	Blacktip sardinella	Toom	توم	Eg	13
675			Sardin	سردين	SA,Su	1,5
676	*Sargocentron caudimaculatum*	Silverspot squirrelfish	Kheha	خيها	Aq	14
677	*Sargocentron diadema*	Crown squirrelfish	Fanas	فانس	Aq	14
678	*Sargocentron rubrum*	Redcoat	Keha	كهة	Aq	14
679			Gagaloom	جاجالوم	Su	6,7
680	*Sargocentron spiniferum*	Sabre squirrelfish	Gehaya	جحية	Er	11
681			Gagaloom	جاجالوم	Su	5,7

(continued)

Table 10.3 (continued)

No.	Scientific name	Common name	Local name	Arabic name	Country[a]	Source[b]
682			Iyya kabeera	ايبا كبيرة	SA	1
683	*Saurida gracilis*	Gracile lizardfish	Macarona	مكرونا	Er	11
684	*Saurida macrolepis*		Haret	حارت	Eg	12
685	*Saurida tumbil*	Greater lizardfish	Baram	برم	Ye	2
686			Dab	دب	Aq	14
687			Macarona	مكرونا	Er,SA,Su	5,9,11
688			Harret	حريت	Eg,SA	9,12
689	*Saurida undosquamis*	Brushtooth lizardfish	Macarona	مكرونا	Er,Su	5,11
690			Dab	دب	Aq	14
691	Scaridae	Parrotfish	Harida	حريدا	Er	10
692	*Scarus collana*	Red Sea parrotfish	Harid	حريد	Aq	14
693			Harida	حريدا	Er	11
694	*Scarus ferrugineus*	Rusty parrotfish	Ghabban	غبان	Aq	14
695	*Scarus frenatus*	Bridled parrotfish	Harida	حريدا	Er	11
696	*Scarus fuscopurpureus*	Purple-brown parrotfish	Ghabban	غابان	Aq	14
697	*Scarus genazonatus*	Sinai parrotfish	Ghabban	غبان	Aq	14
698	*Scarus ghobban*	Blue-barred parrotfish	Harid	حريد	Aq,Er,Su	5,11,14
699	*Scarus gibbus*	Heavybeak parrotfish	Ghabban	غبان	Aq	14
700	*Scarus niger*	Dusky parrotfish	Ghabban	غبان	Aq	14
701	*Scarus psittacus*	Common parrotfish	Ghabban	غبان	Aq	14
702	*Scolopsis ghanam*	Arabian monocle bream	Fanas abiadh	فانس ابيض	Aq	14
703	*Scomber colias*	Atlantic chub mackerel	Shak al-zoor	شك الزور	Eg	13
704	*Scomber japonicus*	Chub mackerel	Scombla	سكمبلا	Aq	14
705			Bagha	باغة	SA	1
706	*Scomber* spp.	Mackerels	Bagha	باغة	SA	9
707	*Scomberoides commersonnianus*	Talang queenfish	Durab	دورب	Er	11
708	*Scomberoides lysan*	Doublespotted queenfish	Todaf	توداف	SA	1
709			Durab	دورب	Er,Ye	2,11
710			Shrow	شرو	Su	5,7
711			Lysan	لسان	SA,Ye	1,2
712	*Scomberomorus commerson*	Narrow-barred Spanish mackerel	Shak abu Isnan	شك ابو اسنان	Aq	14
713			Dirak	ديرك	Er,SA,Su,Ye	1,2,5,10
714	*Scomberomorus guttatus*	Indo-Pacific king mackerel	Dirak	ديرك	Su	5
715	Scombridae	Mackerels, tunas, bonitos	Tun	تن	Ye	2
716			Scherwa	شروة	Ye	2
717	*Scorpaenopsis barbatus*	Bearded scorpionfish	Abu al-Laban	ابو اللبن	Aq	14
718	*Scorpaenopsis diabolus*	False stonefish	Abu al-Laban	ابو اللبن	Aq	14
719	*Scorpaenopsis gibbosa*	Humpbacked scorpionfish	Agrab	عقرب	SA	1
720	*Scorpaenopsis* spp.	Scorpionfishes or rockfishes	Agrab	عقرب	SA	1
721	*Scylla serrata*	Giant mud crab	Hinkakre	هنكاكار	Er	11
722			Abu mekass	ابو مقص	Er	11
723	*Sepia australis*	Southern cuttlefish	Um el hibir	ام الحبر	Er	11
724			Abu midad	ابو مداد	Er	11
725	*Sepia latimanus*	Broadclub cuttlefish	Um el hibir	ام الحبر	Er	11
726			Abu midad	ابو مداد	Er	11
727	*Sepia pharaonis*	Pharaoh cuttlefish	Abu midad	ابو مداد	Er	11
728			Um el hibir	ام الحبر	Er	11

(continued)

Table 10.3 (continued)

No.	Scientific name	Common name	Local name	Arabic name	Country[a]	Source[b]
729	*Sepia* spp.	Cuttlefishes	Fakhd	فكهد	Ye	3
730	Sepiidae	Cuttlefishes	Sebia	سيبيا	Su	5
731	*Seriola dumerili*	Greater amberjack	Hea	هيا	Aq	14
732	*Seriola rivoliana*	Longfin yellowtail	Beyad	بياد	Er	10
733	*Seriola* spp.	Jacks and pompanos	Nazkha	نزخة	SA	1
734	*Seriolina nigrofasciata*	Blackbanded trevally	Beyad	بياد	Er	10
735	Serranidae	Sea basses, groupers and fairy basslets	Kushar	كشر	Er, Ye	2,10
736			Tauwina	طوينا	Ye	3
737	*Siganus argenteus*	Streamlined spinefoot	Sigan	سجان	Er	11
738			Sigan khudhary	سيجان خضاري	Aq	14
739	*Siganus lineatus*	Golden-lined spinefoot	Sigan	سجان	Su	6
740	*Siganus luridus*	Dusky spinefoot	Sigan harafi	سيجان حرفي	Aq	14
741	*Siganus rivulatus*	Marbled spinefoot	Sigan biady	سجان بيادي	Aq	14
742			Sigan	سجان	Er,SA,Su	9,11,15
743	*Siganus* sp.	Rabbitfishes	Sigan	سجان	Su	5
744	*Siganus stellatus*	Brown-spotted spinefoot	Sigan	سجان	Aq	14
745			Sigan al baha	سجان الباحا	SA	1
746	*Sillago sihama*	Silver sillago	Rakad	راكاد	SA	9
747	*Sillago* spp.	Smelt-whitings	Al makhfi	المخفي	Ye	3
748			Al ankood	العنقود	Ye	3
749	*Soleichthys heterorhinos*	Black-tip sole	Tabaq	طبق	Aq	14
750	Sparidae	Porgies	Abyad	ابيض	Er	10
751			Fafal	فافل	Er	10
752			Afsh	عفش	Ye	4
753	*Sparus aurata*	Gilthead seabream	Danees	دنس	Aq	14
754	*Sphyraena barracuda*	Great barracuda	Iqama	عقامه	Aq	14
755			Agam	عقام	Er,SA,Su,Ye	3,5,9,10
756			Todaf	تو داف	Er	11
757	*Sphyraena flavicauda*	Yellowtail barracuda	Malleeta	مليتة	Aq	14
758	*Sphyraena forsteri*	Bigeye barracuda	Todaf	تو داف	Er	10
759			Agam	عقام	Er	10
760	*Sphyraena jello*	Pickhandle barracuda	Agam	عقام	Er,SA	1,10
761			Todaf	تو داف	Er, Ye	2,10
762			Agous	اجوس	Su	5,6
763			Khod	خد	Ye	2
764	*Sphyraena obtusata*	Obtuse barracuda	Zreighan	زريغان	SA,Su	1,5,6
765			Todaf	تو داف	Er	11
766			Agam	عقام	Er	11
767	*Sphyraena putnamiae*	Sawtooth barracuda	Iqama	عقامه	Aq	14
768	*Sphyraena* spp.	Barracudas	Khod	خد	Ye	3
769			Agam	عقام	Ye	3
770	Sphyraenidae	Barracudas	Todaf	تو داف	Er	10
771			Mekazel	مكازل	Er	10
772			Zuran	زوران	Er	10
773			Agam	عقام	Er	10
774			Zuria	زوريا	Er	10
775	*Sphyrna lewini*	Scalloped hammerhead	Gurna	قرن	Er	11
776			Gursh abu burnetta	قرش ابو برنيطة	Aq	14
777			Gurur	جرور	Er	11
778	*Sphyrna zygaena*	Smooth hammerhead	Gursh gurna	قرش قرنا	SA	1
779	*Stolephorus indicus*	Indian anchovy	Figayma	فيجيمة	SA	1

(continued)

Table 10.3 (continued)

No.	Scientific name	Common name	Local name	Arabic name	Country[a]	Source[b]
780	*Stolephorus* spp.	Anchovies	Fagima	فقيمة	Su	5
781	*Stromateus* spp.	Butterfishes	Abu gurz	ابو جرز	SA	9
782	*Sufflamen albicaudatum*	Bluethroat triggerfish	Ajame	عجامة	Er	11
783			Um qaren	ام قرن	Aq	14
784	*Synanceia verrucosa*	Stonefish	Agrab	عقرب	SA	1
785			Abu al-Laban	ابو اللبن	Aq	14
786	Synodontidae	Lizardfishes	Macarona	مكرونا	Er	10
787			Haret	حارت	Er	10
788	*Synodus hoshinonis*	Blackear lizardfish	Dab	دب	Aq	14
789	*Synodus variegatus*	Variegated lizardfish	Dab	دب	Aq	14
790	*Tachysurus* spp.	Sea catfishes	Kumal	كومل	SA	1
791	*Taeniura lymma*	Ribbontail stingray	Halali	هلالى	Er	11
792			Abromis	ابروميس	Er	11
793			Rugtia saghirah	روجتيه صغيرة	SA	1
794			Um qurbal	ام كربل	Aq	14
795	*Terapon jarbua*	Jarbua terapon	Shukrum	شخروم	Er,Su	5,10
796			Henw	شخروم	Aq	14
797	*Terapon* spp.	Terapon	Shukrum	شخروم	SA	9
798			Jaabul	جبول	Ye	3
799	*Tetraodon* spp.	Puffers	Arradh	عراض	SA	9
800	Tetraodontidae	Puffers	Morjan	مرجان	Er	10
801	*Tetrosomus gibbosus*	Humpback turretfish	Sanduk al-bahar	صندوق البحر	Aq	14
802	*Thalassoma klunzingeri*	Klunzinger's wrasse	Muesy	ميسي	Aq	14
803	*Thalassoma lunare*	Moon wrasse	Muesy	ميسي	Aq	14
804	*Thalassoma* spp.	Wrasses	Deek	ديك	SA	1
805	*Thamnaconus modestoides*	Modest filefish	Um qaren	ام قرن	Aq	14
806	*Thenus orientalis*	Flathead lobster	Langus	لانجس	Er	11
807			Sharkha	شركا	Er	10
808			Stacoza	استاكوزا	Er	10
809	*Thryssa baelama*	Baelama anchovy	Belem	بليم	Er	11
810			Aida	عيدة	Er	11
811	*Thunnus alalunga*	Albacore	Tuna	تونة	Su	5
812			Shak	شك	Aq	14
813	*Thunnus albacares*	Yellowfin tuna	Shak zoor	شك زور	Aq	14
814			Tuna	تونة	Su	5
815			Thumad	ثمد	SA	1
816	*Thunnus obesus*	Bigeye tuna	Zeinub	زينوب	Ye	3
817	*Thunnus tonggol*	Longtail tuna	Sherwi	شروي	Er	11
818			Tuna	تونة	Er,Su	5,11
819			Shak Abu thiel	شك ابو ذيل	Aq	14
820			Tomad	تمد	Er	10
821	*Thyrsitoides jordanus*	Black snoek	Saif	سيف	Aq	14
822	*Torpedo panthera*	Panther electric ray	Rugtia kaharabiyyah	روجتيه كهربائية	SA	1
823			Khadala ramlya	خداله رمليه	Aq	14
824	*Torquigener flavimaculatus*		Hadhroom	حضروم	Aq	14
825	*Trachinocephalus myops*	Snakefish	Dab	دب	Aq	14
826			Haret	حارت	Eg	12
827	*Trachinotus baillonii*	Smallspotted dart	Teiman	طيمان	Su	5,6
828	*Trachinotus blochii*	Subnose pompano	Beyad	بياد	Er	10
829			Pompano	بمبانو	Er	11
830			Teiman	طيمان	SA,Su	1,5,6,7
831	*Trachurus indicus*	Arabian scad	Sardina aredha	سردينه عريضة	Aq	14

(continued)

Table 10.3 (continued)

No.	Scientific name	Common name	Local name	Arabic name	Country[a]	Source[b]
832			Bagha	باغة	Su	5
833	*Trachyrhamphus bicoarctatus*	Double-ended pipefish	Masas	ماساس	Aq	14
834	*Triaenodon obesus*	Whitetip reef shark	Sweida	سويدا	Er	11
835	*Trichiurus lepturus*	Largehead hairtail	Saif	سيف	Aq	14
836	*Trichiurus* spp.	Cutlassfishes	Homalan	هوملان	Ye	4
837			Abu saif	ابو سيف	SA,Ye	3,9
838	Trochidae	Trochus shell	Kokian	كوكيان	Er	10
839	*Tylosurus crocodilus crocodilus*	Hound needlefish	Kharam	خرم	Er	11
840			Kombir	كومبير	SA,Su	1,5,6,7
841	*Ulua mentalis*	Longrakered trevally	Beyad	بياد	Er	10
842	*Upeneus moluccensis*	Goldband goatfish	Sabalan	سبلان	Aq	14
843			Abu digin	ابو دجن	Er	10
844			Barbuni	بربونى	Er	10
845			Abu sheneb	ابو شنب	Er	10
846	*Upeneus* spp.	Goatfishes	Abu sheneb	ابو شنب	Su	5,7
847			Barbuni	بربونى	SA	9
848	*Upeneus subvitatus*	Deep-water goatfish	Sabalan	سبلان	Aq	14
849	*Upeneus sulphureus*	Sulphur goatfish	Abu sheneb	ابو شنب	Su	6
850	*Upeneus tragula*	Freckled goatfish	Barbuni	بربونى	Eg	12
851	*Upeneus vittatus*	Yellowstriped goatfish	Barbuni	بربونى	Eg	12
852			Abu sheneb	ابو شنب	Er	11
853	*Uroteuthis duvauceli*	Indian squid	Abu midad	ابو مداد	Er	11
854	*Valamugil cunnesius*	Longarm mullet	Arabi	عربي	Er	10
855	*Valamugil seheli*	Bluespot mullet	Arabi	عربي	Er,Su	5,10
856	*Valamugil* spp.	Mullet	Arabi	عربي	SA	1
857	*Variola louti*	Yellow-edged lyretail	Abu sherif	ابو شريف	SA	9
858			Boosia	بوسيا	Aq	14
859			Louti	لوطى	Er,SA,Su	1,5,6,7,10
860			Rishal	ريشال	Er,Su	5,6,7,10
861	*Xiphias gladius*	Swordfish	Faras al Bahr	فارس البحر	SA	1
862			Abu seif	ابو سيف	Su	5,6,7
863	Xiphiidae	Swordfish	Damriah	دامريه	Ye	2
864	*Xyrichtys melanopus*	Yellowpatch razorfish	Far al-Bahar	فار البحر	Aq	14
865	*Xyrichtys pavo*	Peacock wrasse	Far al-Bahar	فار البحر	Aq	14
866	*Zebrasoma velifer*	Sailfin tang	Zizan	زيزان	Er	11
867			Dajaje el baher	دجاج البحر	Er	11
868	*Zebrasoma xanthurum*	Yellowtail tang	Zizan	زيزان	Er	11
869			Juneh	جنه	Aq	14
870			Gahm	جهم	Er	11

[a]Aq = Aqaba (mainly Jordan), Eg = Egypt, Er = Eritrea, SA = Saudi Arabia, Su = Sudan, Ye = Yemen

[b]1 Barrania et al. (1980), 2 Walczak and Gudmundsson (1975), 3 Walczak (1977), 4 Bonfiglioli and Hariri (2004), 5 MEPI (1993), 6 Abu-Gideiri (1984), 7 Reed (1964), 8 FHAS (1984), 9 El-Saby and Farina (1954), 10 MOF (2012), 11 Tesfamichael and Sebahtu (2006), 12 Bayoumi (1972), 13 Rafail (1972), 14 Khalaf and Disi (1997), 15 Interview

Table 10.4 Common names of Red Sea fish and invertebrates sorted by local names in Arabic

Source[a]	Country[b]	Common name	Scientific name	Local name	Arabic name	ID
11	Er	Honeycomb stingray	*Himantura uarnak*	Abromis	ابروميس	1
11	Er	Ribbontail stingray	*Taeniura lymma*			2
5,6,7	Su	Common bluestripe snapper	*Lutjanus kasmira*	Ableen asfar	ابلين عصفر	3
5	Su	Striped grunter	*Pomadasys striatus*	Ibin skab	ابن سكاب	4
14	Aq	Bearded scorpionfish	*Scorpaenopsis barbatus*	Abu al-Laban	ابو اللبن	5
14	Aq	False stonefish	*Scorpaenopsis diabolus*			6
14	Aq	Stonefish	*Synanceia verrucosa*			7
11	Er	Picnic seabream	*Acanthopagrus berda*	Abu berite	ابو بريت	8
11	Er	Twobar seabream	*Acanthopagrus bifasciatus*			9
9	SA	Mojarras	*Gerres* spp.	Abu gurz	ابو جرز	10
1	SA	Slimys, slipmouths, or ponyfishes	*Leiognathus* spp.			11
9	SA	Butterfishes	*Stromateus* spp.			12
1	SA	Heavybeak parrot fish	*Chlorurus gibbus*	Abu greeyan	ابو جريان	13
1,5,6	SA,Su	Bluespine unicornfish	*Naso unicornis*	Abu garin	ابو جرين	14
5	Su	Flower crab	*Portunus pelagicus*	Abu galambo	ابو جلمبو	15
5,6,7	Su	Humphead wrasse	*Cheilinus undulatus*	Abu jibba	ابو جبة	16
5,6,7	Su	Kawakawa	*Euthynnus affinis*	Abu dam	ابو دم	17
10	Er	Goatfishes	*Mullidae*	Abu digin	ابو دجن	18
5	Su	Yellowstripe goatfish	*Mulloidichthys flavolineatus*			19
1	SA	Red Sea goatfish	*Parupeneus forsskali*			20
10	Er	Rosy goatfish	*Parupeneus rubescens*			21
10	Er	Goldband goatfish	*Upeneus moluccensis*			22
11	Er	Bluestripe herring	*Herklotsichthys quadrimaculatus*	Abu ras	ابو راس	23
15	Ye	Stingrays	Dasyatidae	Abu remis	ابو رميس	24
5	Su	Stingrays	*Dasyatis* spp.	Abu soot	ابو سوط	25
1,5,6	SA,Su	Dorab wolf-herring	*Chirocentrus dorab*	Abu seif	ابو سيف	26
3,9	SA,Ye	Cutlassfishes	Trichiurus spp.			27
5,6,7	Su	Swordfish	*Xiphias gladius*			28
5,6	Su	Indo-Pacific sailfish	*Istiophorus platypterus*	Abu shiraa	ابو شراع	29
14	Aq	Sixblotch hind	*Cephalopholis sexmaculata*	Abu shirni	ابو شرني	30
9	SA	Yellow-edged lyretail	*Variola louti*	Abu sherif	ابو شريف	31
9	SA	Sea catfishes	*Arius* spp.	Abu sheneb	ابو شنب	32
10	Er	Goatfishes	Mullidae			33
11	Er	Cinnabar goatfish	*Parupeneus heptacanthus*			34
10	Er	Rosy goatfish	*Parupeneus rubescens*			35
5	Su	Goatfishes	*Parupeneus* spp.			36
10	Er	Goldband goatfish	*Upeneus moluccensis*			37
5,7	Su	Goatfishes	*Upeneus* spp.			38
6	Su	Sulphur goatfish	*Upeneus sulphureus*			39
11	Er	Yellowstriped goatfish	*Upeneus vittatus*			40
9	SA	Filefishes	*Monacanthus* spp.	Abu shaukah	ابو شوكة	41
1,6	SA,Su	Yellow boxfish	*Ostracion cubicus*	Abu sandoug	ابو صندوق	42
9	SA	Boxfishes	*Ostracion* spp.			43
11	Er	Areolate grouper	*Epinephelus areolatus*	Abu ades	ابو عدس	44
11	Er	Indo-Pacific sailfish	*Istiophorus platypterus*	Abu feres	أبو فرس	45
9	SA	Shrimps	*Penaeus* spp.	Abu gobgab	ابو قبقاب	46
11	Er	Spotted unicornfish	*Naso brevirostris*	Abu karn	ابو قرن	47
8	Su	Unicornfishes	*Naso* spp.			48
6	Su	Picnic seabream	*Acanthopagrus berda*	Abu kuhul	ابو كحل	49

(continued)

Table 10.4 (continued)

Source[a]	Country[b]	Common name	Scientific name	Local name	Arabic name	ID
5,7,10	Er,Su	Twobar seabream	*Acanthopagrus bifasciatus*			50
11	Er	Veined squid	*Loligo forbesii*	Abu midad	ابو مداد	51
11	Er	Southern cuttlefish	*Sepia australis*			52
11	Er	Broadclub cuttlefish	*Sepia latimanus*			53
11	Er	Pharaoh cuttlefish	*Sepia pharaonis*			54
11	Er	Indian squid	*Uroteuthis duvauceli*			55
2,3	Ye	Crab	Brachyura	Abu mokas	ابو مقص	56
11	Er	Flower crab	*Portunus pelagicus*	Abu mekass	ابو مقص	57
11	Er	Giant mud crab	*Scylla serrata*			58
1	SA	Tripletail wrasse	*Cheilinus trilobatus*	Abu mulees	ابو مليس	59
1	SA	Cheeklined wrasse	*Oxycheilinus diagramma*			60
9	SA	Sawfish	*Pristis pectinata*	Abu minshar	ابو منشار	61
10	Er	Tiger shark	*Galeocerdo cuvier*	Abu nebir	ابو نبير	62
5,10	Er,Su	Thumbprint emperor	*Lethrinus harak*	Abu nugta	ابو نقطة	63
11	Er	Atlantic tripletail	*Lobotes surinamensis*	Abu hajar	أبو حجر	64
3	Ye	Grunters	*Pomadasys* spp.			65
10	Er	Twobar seabream	*Acanthopagrus bifasciatus*	Abyad	ابيض	66
10	Er	King soldierbream	*Argyrops spinifer*			67
10	Er	Porgies	Sparidae			68
1	SA	Humpbacked scorpionfish	*Scorpaenopsis gibbosa*	Agrab	عقرب	69
1	SA	Scorpionfishes or rockfishes	*Scorpaenopsis* spp.			70
1	SA	Stonefish	*Synanceia verrucosa*			71
5,6	Su	Pickhandle barracuda	*Sphyraena jello*	Agous	اجوس	72
11	Er	Coral hind	*Cephalopholis miniata*	Ahmer	احمر	73
11	Er	Sandbird octopus	*Octopus aegina*	Akhtebut	اخطبوط	74
11	Er	Big blue octopus	*Octopus cyanea*			75
11	Er	Common octopus	*Octopus vulgaris*			76
5	Su	Rainbow runner	*Elagatis bipinnulata*	Adad	اداد	77
14	Aq	Painted sweetlips	*Diagramma pictum*	Istaf	استاف	78
14	Aq	Minstrel sweetlip	*Plectorhinchus schotaf*			79
5	Su	Spiny lobsters	*Panulirus* spp.	Estakoza	استاكوزا	80
10	Er	Flathead lobster	*Thenus orientalis*	Stacoza	استاكوزا	81
14	Aq	Summan grouper	*Epinephelus summana*	Aqshar	اقشر	82
14	Aq	Greasy grouper	*Epinephelus tauvina*			83
11	Er	Black pomfret	*Parastromateus niger*	Alsa	السا	84
3	Ye	Smelt-whitings	*Sillago* spp.	Al ankood	العنقود	85
3	Ye	Smelt-whitings	*Sillago* spp.	Al makhfi	المخفي	86
14	Aq	Silver-cheeked toadfish	*Lagocephalus sceleratus*	Alnaguem	الناقم	87
11	Er	Southern cuttlefish	*Sepia australis*	Um el hibir	ام الحبر	88
11	Er	Broadclub cuttlefish	*Sepia latimanus*			89
11	Er	Pharaoh cuttlefish	*Sepia pharaonis*			90
14	Aq	Twoband anemonefish	*Amphiprion bicinctus*	Om Al dukhan	ام الدخان	91
2	Ye	Spiny lobsters	Palinuridae	Um ruban	ام ربان	92
14	Aq	False moorish idol	*Heniochus diphreutes*	Um shiraa	ام شراع	93
14	Aq	Red Sea bannerfish	*Heniochus intermedius*			94
1,5,7	SA Su	Common dolphinfish	*Coryphaena hippurus*	Um falloos	ام فلوس	95
14	Aq	Unicorn leatherjacket filefish	*Aluterus monoceros*	Um qaren	ام قرن	96
14	Aq	Broom filefish	*Amanses scopas*			97
14	Aq	Orange-lined triggerfish	*Balistapus undulatus*			98
14	Aq	Honeycomb filefish	*Cantherhines pardalis*			99
14	Aq	Faintstripe filefish	*Paramonacanthus pusillus*			100

(continued)

Table 10.4 (continued)

Source[a]	Country[b]	Common name	Scientific name	Local name	Arabic name	ID
14	Aq		*Pervagor randalli*			101
14	Aq	Bluethroat triggerfish	*Sufflamen albicaudatus*			102
14	Aq	Modest filefish	*Thamnaconus modestoides*			103
14	Aq	Ribbontail stingray	*Taeniura lymma*	Um qurbal	ام كربل	104
2	Ye	Goatfishes	Mullidae	Ambir	امبير	105
11	Er	Sandbird octopus	*Octopus aegina*	Amfesis	امفيسيس	106
11	Er	Big blue octopus	*Octopus cyanea*			107
11	Er	Common octopus	*Octopus vulgaris*			108
14	Aq	Mackerel scad	*Decapterus macarellus*	Amia	اميه	109
1	SA	Pinecone soldierfish	*Myripristis murdjan*	Iyya sagheera	ايبا صغيرة	110
1	SA	Sammara squirrelfish	*Neoniphon sammara*			111
1	SA	Sabre squirrelfish	*Sargocentron spiniferum*	Iyya kabeera	ايبا كبيرة	112
2	Ye	Stingrays	Dasyatidae	Bakhat	باخات	113
2	Ye	Giant guitarfish	*Rhynchobatus djiddensis*			114
14	Aq	Lunar fusilier	*Caesio lunaris*	Bagha	باغة	115
14	Aq	Suez fusilier	*Caesio suevica*			116
5	Su	Japanise scad	*Decapterus maruadsi*			117
13	Eg	Round scad	*Decapterus punctatus*			118
13	Eg	Indian scad	*Decapterus russelli*			119
14	Aq	Goldband fusilier	*Pterocaesio chrysozona*			120
1,2,5,10	Er,SA,Su,Ye	Indian mackerel	*Rastrelliger kanagurta*			121
1	SA	Chub mackerel	*Scomber japonicus*			122
9	SA	Mackerels	*Scomber* spp.			123
5	Su	Arabian scad	*Trachurus indicus*			124
14	Aq	Variable-lined fusilier	*Caesio varilineata*	Bagha hamra	باغة حمراء	125
2	Ye	Golden trevally	*Gnathanodon speciosus*	Bagesh	بجيش	126
10	Er	Goatfishes	Mullidae	Barbuni	بربونى	127
10	Er	Rosy goatfish	*Parupeneus rubescens*			128
10	Er	Goldband goatfish	*Upeneus moluccensis*			129
9	SA	Goatfishes	*Upeneus* spp.			130
12	Eg	Freckled goatfish	*Upeneus tragula*			131
12	Eg	Yellowstriped goatfish	*Upeneus vittatus*			132
2	Ye	Greater lizardfish	*Saurida tumbil*	Baram	برم	133
14	Aq	Whale shark	*Rhincodon typus*	Battan	بطان	134
6,7	Su	Moontail bullseye	*Priacanthus hamrur*	Batel	بطل	135
11	Er	Twobar seabream	*Acanthopagrus bifasciatus*	Butel hammed	بطل حماد	136
11	Er	Spotted sardinella	*Amblygaster sirm*	Belem	بليم	137
14	Aq	Hardyhead silverside	*Atherinomorus lacunosus*			138
11	Er	Rainbow sardine	*Dussumieria acuta*			139
11	Er	Shorthead anchovy	*Encrasicholina heteroloba*			140
11	Er	Bluestripe herring	*Herklotsichthys quadrimaculatus*			141
11	Er	Goldstripe sardinella	*Sardinella gibbosa*			142
11	Er	Baelama anchovy	*Thryssa baelama*			143
11	Er	African pompano	*Alectis ciliaris*	Pompano	بمبانو	144
11	Er	Subnose pompano	*Trachinotus blochii*			145
14	Aq	Slender emperor	*Lethrinus variegatus*	Bunqus khermawi	بنقوس خرماوي	146
1,5,6,7	SA,Su	Bonefish	*Albula vulpes*	Bounouk	بو نوك	147
14	Aq	Yellow-edged lyretail	*Variola louti*	Boosia	بوسيا	148
10	Er	Slimys, slipmouths, or ponyfishes	Leiognathidae	Botatos	بوطاطوس	149

(continued)

Table 10.4 (continued)

Source[a]	Country[b]	Common name	Scientific name	Local name	Arabic name	ID
14	Aq	Starry flying gurnard	*Dactyloptena peterseni*	Boomet al-Bahar	بومة البحر	150
5,6,7	Su	Milkfish	*Chanos chanos*	Bunji	بونجي	151
10	Er	Snappers	Lutjanidae	Bohar	بوهر	152
1,2,5,7,9,10	Aq,Er,SA,Su,Ye	Two-spot red snapper	*Lutjanus bohar*			153
10	Er	Humpback red snapper	*Lutjanus gibbus*			154
10	Er	Common bluestripe snapper	*Lutjanus kasmira*			155
10	Er	Malabar blood snapper	*Lutjanus malabaricus*			156
10	Er	African pompano	*Alectis ciliaris*	Beyad	بياد	157
11	Er	Shrimp scad	*Alepes djedaba*			158
1,11	Er,SA	Yellowtail scad	*Atule mate*			159
11	Er	Longfin trevally	*Carangoides armatus*			160
5,10	Er,Su	Orangespotted trevally	*Carangoides bajad*			161
11	Er	Blue trevally	*Carangoides ferdau*			162
10	Er	Bludger	*Carangoides gymnostethus*			163
11	Er	Malabar trevally	*Carangoides malabaricus*			164
1,5,10	Er,SA,Su	Giant trevally	*Caranx ignobilis*			165
1,5,10	Er,SA,Su	Bigeye trevally	*Caranx sexfasciatus*			166
3,7,9,10	Er,SA,Su,Ye	Jacks and pompanos	*Caranx* spp.			167
10	Er	Rainbow runner	*Elagatis bipinnulata*			168
5,11	Er,Su	Golden trevally	*Gnathanodon speciosus*			169
10	Er	Longfin yellowtail	*Seriola rivoliana*			170
10	Er	Blackbanded trevally	*Seriolina nigrofasciata*			171
10	Er	Subnose pompano	*Trachinotus blochii*			172
10	Er	Longrakered trevally	*Ulua mentalis*			173
1	SA	Indian threadfish	*Alectis indicus*	Beyad abu tabag	بياد ابو تابج	174
1	SA	Yellowspotted trevally	*Carangoides fulvoguttatus*	Beyad gaz	بياد جاز	175
1	SA	Golden trevally	*Gnathanodon speciosus*			176
9	SA	Orangespotted trevally	*Carangoides bajad*	Beyad gazza	بياد جازة	177
9	SA	Coachwhip trevally	*Carangoides oblongus*	Beyad girm	بياد جريم	178
1,10	Er,SA	Bluefin trevally	*Caranx melampygus*			179
6,10	Er,Su	Coastal trevally	*Carangoides coeruleopinnatus*	Beyad goutar	بياد جوتر	180
5,11	Er,Su	Yellowspotted trevally	*Carangoides fulvoguttatus*			181
1	SA	Torpdeo scad	*Megalaspis cordyla*	Beyad turfa	بياد طرفه	182
2	Ye	Giant guitarfish	*Rhynchobatus djiddensis*	Bera	بيرة	183
11	Er	Indian spiny turbot	*Psettodes erumei*	Biedhikabiela	بيضكبيلة	184
1	SA	Blue seachub	*Kyphosus cinerascens*	Tahmal	تحمل	185
6	Su	Freckled tilefish	*Branchiostegus sawakinensis*	Theena	تحينا	186
7	Su	Mullets	*Mugil* spp.	Tadab	تدب	187
10	Er	Emperors or scavengers	Lethrinidae	Terhani	ترهاني	188
7	Su	Jacks and pompanos	*Caranx* spp.	Tak'oi	تكاوي	189
5,6,7	Su	Minstrel sweetlip	*Plectorhinchus schotaf*	Telham	تلهام	190
5	Su	Dogtooth tuna	*Gymnosarda unicolor*	Tomad	تمد	191
10	Er	Longtail tuna	*Thunnus tonggol*			192
2	Ye	Kawakawa	*Euthynnus affinis*	Tun	تن	193
2	Ye	Mackerels, tunas, bonitos	Scombridae			194
1	SA	Doublespotted queenfish	*Scomberoides lysan*	Todaf	توداف	195
11	Er	Great barracuda	*Sphyraena barracuda*			196
10	Er	Bigeye barracuda	*Sphyraena forsteri*			197
2,10	Er, Ye	Pickhandle barracuda	*Sphyraena jello*			198
11	Er	Obtuse barracuda	*Sphyraena obtusata*			199
10	Er	Barracudas	Sphyraenidae			200

(continued)

Table 10.4 (continued)

Source[a]	Country[b]	Common name	Scientific name	Local name	Arabic name	ID
13	Eg	Blacktip sardinella	*Sardinella melanura*	Toom	توم	201
7	Su	Greasy grouper	*Epinephelus tauvina*	Toona	تونا	202
5	Su	Frigate tuna	*Auxis thazard thazard*	Tuna	تونة	203
5	Su	Skipjack tuna	*Katsuwonus pelamis*			204
5	Su	Albacore	*Thunnus alalunga*			205
5	Su	Yellowfin tuna	*Thunnus albacares*			206
5,11	Er,Su	Longtail tuna	*Thunnus tonggol*			207
11	Er	Kawakawa	*Euthynnus affinis*	Tonno	تونو	208
1	SA	Humphead wrasse	*Cheilinus undulatus*	Terbaany	تيرباني	209
1	SA	Yellowfin tuna	*Thunnus albacares*	Thumad	ثمد	210
1,7,11	Er,SA,Su	Blackspotted rubberlip	*Plectorhinchus gaterinus*	Gaterin	جاترين	211
3	Ye	Sweetlips	*Plectorhinchus* spp.			212
6,7	Su	Redcoat	*Sargocentron rubrum*	Gagaloom	جاجالوم	213
5,7	Su	Sabre squirrelfish	*Sargocentron spiniferum*			214
3	Ye	Terapon	*Terapon* spp.	Jaabul	جبول	215
11	Er	Sabre squirrelfish	*Sargocentron spiniferum*	Gehaya	جحية	216
7	Su	Blackspotted rubberlip	*Plectorhinchus gaterinus*	Gadreneb	جدرينيب	217
7	Su	Two-spot red snapper	*Lutjanus bohar*	Garabganat	جرابجانت	218
1	SA	Blackspotted hawkfish	*Cristacirrhitus punctatus*	Jarbua	جربوعه	219
14	Aq	Soldierbream	*Argyrops filamentosus*	Jarbeeden	جربيدن	220
5,7	Su	Hardyhead silverside	*Atherinomorus lacunosus*	Gurgush	جرجوش	221
1	SA	Flyingfishes	*Paraexocoetus* spp.	Jiraad al bahr	جرده البحر	222
11	Er	Scalloped hammerhead	*Sphyrna lewini*	Gurur	جرور	223
1	SA	Threadfin butterflyfish	*Chaetodon auriga*	Gringish	جرينجش	224
1	SA	Pennant coralfish	*Heniochus acuminatus*			225
1	SA	Samoan silverside	*Hypoatherina temminckii*	Gashgoosha	جشجوشه	226
14	Aq	Barred flagtail	*Kuhlia mugil*	Ghlaimeh	جليمه	227
11	Er	Indian white prawn	*Fenneropenaeus indicus*	Gamberi	جمبري	228
11	Er	Kuruma prawn	*Marsupenaeus japonicus*			229
11	Er	Western king prawn	*Melicertus latisulcatus*			230
11	Er	Speckled shrimp	*Metapenaeus monoceros*			231
11	Er	Giant tiger prawn	*Penaeus monodon*			232
11	Er	Green tiger prawn	*Penaeus semisulcatus*			233
2,5	Su,Ye	Shrimps	*Penaeus* spp.			234
1	SA	Live sharksucker	*Echeneis naucrates*	Gamla	جمله	235
14	Aq	Brown surgeonfish	*Acanthurus nigrofuscus*	Juneh	جنه	236
14	Aq	Striated surgeonfish	*Ctenochaetus striatus*			237
14	Aq	Yellowtail tang	*Zebrasoma xanthurum*			238
5	Su	Groupers	*Cephalopholis* spp.	Gahlab	جهلب	239
11	Er	Ringtail surgeonfish	*Acanthurus blochii*	Gahm	جهم	240
1	SA	Black surgeonfish	*Acanthurus gahhm*			241
11	Er	Epaulette surgeonfish	*Acanthurus nigricauda*			242
11	Er	Sohal surgeonfish	*Acanthurus sohal*			243
11	Er	Yellowfin surgeonfish	*Acanthurus xanthopterus*			244
11	Er	Yellowtail tang	*Zebrasoma xanthurum*			245
6	Su	Peacock hind	*Cephalopholis argus*	Ghohlab	جو هلاب	246
6	Su	Coral hind	*Cephalopholis miniata*			247
7	Su	Jacks and pompanos	*Caranx* spp.	Goareit	جواريت	248
10	Er	Slimys, slipmouths, or ponyfishes	Leiognathidae	Gutat	جوتات	249
5	Su	Yellowspotted trevally	*Carangoides fulvoguttatus*	Gutur	جوتر	250

(continued)

Table 10.4 (continued)

Source[a]	Country[b]	Common name	Scientific name	Local name	Arabic name	ID
10	Er	Grunts	Haemulidae	Getran	جيتران	251
10	Er	Silver grunt	*Pomadasys argenteus*			252
11	Er	Javelin grunter	*Pomadasys kaakan*			253
5	Su	Giant trevally	*Caranx ignobilis*	Girim	جيريم	254
7	Su	Jacks and pompanos	*Caranx* spp.			255
5	Su	Brownspotted grouper	*Epinephelus chlorostigma*	Gishir	جيشر	256
5	Su	Camouflage grouper	*Epinephelus polyphekadion*			257
5	Su	Summan grouper	*Epinephelus summana*			258
5,6,7	Su	Greasy grouper	*Epinephelus tauvina*	Gishir tauwina	جيشر توينه	259
5,6,7	Su	Areolate grouper	*Epinephelus areolatus*	Gishir shooni	جيشر شونى	260
7	Su	Mullets	*Mugil* spp.	Jilan	جيلان	261
10	Er	Malabar blood snapper	*Lutjanus malabaricus*	Gehab	جيهاب	262
12	Eg		*Saurida macrolepis*	Haret	حارت	263
10	Er	Lizardfishes	Synodontidae			264
12	Eg	Snakefish	*Trachinocephalus myops*			265
14	Aq	Blackspot snapper	*Lutjanus ehrenbergii*	Hebra	حبرا	266
9,12	Eg,SA	Greater lizardfish	*Saurida tumbil*	Harret	حريت	267
1	SA	Daisy parrotfish	*Cholorurus sordidus*	Hareeth	حريث	268
1	SA	Marbled parrotfish	*Leptoscarus vaigiensis*			269
5,7	Su	Parrotfishes	*Callyodon* spp.	Harid	حريد	270
5,14	Aq,Su	Candelamoa parrotfish	*Hipposcarus harid*			271
14	Aq	Red Sea parrotfish	*Scarus collana*			272
5,11,14	Aq,Er,Su	Blue-barred parrotfish	*Scarus ghobban*			273
11	Er	Green humphead parrotfish	*Bolbometopon muricatum*	Harida	حريدا	274
11	Er	Bicolour parrotfish	*Cetoscarus bicolor*			275
10	Er	Parrotfish	Scaridae			276
11	Er	Red Sea parrotfish	*Scarus collana*			277
11	Er	Bridled parrotfish	*Scarus frenatus*			278
1	SA	Leopard flounder	*Bothus pantherinus*	Hisan al-Bahar	حصان البحر	279
1	SA	Fourlined tonguesole	*Cynoglossus bilineatus*			280
14	Aq	Thorny seahorse	*Hippocampus histrix*			281
14	Aq	Spotted seahorse	*Hippocampus kuda*			282
1	SA	Soles	*Pardachirus* spp.			283
14	Aq	Masked puffer	*Arothron diadematus*	Hadhroom	حضروم	284
14	Aq	White-spotted puffer	*Arothron hispidus*			285
14	Aq	Stellate puffer	*Arothron stellatus*			286
14	Aq	Crowned puffer	*Canthigaster coronata*			287
14	Aq	Pufferfish	*Canthigaster margaritata*			288
14	Aq		*Torquigener flavimaculatus*			289
14	Aq	Spotbase burrfish	*Chilomycterus spilostylus*	Hadhroom abu shouka	حضروم ابو شوكة	290
14	Aq	Spot-fin porcupinefish	*Diodon hystrix*			291
9	SA	Porgies	*Chrysoblephus* spp.	Haffar	حفار	292
1	SA	Karenteen seabream	*Crenidens crenidens*			293
14	Aq	Goldlined seabream	*Rhabdosargus sarba*			294
11	Er	Malabar blood snapper	*Lutjanus malabaricus*	Hamari	حماري	295
5,7	Su	Ruby snapper	*Etelis carbunculus*	Hamaroon	حمرون	296
1	SA	Moontail bullseye	*Priacanthus hamrur*			297
1	SA	Longfin African conger	*Conger cinereus*	Hanish silab	حنيش سيلاب	298
9	SA	Greasy grouper	*Epinephelus tauvina*	Hubog	حوبج	299
1	SA	Dory snapper	*Lutjanus fulviflamma*	Hubr	(حوبر(حبر)	300

(continued)

Table 10.4 (continued)

Source[a]	Country[b]	Common name	Scientific name	Local name	Arabic name	ID
1	SA	John's snapper	*Lutjanus johnii*			301
1	SA	Russell's snapper	*Lutjanus russellii*			302
10	Er	Snappers	Lutjanidae	Huberi	حوبيري	303
10	Er	Humpback red snapper	*Lutjanus gibbus*			304
1,10	Er,SA	Common bluestripe snapper	*Lutjanus kasmira*			305
10	Er	Cobia	*Rachycentron canadum*	Hutian	حوتين	306
2	Ye	Japanese threadfin bream	*Nemipterus japonicus*	Homiara	حوميره	307
15	Er	Sea cucumber	Holothuriidae	Hidra	حيدره	308
14	Aq	Clown coris	*Coris aygula*	Heqab	حيقب	309
1	SA	Silver moony	*Monodactylus argenteus*	Haymaan	حيمان	310
14	Aq	Threespot dascyllus	*Dascyllus trimaculatus*	Kharayeh	خاراية	311
14	Aq	Paletail damsel	*Pomocentrus trichourus*			312
1	SA	Tenpounder	*Elops machnata*	Khanny	خاني	313
2	Ye	Pickhandle barracuda	*Sphyraena jello*	Khod	خد	314
3	Ye	Barracudas	*Sphyraena* spp.			315
14	Aq	Elat electric ray	*Narcine bentuviai*	Khadala	خداله	316
14	Aq	Panther electric ray	*Torpedo panthera*	Khadala ramlya	خداله رمليه	317
5,6,7	Su	Torpdeo scad	*Megalaspis cordyla*	Khurtum	خرطوم	318
11	Er	Hound needlefish	*Tylosurus crocodilus crocodilus*	Kharam	خرم	319
1	SA	Cornetfishes	*Fistularia* spp.	Khurm al baaha	خرم الباحه	320
14	Aq	Striped piggy	*Pomadasys stridens*	Khushrum	خشروم	321
1	SA	Orange-lined triggerfish	*Balistapus undulatus*	Khanzeer	خنزير	322
1	SA	Picasso triggerfish	*Rhinecanthus assasi*			323
4	Ye	Sea cucumber	Holothuriidae	Kheiar albahr	خيار البحر	324
14	Aq	Sammara squirrelfish	*Neoniphon sammara*	Kheha	خيها	325
14	Aq		*Ostichthys hypsipterygion*			326
14	Aq	Silverspot squirrelfish	*Sargocentron caudimaculatum*			327
2	Ye	Swordfish	Xiphiidae	Damriah	دامريه	328
14	Aq	Somali sandperch	*Parapercis somaliensis*	Dab	دب	329
14	Aq	Greater lizardfish	*Saurida tumbil*			330
14	Aq	Brushtooth lizardfish	*Saurida undosquamis*			331
14	Aq	Blackear lizardfish	*Synodus hoshinonis*			332
14	Aq	Variegated lizardfish	*Synodus variegatus*			333
14	Aq	Snakefish	*Trachinocephalus myops*			334
14	Aq	Speckled sandperch	*Parapercis hexophtalma*	Dab ramly	دب رملي	335
1	SA	Radial firefish	*Pterois radiata*	Dujaaja al bahar	دجاج البحر	336
1	SA	Red lionfish	*Pterois volitans*			337
11	Er	Sailfin tang	*Zebrasoma velifer*			338
3	Ye	Shrimp scad	*Alepes djedaba*	Djebbada	دجبدة	339
1	SA	White-spotted puffer	*Arothron hispidus*	Drimma	دريما	340
1	SA	Immaculate puffer	*Arothron immaculatus*			341
1	SA	Stellate puffer	*Arothron stellatus*			342
1	SA	Spot-fin porcupinefish	*Diodon hystrix*			343
14	Aq	Blacktip grouper	*Epinephelus fasciatus*	Daghma	دغمة	344
14	Aq	Comet grouper	*Epinephelus morrhua*			345
14	Aq	Oblique-banded grouper	*Epinephelus radiatus*			346
14	Aq	Gilthead seabream	*Sparus aurata*	Danees	دنس	347
11	Er	Talang queenfish	*Scomberoides commersonnianus*	Durab	دورب	348
2,11	Er,Ye	Doublespotted queenfish	*Scomberoides lysan*			349
14	Aq	Striped bonito	*Sarda orientalis*	Dirak	ديرك	350
1,2,5,10	Er,SA,Su,Ye	Narrow-barred Spanish mackerel	*Scomberomorus commerson*			351

(continued)

Table 10.4 (continued)

Source[a]	Country[b]	Common name	Scientific name	Local name	Arabic name	ID
5	Su	Indo-Pacific king mackerel	*Scomberomorus guttatus*			352
1	SA	Axilspot hogfish	*Bodianus axillaris*	Deek	ديك	353
1	SA	Checkerboard wrasse	*Halichoeres hortulanus*			354
1	SA	Wrasses	*Thalassoma* spp.			355
9	SA	Silver sillago	*Sillago sihama*	Rakad	راكاد	356
1,14	Aq,SA	Twobar seabream	*Acanthopagrus bifasciatus*	Rabaag	رباج	357
14	Aq		*Cheilinus abudjubbe*	Rabadi	ربادي	358
14	Aq	Broomtail wrasse	*Cheilinus lunulatus*			359
14	Aq	Orangespine unicornfish	*Naso lituratus*	Rahu	رحو	360
14	Aq	Bluespine unicornfish	*Naso unicornis*			361
14	Aq	Devil firefish	*Pterois miles*	Rana	رنا	362
14	Aq	Radial firefish	*Pterois radiata*			363
14	Aq	Shortfin turkeyfish	*Dendrochirus brachypterus*	Rani, Abu al-Laban	رنى- ابو اللبن	364
10	Er	Flathead	Platycephalidae	Ruad	رواد	365
2	Ye	Shrimps	*Penaeus* spp.	Rubean	روبين	366
1	SA	Redmouth grouper	*Aethaloperca rogaa*	Ruga	روجا	367
9	SA	Atlantic tripletail	*Lobotes surinamensis*	Rougaah	روجاه	368
1	SA	Stingrays	*Dasyatis* spp.	Rugtia	روجتيه	369
1	SA	Eagle and manta rays	*Mobula* spp.			370
1	SA	Ribbontail stingray	*Taeniura lymma*	Rugtia saghirah	روجتيه صغيرة	371
1	SA	Panther electric ray	*Torpedo panthera*	Rugtia kaharabiyyah	روجتيه كهربائية	372
1	SA	Crocodile flathead	*Cociella crocodilus*	Rugud	روجود	373
1	SA	Eagle and manta rays	*Manta* spp.	Rugia milla	روجي ميله	374
10	Er	Flathead	Platycephalidae	Ro'ed	رويد	375
5,6,7,10	Er,Su	Yellow-edged lyretail	*Variola louti*	Rishal	ريشال	376
14	Aq	Common silver-biddy	*Gerres oyena*	Rishan	ريشان	377
14	Aq	Yellowspotted trevally	*Carangoides fulvoguttatus*	Reem	ريم	378
14	Aq	Orangespotted trevally	*Carangoides bajad*	Reema safra	ريما صفره	379
9	SA	Fairy basslets	*Anthias* spp.	Zargh	زرغ	380
1,5,6	SA,Su	Obtuse barracuda	*Sphyraena obtusata*	Zreighan	زريغان	381
11	Er	Blacktip reef shark	*Carcharhinus melanopterus*	Zingi	زنجى	382
11	Er	Sicklefin lemon shark	*Negaprion acutidens*			383
10	Er	Barracudas	Sphyraenidae	Zuran	زوران	384
10	Er	Barracudas	Sphyraenidae	Zuria	زوريا	385
1	SA	Dussimier's halfbeak	*Hyporhamphus dussumieri*	Zirgaan	زيرجان	386
11	Er	Ringtail surgeonfish	*Acanthurus blochii*	Zizan	زيزان	387
11	Er	Epaulette surgeonfish	*Acanthurus nigricauda*			388
11	Er	Sohal surgeonfish	*Acanthurus sohal*			389
11	Er	Yellowfin surgeonfish	*Acanthurus xanthopterus*			390
11	Er	Sailfin tang	*Zebrasoma velifer*			391
11	Er	Yellowtail tang	*Zebrasoma xanthurum*			392
2	Ye	Kawakawa	*Euthynnus affinis*	Zainub	زينب	393
2	Ye	Shrimps	*Penaeus* spp.	Zinga	زينجا	394
11	Er	Skipjack tuna	*Katsuwonus pelamis*	Zeinub	زينوب	395
3	Ye	Bigeye tuna	*Thunnus obesus*			396
10	Er	Threadfin breams	Nemipteridae	Sare	سار	397
12	Eg	Japanese threadfin bream	*Nemipterus japonicus*			398
1	SA	Yellowtail blue snapper	*Paracaesio xanthura*	Sarra'	ساره	399
1	SA	Snappers	*Pristipomoides* spp.			400
14	Aq	Yellowstripe goatfish	*Mulloidichthys flavolineatus*	Sabalan	سبلان	401

(continued)

Table 10.4 (continued)

Source[a]	Country[b]	Common name	Scientific name	Local name	Arabic name	ID
14	Aq	Yellowfin goatfish	*Mulloidichthys vanicolensis*			402
14	Aq	Cinnabar goatfish	*Parupeneus heptacanthus*			403
14	Aq	Rosy goatfish	*Parupeneus rubescens*			404
14	Aq	Goldband goatfish	*Upeneus moluccensis*			405
14	Aq	Deep-water goatfish	*Upeneus subvitatus*			406
14	Aq	Red Sea goatfish	*Parupeneus forsskali*	Sabalan abu nocta	سبلان ابو نقطة	407
14	Aq	Long-barbel goatfish	*Parupeneus macronemus*	Sabalan ahmar	سبلان احمر	408
14	Aq	Gold-saddle goatfish	*Parupeneus cyclostomus*	Sabalan asfar	سبلان اصفر	409
11	Er	Streamlined spinefoot	*Siganus argenteus*	Sigan	سجان	410
6	Su	Golden-lined spinefoot	*Siganus lineatus*			411
9,11,15	Er,SA,Su	Marbled spinefoot	*Siganus rivulatus*			412
5	Su	Rabbitfishes	*Siganus* sp.			413
14	Aq	Brown-spotted spinefoot	*Siganus stellatus*			414
1	SA	Brown-spotted spinefoot	*Siganus stellatus*	Sigan al baha	سجان الباحا	415
14	Aq	Marbled spinefoot	*Siganus rivulatus*	Sigan biady	سجان بيادي	416
10	Er	Bigeyes	Priacanthidae	Sahr	سحر	417
10	Er	Bigeyes	Priacanthidae	Sahr el Leil	سحر الليل	418
3	Ye	Cobia	*Rachycentron* spp.	Sakhala	سخاله	419
3	Ye	Herrings, shads, sardines etc.	Clupeidae	Sardin	سردين	420
5	Su	Red-eye round herring	*Etrumeus teres*			421
5	Su	Goldstripe sardinella	*Sardinella gibbosa*			422
5	Su	Indian oil sardine	*Sardinella longiceps*			423
1,5	SA,Su	Blacktip sardinella	*Sardinella melanura*			424
14	Aq	Shortfin scad	*Decapterus macrosoma*	Sardina	سردينه	425
14	Aq	Bluestripe herring	*Herklotsichthys quadrimaculatus*			426
14	Aq	Indian mackerel	*Rastrelliger kanagurta*			427
14	Aq	Indian scad	*Decapterus russelli*	Sardina aredha	سردينه عريضة	428
14	Aq	Arabian scad	*Trachurus indicus*			429
13	Eg	Spotted sardinella	*Amblygaster sirm*	Sardina marboum	سردينه مبروم	430
14	Aq	Red-eye round herring	*Etrumeus teres*	Sardina masreya	سردينه مصرية	431
13	Eg	Mauritian sardinella	*Sardinella jussieu*	Sardina mofatar	سردينه مفطر	432
7	Su	Jacks and pompanos	*Caranx* spp.	Safloh	سفلوح	433
5,6,7	Su	Mangrove red snapper	*Lutjanus argentimaculatus*	Safin	سفين	434
13	Eg	Indian mackerel	*Rastrelliger kanagurta*	Scombry	سقمبري	435
14	Aq	Chub mackerel	*Scomber japonicus*	Scombla	سكمبلا	436
3	Ye	Marine turtles	Chelonioidea	Sulhafa	سلحافه	437
1,3,5,6,7,11	Er,SA,Su,Ye	Milkfish	*Chanos chanos*	Salmani	سلماني	438
9	SA	Pacific ladyfish	*Elops affinis*			439
11	Er	Brownspotted grouper	*Epinephelus chlorostigma*	Samman	سمن	440
14	Aq	Sohal surgeonfish	*Acanthurus sohal*	Sahla	سهله	441
10	Er	Bigeyes	Priacanthidae			442
11	Er	Painted sweetlips	*Diagramma pictum*	Sobaity	سوبيتى	443
10	Er	Grunts	Haemulidae			444
11	Er	Blackspotted rubberlip	*Plectorhinchus gaterinus*			445
11	Er	Minstrel sweetlip	*Plectorhinchus schotaf*			446
10	Er	Silver grunt	*Pomadasys argenteus*			447
11	Er	Javelin grunter	*Pomadasys kaakan*			448
1	SA	Sohal surgeonfish	*Acanthurus sohal*	Suhal	سوهل	449
11	Er	Whitetip reef shark	*Triaenodon obesus*	Sweida	سويدا	450

(continued)

Table 10.4 (continued)

Source[a]	Country[b]	Common name	Scientific name	Local name	Arabic name	ID
5	Su	Cuttlefishes	Sepiidae	Sebia	سيبيا	451
7	Su	Areolate grouper	*Epinephelus areolatus*	Seetiati	سيتياتي	452
14	Aq	Dusky spinefoot	*Siganus luridus*	Sigan harafi	سيجان حرفي	453
14	Aq	Streamlined spinefoot	*Siganus argenteus*	Sigan khudhary	سيجان خضاري	454
10	Er	Threadfin breams	Nemipteridae	Ser'a	سيره	455
14	Aq	Common dolphinfish	*Coryphaena hippurus*	Saif	سيف	456
14	Aq	Black snoek	*Thyrsitoides jordanus*			457
14	Aq	Largehead hairtail	*Trichiurus lepturus*			458
10	Er	Cobia	*Rachycentron canadum*	Sikla	سيكلا	459
10	Er	Cobia	*Rachycentron canadum*	Sikin	سيكين	460
6	Su	Dussimier's halfbeak	*Hyporhamphus dussumieri*	Silinti	سيلنتي	461
5,7	Su	Yellowspotted trevally	*Carangoides fulvoguttatus*	Seleikh	سيليخ	462
7	Su	Jacks and pompanos	*Caranx* spp.			463
5	Su	Halfbeaks	*Hemiramphus* spp.	Selenti	سيلينتى	464
10	Er	Grunts	Haemulidae	Shatef	شاتف	465
10	Er	Silver grunt	*Pomadasys argenteus*			466
5,6,7	Su	Indian threadfish	*Alectis indicus*	Shawish	شاويش	467
14	Aq	Scissortail sergeant	*Abudefduf sexfasciatus*	Shabbar	شبار	468
14	Aq	Blackspot sergeant	*Abudefduf sordidus*			469
14	Aq	Indo-Pacific sergeant	*Abudefduf vaigiensis*			470
5,6,7	Su	Tenpounder	*Elops machnata*	Shagool	شجول	471
5	Su	Spotback herring	*Herklotsichthys punctatus*			472
5	Su	Bigeye snapper	*Lutjanus lutjanus*	Shukrum	شخروم	473
12	Eg	Striped piggy	*Pomadasys stridens*			474
5,10	Er,Su	Jarbua terapon	*Terapon jarbua*			475
9	SA	Terapon	*Terapon* spp.			476
5,6,7	Su	Indian scad	*Decapterus russelli*	Shadba	شدبا	477
1	SA	Scads	*Decapterus* spp.	Shaduba	شدوبا	478
10	Er	Spiny lobsters	Palinuridae	Sharkha	شركا	479
11	Er	Scalloped spiny lobster	*Panulirus homarus*			480
10	Er	Ornate spiny lobster	*Panulirus ornatus*			481
10	Er	Pronghorn spiny lobster	*Panulirus penicillatus*			482
10	Er	Painted spiny lobster	*Panulirus versicolor*			483
10	Er	Flathead lobster	*Thenus orientalis*			484
5,7	Su	Doublespotted queenfish	*Scomberoides lysan*	Shrow	شرو	485
2,11	Er,Ye	Kawakawa	*Euthynnus affinis*	Sherwa	شروة	486
2	Ye	Mackerels, tunas, bonitos	Scombridae	Scherwa	شروة	487
14	Aq	Red-toothed triggerfish	*Odonus niger*	Shuroma	شرومة	488
1	SA	Frigate tuna	*Auxis thazard thazard*	Sherwi	شروي	489
11	Er	Skipjack tuna	*Katsuwonus pelamis*			490
11	Er	Torpdeo scad	*Megalaspis cordyla*			491
11	Er	Longtail tuna	*Thunnus tonggol*			492
1	SA	Moray eels	*Gymnothorax* spp.	Shaaga	شعاجه	493
1	SA	Undulated moray	*Gymnothorax undulatus*			494
8,7	Su	Mullets	*Mugil* spp.	Sha'aboi	شعبوي	495
10	Er	Snappers	Lutjanidae	Shaefen	شعفن	496
1,11	Er,SA	Mangrove red snapper	*Lutjanus argentimaculatus*			497
11	Er	Blackspot snapper	*Lutjanus ehrenbergii*			498
11	Er	Dory snapper	*Lutjanus fulviflamma*			499
11	Er	Blubberlip snapper	*Lutjanus rivulatus*			500
11	Er	Black and white snapper	*Macolor niger*			501
8,10	Er,Su	Emperors or scavengers	Lethrinidae	Shu'ur	شعور	502

(continued)

Table 10.4 (continued)

Source[a]	Country[b]	Common name	Scientific name	Local name	Arabic name	ID
5,10	Er,Su	Thumbprint emperor	*Lethrinus harak*			503
5,10	Er,Su	Pink ear emperor	*Lethrinus lentjan*			504
5,10	Er,Su	Sky emperor	*Lethrinus mahsena*			505
10	Er	Smalltooth emperor	*Lethrinus microdon*			506
5,10,14	Aq,Er,Su	Spangled emperor	*Lethrinus nebulosus*			507
5,7,9	SA,Su	Emperors or scavengers	*Lethrinus* spp.			508
10	Er	Spotted coralgrouper	*Plectropomus maculatus*			509
1	SA	Emperors or scavengers	*Lethrinus* spp.	Shu'ur abu zahwa	شعور ابو زهوه	510
1	SA	Humpnose big-eye bream	*Monotaxis grandoculis*	Shu'ur abu'ayn	شعور ابو عين	511
1	SA	Emperors or scavengers	*Lethrinus* spp.	Shu'ur khirmiya	شعور خرمية	512
9	SA	Trumpet emperor	*Lethrinus miniatus*	Shu'ur dibi	شعور دبي	513
1	SA	Emperors or scavengers	*Lethrinus* spp.	Shu'ur deeb	شعور ديب	514
9	SA	Spangled emperor	*Lethrinus nebulosus*	Shu'ur ramaka	شعور رامكا	515
1	SA	Emperors or scavengers	*Lethrinus* spp.			516
1	SA	Spangled emperor	*Lethrinus nebulosus*	Shu'ur mehseny	شعور محيسنى	517
14	Aq	Albacore	*Thunnus alalunga*	Shak	شك	518
14	Aq	Narrow-barred Spanish mackerel	*Scomberomorus commerson*	Shak abu Isnan	شك ابو اسنان	519
14	Aq	Longtail tuna	*Thunnus tonggol*	Shak Abu thiel	شك ابو ذيل	520
14	Aq	Dogtooth tuna	*Gymnosarda unicolor*	Shak abu ein	شك ابو عين	521
13	Eg	Atlantic chub mackerel	*Scomber colias*	Shak al-zoor	شك الزور	522
14	Aq	Yellowfin tuna	*Thunnus albacares*	Shak zoor	شك زور	523
14	Aq	Cobia	*Rachycentron canadum*	Shakan	شكان	524
5,6,7	Su	Painted sweetlips	*Diagramma pictum*	Shakfa	شكفا	525
2,11	Er,Ye	Painted sweetlips	*Diagramma pictum*	Shutaf	شوطاف	526
9	SA	Haemulidae	*Diagramma* spp.			527
1	SA	Minstrel sweetlip	*Plectorhinchus schotaf*			528
7	Su	Mullets	*Mugil* spp.	Shole	شول	529
11	Er	Leopard flounder	*Bothus pantherinus*	Shebet al bahir	شيبت البحر	530
11	Er	Largetooth flounder	*Pseudorhombus arsius*			531
10	Er	Snappers	Lutjanidae	Sheik ali	شيخ على	532
9	SA	Triggerfishes	*Balistes* spp.	Schiyram shiram	شيرام شيرام	533
14	Aq	Coral hind	*Cephalopholis miniata*	Shirni	شيرنى	534
10	Er	Grunts	Haemulidae	Shefsh	شيفش	535
10	Er	Silver grunt	*Pomadasys argenteus*			536
10	Er	Giant sea catfish	*Netuma thalassina*	Shilan	شيلان	537
14	Aq	Areolate grouper	*Epinephelus areolatus*	Shelwa	شيلوة	538
15	Ye	Yellowtail scad	*Atule mate*	Saibariya	صعباريا	539
3	Ye	Whitefin trevally	*Carangoides equula*	Subaria	صعباريا	540
14	Aq	Spotted guitarfish	*Rhinobatos punctifer*	Salfooh	صلفوه	541
14	Aq	Yellow boxfish	*Ostracion cubicus*	Sanduk al-bahar	صندوق البحر	542
14	Aq	Bluetail trunkfish	*Ostracion cyanurus*			543
14	Aq	Humpback turretfish	*Tetrosomus gibbosus*			544
11	Er	Pink ear emperor	*Lethrinus lentjan*	Suli	صولي	545
10	Er	Smalltooth emperor	*Lethrinus microdon*			546
11	Er	Spangled emperor	*Lethrinus nebulosus*			547
10	Er	Flathead	Platycephalidae	Sumar	صومر	548
2	Ye	Requiem sharks	Carcharhinidae	Dohoosh	دهوش	549
14	Aq	Leopard flounder	*Bothus pantherinus*	Tabaq	طبق	550
14	Aq	Finless sole	*Pardachirus marmoratus*			551

(continued)

Table 10.4 (continued)

Source[a]	Country[b]	Common name	Scientific name	Local name	Arabic name	ID
14	Aq	Black-tip sole	*Soleichthys heterorhinos*			552
5,6,7,11	Er,Su	Greasy grouper	*Epinephelus tauvina*	Tauwina	طوينا	553
3	Ye	Sea basses, groupers and fairy basslets	Serranidae			554
3	Ye	Stingrays	Dasyatidae	Taira	طيرة	555
5,6	Su	Smallspotted dart	*Trachinotus baillonii*	Teiman	طيمان	556
1,5,6,7	SA,Su	Subnose pompano	*Trachinotus blochii*			557
11	Er	Orange-lined triggerfish	*Balistapus undulatus*	Ajame	عجامة	558
11	Er	Titan triggerfish	*Balistoides viridescens*			559
11	Er	Yellowmargin triggerfish	*Pseudobalistes flavimarginatus*			560
11	Er	Yellow-spotted triggerfish	*Pseudobalistes fuscus*			561
11	Er	Picasso triggerfish	*Rhinecanthus assasi*			562
11	Er	Bluethroat triggerfish	*Sufflamen albicaudatum*			563
9	SA	Guitarfishes	*Rhinobatos* spp.	Orab	عراب	564
9	SA	Puffers	*Tetraodon* spp.	Arradh	عراض	565
14	Aq	Fringelip mullet	*Crenimugil crenilabis*	Arabi	عربي	566
1,10	Er,SA	Mullet	*Crenimugil* spp.			567
1,11	Er,SA	Largescale mullet	*Liza macrolepis*			568
10	Er	Squaretail mullet	*Liza vaigiensis*			569
1,10	Er,SA	Flathead grey mullet	*Mugil cephalus*			570
5,7,8	Su	Mullets	*Mugil* spp.			571
10,15	Er,Ye	Mullets	Mugilidae			572
10	Er	Hornlip mullet	*Oedalechilus labiosus*			573
10	Er	Longarm mullet	*Valamugil cunnesius*			574
5,10	Er,Su	Bluespot mullet	*Valamugil seheli*			575
1	SA	Mullet	*Valamugil* spp.			576
14	Aq	Leopard blenny	*Exallias brevis*	Arfaj	عرفج	577
14	Aq	Rippled rockskipper	*Istiblennius edentulus*			578
11	Er	Abudjubbe wrass	*Cheilinus abudjubbe*	Arousset el baher	عروسة البحر	579
11	Er	Humphead wrasse	*Cheilinus undulatus*			580
5	Su	Slimys, slipmouths, or ponyfishes	*Leiognathus* spp.	Arian	عريان	581
1	SA	Goldlined seabream	*Rhabdosargus sarba*	Areedh	عريض	582
11	Er	Abudjubbe wrass	*Cheilinus abudjubbe*	Esha mer'e	عش ميري	583
11	Er	Humphead wrasse	*Cheilinus undulatus*			584
5,6,11	Er,Su	Humpback red snapper	*Lutjanus gibbus*	Asmoot	عصموت	585
8	Su	Snappers	*Lutjanus* spp.			586
1	SA	Humpback red snapper	*Lutjanus gibbus*	Asmoodi	عصمودي	587
9	SA	Mojarras	*Gerres* spp.	Afs	عفس	588
10	Er	Slimys, slipmouths, or ponyfishes	Leiognathidae	Afsh	عفش	589
2	Ye	Emperors or scavengers	Lethrinidae			590
2	Ye	Spangled emperor	*Lethrinus nebulosus*			591
3	Ye	Emperors or scavengers	*Lethrinus* spp.			592
4	Ye	Porgies	Sparidae			593
3,5,9,10	Er,SA,Su,Ye	Great barracuda	*Sphyraena barracuda*	Agam	عقام	594
10	Er	Bigeye barracuda	*Sphyraena forsteri*			595
1,10	Er,SA	Pickhandle barracuda	*Sphyraena jello*			596
11	Er	Obtuse barracuda	*Sphyraena obtusata*			597
3	Ye	Barracudas	*Sphyraena* spp.			598
10	Er	Barracudas	Sphyraenidae			599
14	Aq	Great barracuda	*Sphyraena barracuda*	Iqama	عقامه	600
14	Aq	Sawtooth barracuda	*Sphyraena putnamiae*			601

(continued)

Table 10.4 (continued)

Source[a]	Country[b]	Common name	Scientific name	Local name	Arabic name	ID
1	SA	Orangespine unicornfish	*Naso lituratus*	Akra abu garn	عكرا ابيو جرم	602
12	Eg	Red Sea goatfish	*Parupeneus forsskali*	Inber baladi	عنبر بلدي	603
10	Er	Small toothed jobfish	*Aphareus furca*	Anteg	عنتق	604
10	Er	Snappers	Lutjanidae			605
10	Er	Goldband jobfish	*Pristipomoides multidens*			606
15	Ye	Smallscaled grouper	*Epinephelus polylepis*	Angar	عنقر	607
10	Er	Requiem sharks	Carcharhinidae	Autat	عوتات	608
11	Er	Blacktip shark	*Carcharhinus limbatus*			609
11	Er	Blacktip reef shark	*Carcharhinus melanopterus*			610
11	Er	Sandbar shark	*Carcharhinus plumbeus*			611
11	Er	Sicklefin lemon shark	*Negaprion acutidens*			612
11	Er	Milk shark	*Rhizoprionodon acutus*			613
11	Er	Halavi ray	*Glaucostegus halavi*	O'ud	عود	614
5,6,7	Su	Goldlined seabream	*Rhabdosargus sarba*	Eibad	عيبد	615
11	Er	Spotted sardinella	*Amblygaster sirm*	Aida	عيدة	616
3	Ye	Herrings, shads, sardines etc.	Clupeidae			617
11	Er	Rainbow sardine	*Dussumieria acuta*			618
11	Er	Shorthead anchovy	*Encrasicholina heteroloba*			619
11	Er	Bluestripe herring	*Herklotsichthys quadrimaculatus*			620
11	Er	Goldstripe sardinella	*Sardinella gibbosa*			621
11	Er	Baelama anchovy	*Thryssa baelama*			622
14	Aq	Purple-brown parrotfish	*Scarus fuscopurpureus*	Ghabban	غابان	623
14	Aq	Viridescent parrotfish	*Calotomus viridescens*			624
14	Aq	Bicolour parrotfish	*Cetoscarus bicolor*			625
14	Aq	Daisy parrotfish	*Chlorurus sordidus*			626
14	Aq	Rusty parrotfish	*Scarus ferrugineus*			627
14	Aq	Sinai parrotfish	*Scarus genazonatus*			628
14	Aq	Heavybeak parrotfish	*Scarus gibbus*			629
14	Aq	Dusky parrotfish	*Scarus niger*			630
14	Aq	Common parrotfish	*Scarus psittacus*			631
11	Er	Red Sea halfbeak	*Hyporhamphus gamberur*	Far	فار	632
14	Aq	Robust tuskfish	*Choerodon robustus*	Far al-Bahar	فار البحر	633
14	Aq	Fivefinger wrasse	*Iniistius pentadactylus*			634
14	Aq	Yellowpatch razorfish	*Xyrichtys melanopus*			635
14	Aq	Peacock wrasse	*Xyrichtys pavo*			636
14	Aq	Indo-Pacific sailfish	*Istiophorus platypterus*	Faras	فارس	637
12	Eg	Red filament threadfin bream	*Nemipterus marginatus*			638
1	SA	Black marlin	*Istiompax indica*	Faras al Bahr	فارس البحر	639
1,5,6	SA,Su	Indo-Pacific sailfish	*Istiophorus platypterus*			640
1	SA	Blue marlin	*Makaira nigricans*			641
1	SA	Swordfish	*Xiphias gladius*			642
5,7	Su	Snappers	*Aprion* spp.	Farsi	فارسي	643
6	Su	Green jobfish	*Aprion virescens*			644
8	Su	Snappers	Lutjanidae			645
1	SA	Seahorses	*Hippocampus* spp.	Fara al bahr	فاره البحر	646
7	Su	Mullets	*Mugil* spp.	Fasekh	فاسخ	647
10	Er	Porgies	Sparidae	Fafal	فافل	648
5,7	Su	Titan triggerfish	*Balistoides viridescens*	Faki sharam	فاكي شرام	649
14	Aq	Ring-tailed cardinalfish	*Apogon aureus*	Fanas	فانس	650
14	Aq	Twobelt cardinal	*Apogon bifasciatus*			651
14	Aq	Yellowstriped cardinalfish	*Apogon cyanosoma*			652

(continued)

Table 10.4 (continued)

Source[a]	Country[b]	Common name	Scientific name	Local name	Arabic name	ID
14	Aq	Narrowstripe cardinalfish	*Apogon exostigma*			653
14	Aq	Iridescent cardinalfish	*Apogon kallopterus*			654
14	Aq	Blackstripe cardinalfish	*Apogon nigrofasciatus*			655
14	Aq		*Cheilodipterus lachneri*			656
14	Aq	Large toothed cardinalfish	*Cheilodipterus macrodon*			657
14	Aq	Indian Ocean twospot cardinalfish	*Cheilodipterus novemstriatus*			658
14	Aq	Pinecone soldierfish	*Myripristis murdjan*			659
14	Aq	Eightspine cardinalfish	*Neamia octospina*			660
14	Aq	Vanikoro sweeper	*Pempheris vanicolensis*			661
14	Aq	Crown squirrelfish	*Sargocentron diadema*			662
14	Aq	Arabian monocle bream	*Scolopsis ghanam*	Fanas abiadh	فانس ابيض	663
5,6,7	Su	Painted sweetlips	*Diagramma pictum*	Fataleeta	فتاليتا	664
14	Aq	Kawakawa	*Euthynnus affinis*	Fatla	فتلة	665
10	Er	Cobia	*Rachycentron canadum*			666
14	Aq	Skipjack tuna	*Katsuwonus pelamis*	Fatleh	فتله	667
14	Aq	Sailfin flyingfish	*Parexocoetus brachypterus*	Farash	فراش	668
1	SA	Snappers	*Pristipomoides* spp.	Fars	فرس	669
14	Aq	Blueskin seabream	*Polysteganus coeruleopunctatus*	Fareedin	فريدين	670
11	Er	Coral hind	*Cephalopholis miniata*	Ferek	فريق	671
5,7	Su	Anchovies	*Anchoviella* spp.	Fagima	فقيمة	672
5	Su	Anchovies	*Stolephorus* spp.			673
3	Ye	Cuttlefishes	*Sepia* spp.	Fakhd	فكهد	674
2	Ye	Giant guitarfish	*Rhynchobatus djiddensis*	Fakhadoo	فكهدو	675
14	Aq	Moontail bullseye	*Priacanthus hamrur*	Fanas abu-ein	فناس ابو عين	676
2	Ye	Stingrays	Dasyatidae	Fahodoo	فهودو	677
5	Su	Twobar seabream	*Acanthopagrus bifasciatus*	Fogil	فوجيل	678
5,7,10	Er,Su	King soldierbream	*Argyrops spinifer*	Fofal	فوفل	679
10	Er	Cobia	*Rachycentron canadum*	Fitle	فيتل	680
1	SA	Indian anchovy	*Stolephorus indicus*	Figayma	فيجيمة	681
2	Ye	Billfishes	Istiophoridae	Feraz	فيرز	682
14	Aq	Snubnose emperor	*Lethrinus borbonicus*	Qeda	قدة	683
14	Aq	Sky emperor	*Lethrinus mahsena*			684
14	Aq	Sandbar shark	*Carcharhinus plumbeus*	Gursh	قرش	685
1	SA	Great white shark	*Carcharodon carcharias*			686
14	Aq	Tiger shark	*Galeocerdo cuvieri*			687
14	Aq	Arabian smooth-hound	*Mustelus mosis*			688
1	SA	Blue shark	*Prionace glauca*			689
14	Aq	Scalloped hammerhead	*Sphyrna lewini*	Gursh abu burnetta	قرش ابو برنيطة	690
1	SA	Guitarfishes	*Rhinobatos* spp.	Gursh abu halawa	قرش ابو حلاوة	691
1	SA	Sawfishes	*Pristis* spp.	Gursh abu minshaar	قرش ابو منشار	692
1	SA	Grey reef shark	*Carcharhinus amblyrhynchos*	Gursh al baba	قرش البابا	693
1	SA	Blacktip shark	*Carcharhinus limbatus*	Gursh al sahl	قرش السهل	694
1	SA	Whale shark	*Rhincodon typus*	Gursh bitaan	قرش بيتان	695
1	SA	Thintail thresher	*Alopias vulpinus*	Gursh husseni	قرش حصينى	696
1	SA	Mackerel sharks or white shark	*Isurus* spp.	Gursh deeba	قرش ديبا	697
1	SA	Smooth hammerhead	*Sphyrna zygaena*	Gursh gurna	قرش قرنا	698
1	SA	Tawny nurse shark	*Nebrius ferrugineus*	Gursh massassa	قرش مصاصا	699
1	SA	Tiger shark	*Galeocerdo cuvier*	Gursh nimrany	قرش نمراني	700
5	Su	Giant seacatfish	*Netuma thalassina*	Garmout	قرموط	701

(continued)

Table 10.4 (continued)

Source[a]	Country[b]	Common name	Scientific name	Local name	Arabic name	ID
11	Er	Scalloped hammerhead	*Sphyrna lewini*	Gurna	قرن	702
11	Er	Spotted unicornfish	*Naso brevirostris*	Kurnjal	قرنجل	703
11	Er	Bluespine unicornfish	*Naso unicornis*			704
5,8,9	SA,Su	Requiem sharks	Carcharhinidae	Girish	قريش	705
9	SA	Mojarras	*Gerres* spp.	Gash	قش	706
3	Ye	Emperors or scavengers	*Lethrinus* spp.			707
6	Su	Honeycomb grouper	*Epinephelus merra*	Ghoshar	قشار	708
2	Ye	Painted sweetlips	*Diagramma pictum*	Caterin	قطرين	709
14	Aq	Blackspotted rubberlip	*Plectorhinchus gaterinus*	Qatran, staff	قطرين, ستاف	710
14	Aq	Blue-lined large-eye bream	*Gymnocranius grandoculis*	Qamar	قمر	711
14	Aq	Humpnose big-eye bream	*Monotaxis grandoculis*	Qamar Abu ein	قمر ابو عين	712
9	SA	Remoras	*Echeneis* spp.	Kamlet al darfil	قملة الدرفيل	713
14	Aq	Live sharksucker	*Echeneis naucrates*	Qamlet alqersh	قملة القرش	714
14	Aq	Shark sucker	*Remora remora*			715
14	Aq	Whitespotted moray	*Gymnothorax johnsoni*	Qmum	قموم	716
14	Aq	Starry moray	*Gymnothorax nudivomer*			717
14	Aq	Snowflake moray	*Echidna nebulosa*	Qmum muraqata	قموم مرقطة	718
5,7	Su	Jacks and pompanos	*Caranx* spp.	Goutar	قوتر	719
11	Er	Red Sea halfbeak	*Hyporhamphus gamberur*	Korom	قورم	720
7	Su	Mullets	*Mugil* spp.	Ka'oi	كاوي	721
5	Su	Flower crab	*Portunus pelagicus*	Kaboriah	كبوريا	722
7	Su	Black and white snapper	*Macolor niger*	Kut	كت	723
5,7	Su	Redmouth grouper	*Aethaloperca rogaa*	Katarban	كتربان	724
10	Er	Indian spiny turbot	*Psettodes erumei*	Kutian	كتيان	725
7	Su	Areolate grouper	*Epinephelus areolatus*	Kodad	كداد	726
11	Er	Striated fusilier	*Caesio striata*	Kourab el bahr	كراب البحر	727
11	Er	Suez fusilier	*Caesio suevica*			728
5,7	Su	Jacks and pompanos	*Caranx* spp.	Karb	كرب	729
10	Er	Redmouth grouper	*Aethaloperca rogaa*	Karban	كربان	730
1,6,7,11	Er,SA,Su	Common silver-biddy	*Gerres oyena*	Kass	كس	731
14	Aq	Bluespotted cornetfish	*Fistularia commersonii*	Qasaba	قصبة	732
5	Su	Black and white snapper	*Macolor niger*	Kust	كست	733
2	Ye	Slimys, slipmouths,or ponyfishes	Leiognathidae	Kash	كش	734
11	Er	Redmouth grouper	*Aethaloperca rogaa*	Kushar	كشر	735
1	SA	Coral hind	*Cephalopholis miniata*			736
8	Su	Sea basses, groupers and fairy basslets	*Epinephelus* spp.			737
11	Er	Blacktip grouper	*Epinephelus fasciatus*			738
11	Er	Brown-marbled grouper	*Epinephelus fuscoguttatus*			739
11	Er	Camouflage grouper	*Epinephelus polyphekadion*			740
9,10	Er,SA	Sea basses, groupers and fairy basslets	*Epinephelus* spp.			741
2,11	Er,Ye	Greasy grouper	*Epinephelus tauvina*			742
11	Er	Spotted coralgrouper	*Plectropomus maculatus*			743
2,10	Er,Ye	Sea basses, groupers and fairy basslets	Serranidae			744
1,11	Er,SA	Peacock hind	*Cephalopholis argus*	Kushar abu blaha	كشر ابو بلحة	745
9	SA	Coral hind	*Cephalopholis miniata*	Kushar abu adas	كشر ابو عد س	746
1	SA	Fairy basslets	*Grammistes* spp.			747
1	SA	Comet grouper	*Epinephelus morrhua*	Kushar abu lulu	كشر ابو لولو	748
1,12	Eg,SA	Greasy grouper	*Epinephelus tauvina*	Kushar tauwina	كشر توينه	749

(continued)

Table 10.4 (continued)

Source[a]	Country[b]	Common name	Scientific name	Local name	Arabic name	ID
9	SA	Giant grouper	*Epinephelus lanceolatus*	Kushar twini	كشر طويني	750
9,10	Er,SA	Malabar grouper	*Epinephelus malabaricus*			751
1,11	Er,SA	Summan grouper	*Epinephelus summana*	Kushar mubal'at	كشر موبالات	752
9,11	Er,SA	Areolate grouper	*Epinephelus areolatus*	Kushar nagel	كشر ناجل	753
5	Su	Squids	Loliginidae	Kalamari	كلاماري	754
2	Ye	Indian spiny turbot	*Psettodes erumei*	Kelb	كلب	755
1	SA	Cobia	*Rachycentron canadum*	Kuml nu'aakhr	كمل نواخر	756
5,7	Su	Orbicular batfish	*Platax orbicularis*	Kanaf	كناف	757
1	SA	Dusky batfish	*Platax pinnatus*			758
10	Er	Triggerfishes	*Balistes* spp.	Canzir	كنزير	759
14	Aq	Redcoat	*Sargocentron rubrum*	Keha	كهة	760
0	Su	Scribbled leatherjacket filefish	*Aluterus scriptus*	Kotub	كوتب	761
5,7	Su	Crimson jobfish	*Pristipomoides filamentosus*	Koreib	كوريب	762
10	Er	Grunts	Haemulidae	Koko	كوكو	763
5,10	Er,Su	Silver grunt	*Pomadasys argenteus*			764
5	Su	Smallspotted grunter	*Pomadasys commersonnii*			765
10	Er	Trochus shell	Trochidae	Kokian	كوكيان	766
1,5,6,7	SA,Su	Hound needlefish	*Tylosurus crocodilus crocodilus*	Kombir	كومبير	767
11	Er	Giant seacatfish	*Netuma thalassina*	Kumal	كومل	768
2	Ye	Blacktip catfish	*Plicofollis dussumieri*			769
1	SA	Sea catfishes	*Tachysurus* spp.			770
5,7	Su	Black surgeonfish	*Acanthurus gahhm*	Kohom	كوهم	771
11	Er	Scalloped spiny lobster	*Panulirus homarus*	Langus	لانجس	772
11	Er	Ornate spiny lobster	*Panulirus ornatus*			773
11	Er	Painted spiny lobster	*Panulirus versicolor*			774
11	Er	Flathead lobster	*Thenus orientalis*			775
7	Su	Jacks and pompanos	*Caranx* spp.	Lamenab	لمناب	776
5	Su	Drums or croakers	*Otolithes* spp.	Lut	لت	777
2	Ye	Requiem sharks	Carcharhinidae	Lokhem	لخام	778
1,2	SA,Ye	Doublespotted queenfish	*Scomberoides lysan*	Lysan	لسان	779
7	Su	Two-spot red snapper	*Lutjanus bohar*	Lolab	لو لاب	780
1,5,6,7,10	Er,SA,Su	Yellow-edged lyretail	*Variola louti*	Louti	لوطي	781
7	Su	Humphead wrasse	*Cheilinus undulatus*	Limalima	ليما ليما	782
14	Aq	Double-ended pipefish	*Trachyrhamphus bicoarctatus*	Masas	ماساس	783
1,5,6,7,10	Er,SA,Su	Kawakawa	*Euthynnus affinis*	Ma'agab	ماعجب	784
1	SA	Skipjack tuna	*Katsuwonus pelamis*			785
11	Er	Sky emperor	*Lethrinus mahsena*	Mahsena	محسنه	786
11	Er	Dorab wolf-herring	*Chirocentrus dorab*	Mekhlef	مخلف	787
7,12	Eg,Su	King soldierbream	*Argyrops spinifer*	Morjan	مرجان	788
2	Ye	Humphead snapper	*Lutjanus sanguineus*			789
9	SA	Japanese threadfin bream	*Nemipterus japonicus*			790
9	SA	Threadfin breams	Nemipterus spp.			791
10	Er	Puffers	Tetraodontidae			792
14	Aq	Network pipefish	*Corythoichthys flavofasciatus*	Masas ramli	مساس رملي	793
14	Aq	Schultz's pipefish	*Corythoichthys schultzi*			794
1,11	Er,SA	Rainbow runner	*Elagatis bipinnulata*	Muslabah	مسلابه	795
14	Aq	Pomfret	*Ariomma brevimanus*	Maslimani	مسلماني	796
14	Aq	Yellow-ear angelfish	*Apolemichthys xanthotis*	Moscht	مشط	797
14	Aq	Dusky angelfish	*Centropyge multispinus*			798
14	Aq	Threadfin butterflyfish	*Chaetodon auriga*			799
14	Aq	Blacktail butterflyfish	*Chaetodon austriacus*			800

(continued)

Table 10.4 (continued)

Source[a]	Country[b]	Common name	Scientific name	Local name	Arabic name	ID
14	Aq	Diagonal butterflyfish	*Chaetodon fasciatus*			801
14	Aq	Blackback butterflyfish	*Chaetodon melannotus*			802
14	Aq	Eritrean butterflyfish	*Chaetodon paucifasciatus*			803
14	Aq	Bluecheek butterflyfish	*Chaetodon semilarvatus*			804
14	Aq	Chevron butterflyfish	*Chaetodon trifascialis*			805
14	Aq	Zebra angelfish	*Genicanthus caudovittatus*			806
14	Aq	Emperor angelfish	*Pomacanthus imperator*			807
14	Aq	Yellowbar angelfish	*Pomacanthus maculosus*			808
14	Aq	Regal angelfish	*Pygoplites diacanthus*			809
11	Er	Strongspine silver-biddy	*Gerres longirostris*	Mukeresh	مكارش	810
10	Er	Barracudas	Sphyraenidae	Mekazel	مكازل	811
11	Er	Gracile lizardfish	*Saurida gracilis*	Macarona	مكرونا	812
5,9,11	Er,SA,Su	Greater lizardfish	*Saurida tumbil*			813
5,11	Er,Su	Brushtooth lizardfish	*Saurida undosquamis*			814
10	Er	Lizardfishes	Synodontidae			815
14	Aq	Yellowtail barracuda	*Sphyraena flavicauda*	Malleeta	مليتة	816
15	Ye	Common silver-biddy	*Gerres oyena*	Mehara	مهارا	817
13	Eg	Smooth-belly sardinella	*Amblygaster leiogaster*	Moza	موزة	818
10	Er	Indian spiny turbot	*Psettodes erumei*	Muse	موس	819
10	Er	Indian spiny turbot	*Psettodes erumei*	Mousa	موسى	820
14	Aq	Yellowfin hind	*Cephalopholis hemistiktos*	Mumen	مومن	821
14	Aq	Bluespotted wrasse	*Anampses caeruleopunctatus*	Muesy	ميسي	822
14	Aq	Lined wrasse	*Anampses lineatus*			823
14	Aq	Spotted wrasse	*Anampses meleagrides*			824
14	Aq	Yellowbreasted wrasse	*Anampses twistii*			825
14	Aq	Lyretail hogfish	*Bodianus anthioides*			826
14	Aq	Mental wrasse	*Cheilinus mentalis*			827
14	Aq	Spottail coris	*Coris caudimacula*			828
14	Aq	Dapple coris	*Coris variegata*			829
14	Aq	Green birdmouth wrasse	*Gomphosus caeruleus*			830
14	Aq	Checkerboard wrasse	*Halichoeres hortulanus*			831
14	Aq	Zigzag wrasse	*Halichoeres scapularis*			832
14	Aq	Klunzinger's wrasse	*Thalassoma klunzingeri*			833
14	Aq	Moon wrasse	*Thalassoma lunare*			834
1	SA	Molas or ocean sunfishes	*Mola* spp.	Milla	ميله	835
1,5,7,10	Er,SA,Su	Spotted coralgrouper	*Plectropomus maculatus*	Najil	ناجل	836
14	Aq	Roving coralgrouper	*Plectropomus pessuliferus*			837
8	Su	Groupers	*Plectropomus* spp.			838
2	Ye	Grunts	Haemulidae	Nakem	ناكم	839
1	SA	Smallspotted grunter	*Pomadasys commersonnii*			840
3	Ye	Grunters	*Pomadasys* spp.			841
11	Er	Tiger shark	*Galeocerdo cuvier*	Nebrawi	نبراوي	842
1,9	SA	King soldierbream	*Argyrops spinifer*	Najar	نجار	843
14	Aq	King soldier bream	*Argyrops spinifer*			844
1	SA	Jacks and pompanos	*Seriola* spp.	Nazkha	نزخة	845
3	Ye	Dorab wolf-herring	*Chirocentrus dorab*	Nakanaf	نكاناف	846
3	Ye	Tiger shark	*Galeocerdo cuvier*	Numrani	نمراني	847
9	SA	Threadfin breams	*Nemipterus* spp.	Nofrah	نوفره	848
14	Aq	Red Sea seabream	*Diplodus noct*	Noct	نوكت	849
2	Ye	Humphead snapper	*Lutjanus sanguineus*	Nirjan	نيرجان	850
10	Er	Snappers	Lutjanidae	Naisarah	نيصاره	851

(continued)

Table 10.4 (continued)

Source[a]	Country[b]	Common name	Scientific name	Local name	Arabic name	ID
5,7	Su	Yellowtail scad	*Atule mate*	Haboot	هابوت	852
7	Su	Jacks and pompanos	*Caranx* spp.			853
5,6,7	Su	Dory snapper	*Lutjanus fulviflamma*	Habair	هبير	854
14	Aq	Starry triggerfish	*Abalistes stellatus*	Hijma	هجمه	855
14	Aq	Yellow-spotted triggerfish	*Pseudobalistes fuscus*			856
11	Er	Honeycomb stingray	*Himantura uarnak*	Halali	هلالى	857
11	Er	Ribbontail stingray	*Taeniura lymma*			858
11	Er	Flower crab	*Portunus pelagicus*	Hinkakre	هنكاكار	859
11	Er	Giant mud crab	*Scylla serrata*			860
14	Aq	Jarbua terapon	*Terapon jarbua*	Henw	هنو	861
9	SA	Dory snapper	*Lutjanus fulviflamma*	Hobara	هوبارا	862
9	SA	Snappers	*Lutjanus* spp.			863
4	Ye	Cutlassfishes	*Trichiurus* spp.	Homalan	هوملان	864
14	Aq	Greater amberjack	*Seriola dumerili*	Hea	هيا	865
11	Er	Humpback red snapper	*Lutjanus gibbus*	Himbuk	هيمبوك	866
11	Er	Emperor red snapper	*Lutjanus sebae*			867
5,7	Su	Karenteen seabream	*Crenidens crenidens*	Hindook	هيندوك	868
2	Ye	Herrings, shads, sardines etc.	Clupeidae	Wasif	وزف	869
11	Er	Shorthead anchovy	*Encrasicholina heteroloba*			870

[a]A Barrania et al. (1980), 2 Walczak and Gudmundsson (1975), 3 Walczak (1977), 4 Bonfiglioli and Hariri (2004), 5 MEPI (1993), 6 Abu-Gideiri (1984), 7 Reed (1964), 8 FHAS (1984), 9 El-Saby and Farina (1954), 10 MOF (2012), 11 Tesfamichael and Sebahtu (2006), 12 Bayoumi (1972), 13 Rafail (1972), 14 Khalaf and Disi (1997), 15 Interview

[b]1 Aq = Aqaba (mainly Jordan), Eg = Egypt, Er = Eritrea, SA = Saudi Arabia, Su = Sudan, Ye = Yemen

Appendix 2: Some of the most common Fishes in the Red Sea fisheries

Acanthopagrus bifasciatus
Twobar seabream
Photo: Ministry of Fisheries, Eritrea

Acanthurus sohal
Sohal surgeonfish
Photo: Andrew Bruckner

Aethaloperca rogaa
Redmouth grouper
Photo: Tana S. Taylor

Argyrops spinifer
King soldierbream
Photo: Ministry of Fisheries, Eritrea

Atule mate
Yellowtail scad
Photo: Wikipedia commons, Kare Kare

Carangoides bajad
Orangespotted trevally
Photo: Andrew Bruckner

Carangoides fulvoguttatus
Yellowspotted trevally
Photo: Wikipedia commons, Kare Kare

Caranx ignobilis
Giant trevally
Photo: Andrew Bruckner

Carcharhinus melanopterus
Blacktip reef shark
Photo: Tana S. Taylor

Cephalopholis miniata
Coral hind
Photo: Andrew Bruckner

Cheilinus abudjubbe
Abudjubbe wrass
Photo: Anat Lynn

Cheilinus undulatus
Humphead wrasse
Photo: Andrew Bruckner

Decapterus russelli
Indian scad
Photo: Wikipedia commons, Robbie Cada

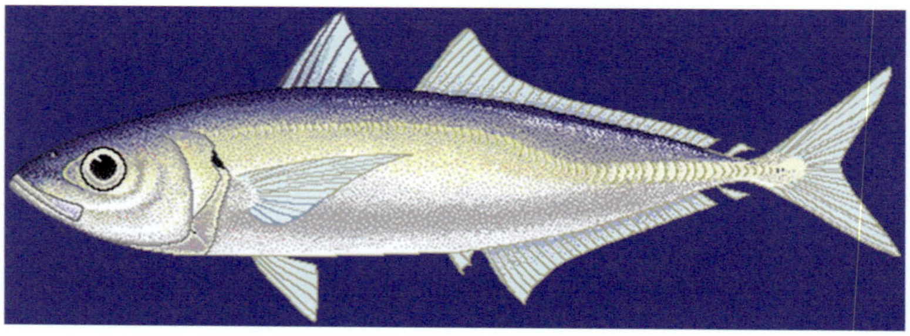

Diagramma pictum
Painted sweetlips
Photo: Ministry of Fisheries, Eritrea

Epinephelus areolatus
Areolate grouper
Photo: Ministry of Fisheries, Eritrea

Epinephelus tauvina
Greasy grouper
Photo: Andrew Bruckner

Euthynnus affinis
Kawakawa
Photo: Wikipedia commons, Robbie Cada

Galeocerdo cuvier
Tiger shark
Photo: Wikipedia commons, Albert Kok

Gerres oyena
Common silver-biddy
Photo: Ministry of Fisheries, Eritrea

Herklotsichthys quadrimaculatus
Bluestripe herring
Photo: Jonathan A. Anticamara

Himantura uarnak
Honeycomb stingray
Photo: Wikipedia commons, Aurimas Mikalauskas

Holothuroidea
Sea cucumber
Photo: Steffan Howe

Istiophorus platypterus
Indo-Pacific sailfish
Photo: Ministry of Fisheries, Eritrea

Leiognathus equulus
Common ponyfish
Photo: Ministry of Fisheries, Eritrea

Lethrinus mahsena
Sky emperor
Photo: Ministry of Fisheries, Eritrea

Lethrinus nebulosus
Spangled emperor
Photo: Tana S. Taylor

Lutjanus bohar
Two-spot red snapper
Photo: Ministry of Fisheries, Eritrea

Lutjanus fulviflamma
Dory snapper
Photo: Andrew Bruckner

Lutjanus gibbus
Humpback red snapper
Photo: Tana S. Taylor

Mugil cephalus
Flathead grey mullet
Photo: Ministry of Fisheries, Eritrea

Mulloidichthys flavolineatus
Yellowstripe goatfish
Photo: Tana S. Taylor

Naso unicornis
Bluespine unicornfish
Photo: Wikipedia commons, BS Thurner Hof

Nemipterus japonicus
Japanese threadfin bream
Photo: Ministry of Fisheries, Eritrea

Parupeneus forsskali
Red Sea goatfish
Photo: Ministry of Fisheries, Eritrea

Penaeus semisulcatus
Green tiger prawn
Photo: Ministry of Fisheries, Eritrea

Plectorhinchus gaterinus
Blackspotted rubberlip
Photo: Tana S. Taylor

Plectorhinchus schotaf
Minstrel sweetlip
Photo: Ministry of Fisheries, Eritrea

Pomadasys argenteus
Silver grunt
Photo: Ministry of Fisheries, Eritrea

Pomadasys kaakan
Javelin grunter
Photo: Ministry of Fisheries, Eritrea

Portunus pelagicus
Flower crab
Photo: Wikipedia commons, Tanaka Juuyoh

Psettodes erumei
Indian spiny turbot
Photo: Ministry of Fisheries, Eritrea

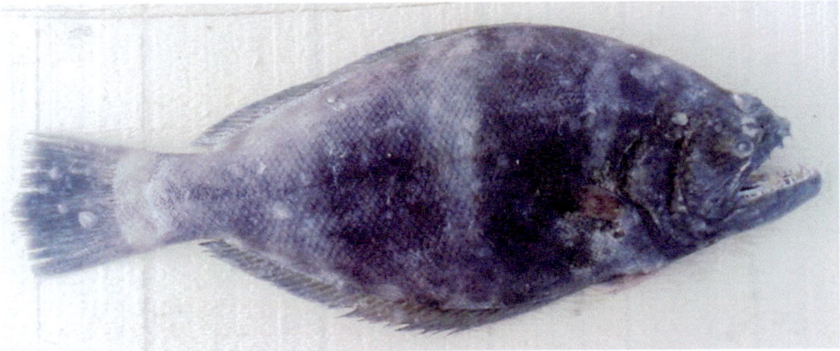

Rachycentron canadum
Cobia
Photo: Ministry of Fisheries, Eritrea

Rastrelliger kanagurta
Indian mackerel
Photo: Ministry of Fisheries, Eritrea

Rhincodon typus
Whale shark
Photo: Andrew Bruckner

Saurida tumbil
Greater lizardfish
Photo: Ministry of Fisheries, Eritrea

Scomberoides lysan
Doublespotted queenfish
Photo: Jonathan A. Anticamara

Sphyraena jello
Pickhandle barracuda
Photo: Ministry of Fisheries, Eritrea

Taeniura lymma
Ribbontail stingray
Photo: Tana S. Taylor

Thunnus tonggol
Longtail tuna
Photo: Wikipedia commons, Robbie Cada

Upeneus moluccensis
Goldband goatfish
Photo: Ministry of Fisheries, Eritrea

Variola louti
Yellow-edged lyretail
Photo: Tana S. Taylor

References

Abu-Gideiri YB (1984) Fishes of the Sudan. University Press, Khartoum, 66 p

Ammon U (ed) (2001) The dominance of English as a language of science: effects on other languages and language communities. Walter de Gruyter, Berlin

Barrania A, Bringi MR, Saleh M (1980) Socio-economic aspects of the Saudi Arabian fisheries in the Red Sea. FAO/UNDP Project for Development of Fisheries in Areas of the Red Sea and Gulf of Aden, Suez, 54 p

Bayoumi AR (1972) Recent biological investigation in the Red Sea along the A.R.E. coasts. I – on some demersal fishes of economic importance from the Red Sea with notes on migration of fish through the Suez Canal. Bull Inst Oceanogr Fish, UAR 2:159–183

Becker CD, Ghimire K (2003) Synergy between traditional ecological knowledge and conservation science supports forest preservation in Ecuador. Conserv Ecol 8(1):1

Berlin B (1992) Ethnobiological classification: principles of categorization of plants and animals in traditional societies. Princeton University Press, Princeton, 335 p

Bonfiglioli A, Hariri KI (2004) Small-scale fisheies in Yemen: social assessment and development prospects. FAO/The World Bank, 111 p

Burns TR, Engdahl E (1998) The social construction of consciousness. Part 1: collective consciousness and its socio-cultural foundations. J Conscious Stud 5(1):67–85

El-Saby MK, Farina L (1954) Report to the government of Saudi Arabia on possibility for development of marine fisheries. FAO, Rome, 29 p

FHAS (1984) Fish production in Sudan – a study on the methodology and the data collected during 1979 and 1980. FAO/UNDP, Suez, 61 p

Forsskål P (1775) Descriptiones animalium, avium, amphibiorum, piscium, insectorum, vermium; quae in itinere orientali observavit. Hauniae, Copenhagen

Freire KMF, Pauly D (2003) What's in there: common name of Brazilian marine fishes. In: Haggan N, Brignall C, Wood L (eds) Putting fisher's knowledge to work. Fisheries Centre Research Reports, vol 11(1). Fisheries Centre, University of British Columbia, Vancouver, pp 439–444

Freire KMF, Pauly D (2005) Richness of common names of Brazilian marine fishes and its effect on catch statistics. J Ethnobiol 25(2):279–296

Froese R, Pauly D (2012) FishBase. Available at: www.fishbase.org.

Johannes RE (1981) Words of the lagoon: fishing and marine lore in the Palau District of Micronesia. University of California, Berkeley

Johannes RE, Freeman MMR, Hamilton RJ (2000) Ignore fishers' knowledge and miss the boat. Fish Fish 1(3):257–271

Khalaf MA, Disi AM (1997) Fishes of the Gulf of Aqaba. Marine Science Station, Aqaba, 252 p

MEPI (1993) Identification study for Sudan Red Sea fisheries. Ministry of Economic Planning and Investment, Khartoum, 213 p

MOF (2012) Eritrean fisheries database system. Ministry of Fisheries, Massawa

Palomares MLD, Garilao CV, Pauly D (1999) On the biological information content of common names: a quantitative case study of Philippine fishes. In: Séret B, Sire J-Y (eds) Proceedings of the 5th Indo-Pacific fish conference, Nouméa, 3–8 November 1997, Paris, pp 861–866

Palomares MLD, Pauly D (2012) SeaLifeBase. Available at: www.sealifebase.org .

Rafail SZ (1972) Studies of Red Sea fisheries by light and purse-seine near Al-Ghardaqa. Bull Inst Oceanogr Fish, UAR 2:25–49

Reed W (1964) Red Sea fisheries of Sudan. Game and Fisheries Section, Ministry of Animal Resources, Sudan

Rüppell E (1826) Atlas zu der Reise im nördlichen Afrika1826–1828. Gedruckt und in Commission bei Heinr. Ludw. Brönner, Frankfurt am Main

Tesfamichael AG, Sebahtu SH (2006) Commercial fish of the Eritrean Red Sea. Apex publishing Ltd, Essex

Tesfamichael D (2005) Reconstructing the John Murray/Mabahiss expedition. The Sea Around Us Project Newsletter 29:4

Tesfamichael D, Pitcher TJ, Pauly D (2014) Assessing changes in fisheries using fishers' knowledge to generate long time series of catch rates: a case study from the Red Sea. Ecol Soc 19(1)

Tonkin E (1986) Investigating oral tradition. J Afr Hist 27(2):203–213

Walczak P (1977) The Yemen Arab republic. A study of the marine resources of the Yemen Arab republic. A report prepared for the fisheries development project. FAO, Rome, 67 p

Walczak P, Gudmundsson J (1975) Yemen Arab republic – summary of fish stock assessment work. FAO, Rome, 14 p

Acronyms and Glossary[1]

Acoustic survey Surveys of actual potential fishing grounds by one or several vessels equipped with a sound emitting device (i.e., an echo sounder), producing sounds which are reflected by fish schools and even single fish (especially if they have a gas bladder), and whose abundance can thus be estimated.

Aquaculture According to *FAO, aquaculture is the farming of aquatic organisms including fish, molluscs, crustaceans and aquatic plants, with some sort of intervention in the rearing process to enhance production, such as regular stocking, feeding, protection from predators, etc. See also *Mariculture.

Areas beyond national jurisdiction See *High Sea(s).

Artisanal Referring to small-scale fishers who are catching fish that is predominantly sold, i.e., small-scale commercial fishers

Atoll An island (of groups of islands), mostly surrounded by deep water, and consisting of a ring-like perimeter of shallow coral reefs enclosing a shallow *lagoon, the entire structure sitting on the tip of a sunken volcano.

B_{MSY} Biomass at which an exploited stock generates *Maximum Sustainable Yield (*MSY), i.e., generally half the unexploited biomass.

Bait/baitfish Fish used to catch other fishes, e.g., in pole and line fishing, or in longlining for tuna.

Beach seining A fishing method where a net and a length of rope are laid out from and back to the shore and retrieved by hauling the net on to the shore. Often, the hauling is performed by a large group of people (e.g., from a village community), with the fish that are caught then shared between them. Beach seines are problematic in that they catch juvenile fish and thus contribute to *growth overfishing.

Benthos The community of organisms which live on the bottom of a water body, in it or near it.

Biodiversity A term used to refer to the full range of living organisms, including, inter alia, terrestrial, marine and other aquatic ecosystems and the ecological complexes of which they are part; this includes diversity within species, between species and of ecosystems.

Biomass Weight of a *stock or of one of its components; thus, e.g., 'spawning biomass' is the combined weight of all sexually mature animals in a stock. *Standing stock is an alternative term for biomass. Also, the mass of living tissues across organisms in a *population or *ecosystem. Used as a measure of population abundance.

By-catch/bycatch That part of a fish catch that is caught in addition to the *target species because the fishing gear (e.g., a *trawl) is not selective. By-catch may be retained, landed and sold or used, or may be dumped at sea (see *discard).

Canoe The smallest boat used by artisanal and subsistence fisheries, usually on a day trip.

Capacity (fleet) In input terms, fleet capacity can be considered as the minimum fleet size and effort required to produce a given catch. In output terms, capacity can be considered as the maximum catch that a fisher or a fleet can produce with given levels of inputs, such as fuel, amount of fishing gear, ice, bait, engine horsepower and vessel size.

Cascade (trophic) Trophic cascades occur when predators in a *food web suppress the abundance and/or alter the behavior of their prey, such that the next lower *trophic level is released from predation (or *herbivory if the intermediate trophic level is a herbivore). For example, if the abundance of large piscivorous fish is increased, the abundance of their prey, zooplanktivorous fish, should decrease, large *zooplankton abundance should increase, and *phytoplankton biomass should decrease. This concept has stimulated research in many areas of marine ecology and fisheries biology.

Catch The number or weight of fish or other animals caught or killed by a fishery, including fishes that are landed (whether reported in statistics or not), discarded at sea, or killed by lost gear (*ghost fishing').

Catch composition Refers to the different taxa (species, genera, family) making up the catch of a fishery. The more detailed a catch composition is, the more useful it is. When their composition is unavailable, catches are often labelled as 'miscellaneous fish' or similarly uninformative labels.

[1]Adapted from the glossary in FishBase (www.fishbase.org), from Wikipedia and other sources including Holt, S.J. 1960. Multilingual vocabulary and notation for fishery dynamics. FAO, Rome. The symbol * refers to an entry elsewhere in this list of acronyms and glossary.

D. Tesfamichael, D. Pauly (eds.), *The Red Sea Ecosystem and Fisheries*, Coral Reefs of the World 7,
DOI 10.1007/978-94-017-7435-2

Catch per unit of effort (CPUE; or catch/effort) A measure of relative abundance, obtained by dividing the catch by a measure of the fishing effort required to realize this catch. Generally proportional to *biomass.

Climate change A lasting change in the statistical distribution of weather patterns over periods ranging from decades to millions of years. It may be a change in average weather conditions, or in the distribution of weather around the average conditions (i.e., more or fewer extreme weather events such as storms or heat waves). Climate change is caused by factors such as variations in solar radiation received by Earth, plate tectonics, and volcanic eruptions. The release of greenhouse gases (notably carbon dioxide and methane) by human activities have also been identified as the cause of recent climate change, often referred to as 'global warming'.

Coastal zone The region where interactions of the sea and land processes occur.

Collapse(d) Rapid decline in the abundance (*biomass) of a stock, generally reflected in a rapid decline of catches from that stock, either because there are fewer fish to be caught than previously, or less often, because the fishery for this stock is closed or strongly reduced. In *SSPs, collapsed stock are defined as stocks with catches of less than 10 % of the historically maximum catch.

Commercial Refers to a fishery whose catch is sold. This means that both large-scale (or industrial) and small-scale (i.e., artisanal) are commercial fisheries, and that the term 'commercial fisheries' should not be considered synonymous with industrial or large-scale fisheries.

Commercial fisher A person who fishes for a living and sells the vast majority of his/her catch.

Common names Here, the locally variable names of fishes and other Red Sea animals in vernacular languages, notably Arabic and English and which can be written with any characters, in contrast to *Scientific names.

Compliance Adherence of individual fishers to the harvest strategies of official fisheries management bodies.

CPUE *Catch per Unit of Effort, or catch/effort.

Demersal Organisms swimming just above or lying on the seafloor and usually feeding on *benthic organisms.

Discard Portion of catch that is thrown overboard, but which may be of important ecological or commercial value. Discard typically consists of 'non-target' species or undersized specimens of the target species. High-grading is a special form of (mostly illegal) discarding where a catch of target species is thrown overboard to make space in the hull (or accommodate under a quota) fresher, larger or otherwise more valuable catch of the same species.

Distant-water fleet/fishery The fleet of a country that is fishing in the *EEZ of another country (or the EEZs of other countries), or in *High Sea regions not adjacent to it own EEZ. Under *UNCLOS, a distant-water fishery can be conducted in EEZ of a coastal state only with its explicit *access agreement, generally in exchange for a compensation.

Domestic Here, pertaining to a country's or territory's own *EEZ.

Driftnet Nets hanging vertically in the water column, without being anchored to the bottom. The nets are keep vertical in the water by floats attached to a rope along the top of the net and weights attached to another rope along the bottom of the net. Driftnets generally rely on the entanglement properties of loosely affixed netting. Folds of loose netting, much like a window drapery, snag on a fish's fins and tail and wrap it up in loose netting as it struggles to escape. However the nets can also function as gill nets if fish are captured when their head get stuck in the net. Drift net are unselective, and thus kill thousands of marine mammals, turtles and seabirds, beside the fish *bycatch. Prior to the 1960s, the size of drift nets was not limited, and they grew to lengths in excess of 50 km. In 1992, the UN banned the use of drift nets longer than 2.5 km long in the *High Seas.

DWF *Distant-water fleet or fishery.

Ecopath An approach and software package allowing for the straightforward construction a mass-balance models (=quantified representations) of the trophic linkages in (aquatic) ecosystems at a given time, or during a given period.

Ecopath with Ecosim An ecosystem modeling software currently integrating *Ecopath, *Ecosim and *Ecospace.

Ecosim An add-on to *Ecopath, which uses its parametrization to define a system of differential equation allowing changes in e.g., fishing tactics or environmental forcing to be evaluated in term of their effects on the ecosystem as a whole.

Ecospace An add-on to *Ecosim which allows the processes it simulate to be represented spatially.

Ecosystem Community of plants, animals and other living organisms, together with the non-living components of their environment, found in a particular habitat and interacting with each other.

EEZ *Exclusive Economic Zone

Effort (fishing) Any activity or devices deployed to catch fish, and which can be quantified. Thus, the number of nets of a certain type deployed in a set period is a measure of effort, as is the amount of fuel used by a fishing fleetor days fished per year, etc.

Endemic Native and restricted to a particular area, e.g., an island, a country, a continent, an ocean.

EwE *Ecopath with Ecosim (and *Ecospace).

Exclusive Economic Zone (EEZ) Generally, all waters within 200 nautical miles (370 km) of a country and its outlying islands, unless such areas would overlap because neighboring countries are less than 400 nautical miles

(740 km) apart. If an overlap exists, it is up to countries to negotiate a delineation of the actual maritime boundary. Under *UNCLOS, a country has special rights regarding the exploration and use of marine resources inside its EEZ, such as the power to control and manage all fishery resources in this zone. Not until 1982, with the adoption of UNCLOS, did 200 nm EEZs become formally adopted; a country needs to formally declare its EEZ to have one.

F_{MSY} The value of fishing mortality *F* which produces the maximum yield in the long-term, i.e., *Maximum Sustainable Yield (MSY).

FAO Food and Agriculture Organization of the United Nations.

Food and Agriculture Organization of the United Nations (FAO) The only agency in the world tasked with annually assembling global fisheries statistics, and generally assisting member countries in managing their fisheries.

Feeding (or trophic) interactions Linkages between (groups of) species dues to grazing (in herbivores) or predation (in other animals), and whose strength defines the dependencies of these species upon each others. Feeding interactions are stronger *within* *ecosystems than *between* them, and this in fact, largely defines their borders.

Finfish Members of class *Pisces*, i.e., aquatic animal with fins, as distinguished from *shellfish.

Finning Removing (by cutting) the valuable fins off freshy caught sharks (i.e., while they are alive), often with subsequent discarding of the carcasses.

Fish/Fishes The term 'fish' *sensu stricto* refers to the taxonomic (*Taxon) class *Pisces* (*Finfish) in the Subphylum Vertebrata, Phylum Chordata. In the wider sense, 'fish' refer to aquatic animals sought by fisheries, i.e., *Finfish + invertebrate *shellfish; the plural 'fishes' is used when explicitly referring to more than one species of *Finfish.

Fish Aggregating Devices (FADs) Floating objects made of vegetable matter (e.g., palm fronds) or artificial materials (e.g., plastic, steel and even concrete), some anchored on the sea floor, which attract *pelagic fishes, mainly tuna, based on their propensity to congregate under floating debris. Some FADs are now equipped with sensors (linked to satellites), to assess when whey can be fished. FADs attract juvenile fish and thus can contribute to *growth overfishing

Fisheries-independent data Information about a fishery resources not based on catches and derived statistics such as *catch/effort (or *CPUE). Typically, fisheries-independent data are obtained from dedicated research vessels performing *trawling or *acoustic surveys, and from fish *tagging operations. This can, however, involve remote sensing (satellite) data, or shore based surveys, e.g., of human fish consumption.

Fisherman A person who fishes professionally (or for *subsistence or recreation). Nowadays commonly replaced by 'fisher', to allow for the many women who are also engaged in fishing.

Fishery A set of persons and gear interacting with an aquatic resource (one or several species of fish) for the purpose of generating a *catch.

Fishing down (marine food webs) The process whereby fisheries in a given ecosystem, having depleted the large predatory fish on top of the food web, turn to increasingly smaller species, finally ending up with previously spurned small fish and invertebrates. See also see Wikipedia entry on this, and www.fishingdown.org.

Fishmeal Protein-rich animal feed product based on ground up fish, usually small *pelagic fishes such as anchovies and sardines, which are also directly consumed by people.

Food security According to the FAO, this occurs "when all people, at all times, have physical and economic access to sufficient, safe and nutritious food to meet their dietary needs and food preferences for an active and healthy life". *Seafood contributes crucially to food security in numerous countries where alternative sources of animal protein and micronutrients are lacking.

Food web The ensemble of *feeding (or trophic) interactions connecting the elements of (and defining) an *ecosystem.

Growth overfishing The catching of *fish that are too small relative to their growth potential, even when account is taken of their natural mortality. Growth overfishing is caused by the excessive deployment of gear catching undersize fish.

Herbivory Feeding only on organisms with a *trophic level of 1, i.e., plants.

High Sea(s) The areas of the world ocean that is outside of the 200 mile Exclusive Economic Zone of coastal states; the High Seas cover about 60 % of the world ocean.

Huri One of the commonly used wooden boats by artisanal fishers in the Red Sea. It is smaller than *Sambuk usually with outboard engine.

IFA Inshore Fishing Area, i.e., up to 50 km or a depth of 200 m from an exploited coastline.

Illegal, Unreported and Unregulated (IUU) Illegal, Unreported and Unregulated; an acronym proposed by *FAO to describe fisheries and catch that have issues. This acronym has become synonymous with 'illegal' in practice, and thus confuses people.

Industrial Referring to *large-scale fisheries which are catching fish for commercial marketing or global export, i.e., large-scale commercial fisheries. The distinction between large-scale and *small-scale (i.e., *artisanal) is, for almost all countries, the definition prevailing in those very countries, and is usually related to vessel size and gear type used (For countries without such definition, the definitions used in similar countries were used).

Kusar A local term used to describe a tradition where part of the catch (usually from artisanal) is given freely to family, friends and people with need (e.g., widows, the elderly). It is an informal food security social network that fishers are expected to share part of their catch before they market the remaining. Not to participate in the norm can result in social ostracism: not allowed to sell and buy in the local market or marriage prohibition.

Lagoon Smaller water bodies, associated with either coastlines or coral reefs. Costal lagoons are formed behind permanent of occasionally (or seasonally) occurring sand bars, and their salinity (and hence the flora and fauna of costal lagoons) depend on their water exchange with the land interior (rivers) and the sea (breaks in the sand bars). Coastal lagoons can be very productive, e.g., along the cast of the Gulf of Guinea. The shallow water body whose periphery is defined by *atolls are also called lagoons, particularly if the water exchanges with the outside can also be high. Lagoon of both types are vulnerable to sudden drops in salinity, which can kill the resident organisms.

Landings Weight of the catch landed at a wharf or beach. Also: The number or weight of fish unloaded at a dock by *commercial fishers or brought to shore by *subsistence and *recreational fishers for personal use. Landings are reported at the points at which fish are brought to shore. Note that here, catch = landing + *discards.

Large Marine Ecosystem Large Marine Ecosystems (*LMEs) are large areas of ocean space (approximately 200,000 km^2 or greater), adjacent to the continents in coastal waters where *primary productivity is generally higher than in open ocean areas. The delineation and definition of individual LMEs and their boundaries are based on four ecological, rather than political (see *EEZ) or economic, criteria. These are: (i) bathymetry, (ii) hydrography, (iii) productivity, and (iv) trophic relationships. Based on these ecological criteria, to-date 66 distinct LMEs have been delineated around the coastal margins of the Atlantic, Pacific and Indian Oceans.

Lessepsian migrants Organisms that have migrated from the Red Sea into the Mediterranean through the Suez Canal, or much more rarely, from the Mediterranean into the Red Sea.

LME *Large Marine Ecosystem.

Longline/Longlining A line of considerable length, bearing numerous baited hooks, and which is usually set horizontally in the water column; used, e.g., in snapper, grouper, ling and tuna fisheries. The line is set for varying periods up to several hours on the seafloor, or in the case of tuna, in mid-water water at various depths. Also a fishing line with baited hooks set at intervals on branch lines; it may be 150 km long and have several thousand hooks and can be on the sea bed or above it supported by floats. It may be anchored or drift free and is marked by floats.

Mangrove Trees and shrubs that grow in saline coastal sediment habitats in the tropics and subtropics, mainly between latitudes 25°N and 25°S. Mangroves are saline wood- or shrub-land habitats that are characterized by depositional coastal environments, where fine sediments (often with high organic content) collect in areas protected from high-energy wave action. The saline conditions tolerated by various mangrove species range from brackish water, through pure seawater, to water concentrated by evaporation to over twice the salinity of ocean seawater. Mangroves are crucial juvenile *nursery habitats for many fisheries species, and also fulfill a very important coastal protection function, where they, similar to healthy coral reefs, shield coastlines from the impacts of high energy ocean surges and tropical storm damage.

Mariculture The farming of aquatic organisms in seawater, such as fjords, inshore and open waters in which the salinity generally exceeds 20 *psu. Earlier stages in the life cycle of these aquatic organisms may be spent in *brackish water or freshwater.

Marine Protected Area (MPA) Areas of the ocean within which fishing and other extractive activities are limited. Often used to mean 'no-take area', where no fishing is allowed, but for which the term *marine reserve is more appropriate.

Marine reserve A version of *marine protected area that has legal protection against fishing (i.e., is designated a 'no-take area'), i.e., an area of ocean space where all fishing is prohibited. Benefits include increases in the biodiversity, abundance, biomass, body size and reproductive output of fisheries populations.

Maximum Sustainable Yield (MSY) The maximum amount that can be taken (caught) over the long term from a fisheries resource. MSY is best considered an upper limit for fishery management, as opposed to a *target level.

Midwater Refers to trawling, net or line fishing at a water depth that is higher in the water column than the bottom of the ocean. It is contrasted with bottom (or *benthic) fishing. Also known as *pelagic fishing.

Monitoring, Control and Surveillance (MCS) Term used for the broadening of traditional means of enforcing national rules over fishing, to the support of the broader problem of fisheries management. MCS has aspects distinct from fisheries management, although there is overlap, and increasingly management cannot be effective without MCS. While MCS, in the traditional definition, does not include enforcement, that category will need to be included as part of successful implementation of MCS operations. There is a strong emphasis that the success of MCS is not to be measured in number of arrests, but in the level of compliance with reasonable management frameworks.

Nei FAO acronym for 'not elsewhere included'; a synonym for 'other fish species' or 'other invertebrate species' not specifically identified.

Neritic The shallow pelagic zone over the continental shelf, down to a depth of 200 m.

NPP Net primary *production; the fraction of primary production that is available to grazers, e.g., *zoooplankton.

Nursery The area where fish larvae metamorphose into juveniles and where the latter remain until they 'recruit' to the adult stock. *Mangroves, in the tropics, often serve as nurseries, as do *estuaries, coastal *lagoons and generally, shallow areas. However, contrary to a widespread belief, mangroves, and the other nearshore habitats are *not* the places where most marine fish spawn (which generally occurs offshore).

Nutrients Those nutritional constituents required by organisms for body maintenance and growth; for marine primary production, the key nutrients are nitrates, silicates and phosphates, generally supplied by *tidal mixing and *upwellings.

Omnivory This occurs when consumers feed at several trophic level, such as a bear feeding on both berries (TL = 1) and salmon (TL = 3.5 – 4.0); also see *Herbivory.

Open-access fisheries Fisheries operated by fishers who do not have exclusive rights to the resource they exploit, i.e., anyone can enter the fishery. Many fisheries which appear to be open-access upon superficial examination are actually limited-entry fisheries with customary limits to entry. Examples are some of the coral reef fisheries of the tropical Pacific, where traditional tenure systems persist.

Overfishing Applying a level of fishing effort beyond which will generate a desirable, sustainable, or 'safe' population or stock level. The level of effort can be in excess of that required to generate *Maximum Sustainable Yield (biological overfishing), *Maximum Economic Yield (economic overfishing), maximum yield per recruit (growth overfishing), or maximum *recruitment (recruitment overfishing)

P/B ratio The ratio of *production to biomass of an organism (or roughly: its 'turnover rate'), of dimension $time^{-1}$, is a measure of its productivity. It is usually equivalent to its observed instantaneous rate of total *mortality, because any population of organisms, if it is going to persist, must compensate the losses it experiences by *recruitment and growth, which jointly define productivity.

Pelagic Living and feeding in the open sea; associated with the surface or middle depths of a body of water; free swimming in the seas, oceans or open waters; not in association with the bottom (*benthic). Many pelagic fish feed on plankton.

Plankton Community of living plants (microscopic *phytoplankton) and animals (*zooplankton) whose lack of powerful propulsive organs force them to drift with the water body in which the vagaries of turbulence, *currents or *upwelling have placed them.

Population A set of interacting organisms of the same species that live in the same geographical area, and have the capability of interbreeding. Roughly corresponds to the concept of *stock as used by fishery scientists.

Practical Salinity Unit The ratio K of the electrical conductivity of a sea water sample of 15 °C and the pressure of one standard atmosphere, to that of a potassium chloride (KCl) solution, in which the mass fraction of KCl is 0.0324356, at the same temperature and pressure. The K value exactly equal to one corresponds, by definition, to a practical salinity equal to 35. In this definition, salinity is a ratio and parts per thousand (‰) is therefore no longer used, but an old value of 35‰ corresponds to a value of 35 practical salinity unit. Practical salinity is a ratio and strictly no units should be used but often 'psu' is added to the value.

Production In Ecology and Fisheries biology, production refers to the sum of all growth increments of the animals or plants of a population over a given time period, including the growth of individual that may not have survived to the end of that period. Most of the *primary production of the ocean is due to *phytoplankton, while secondary production is due to herbivorous *zooplankton. The term 'production', which is appropriate in agriculture, may be applied to *aquaculture (incl. *mariculture), but should never be applied to fisheries *catches (or even *landings) which are not 'produced' by fishing.

Psu *Practical Salinity Unit.

Purse seine A fishing net used to encircle *pelagic fish. The net may be of up to 1 km length and 300 m depth and is used to encircle surface schooling fish such as anchovies, mackerel or tuna. It is usually set at speed from a larger vessel while the other end is anchored by a small boat. During retrieval, the bottom of the net is closed or pursed by drawing a purse line through a series of rings to prevent the fish escaping.

Q/B ratio The ratio of consumption to biomass of an organism (dimension $time^{-1}$) which is a measure of its food intake.

Quota The amount of fish that a country, enterprise or fisher is allowed to take in a given year. Also refers to a constant fraction of a variable *Total Allowable Catch or TAC. Quotas can also be given to individual fisheries; see *ITQ.

Rebuilding To reduce fishing (e.g., via low to zero quotas) until the natural processes of recruitment and individual growth cause the biomass of a stock to increase to some pre-set level, e.g., *B_{MSY}.

Reconstruction (catch) A set of procedures to derive a coherent time series of likely total catches for the fisheries of a country or area from various sources, not necessarily including official catch statistics; also, the product of this procedures. The word and concept are derived from the science of linguistics, which 'reconstructs' extinct words (and/or languages) from words in daughter languages.

Recreational fishing This form of fishing, also called 'sport' fishing, is fishing for pleasure or in competition. This differs from *commercial fishing (both *artisanal and *industrial), where the main motivation is to catch fish for eventual sale, and from *subsistence fishing, where fish is mainly caught for personal of family and friends' consumption.

Recruitment The process by which young fish enter a fish population. Recruitment is distinguished from 'reproduction' because the high mortality experienced by fish eggs and larvae usually precludes the prediction of population sizes from the abundance of these early stages (while the number of recruits can be used to predict the number of adults).

Reduction fisheries Fisheries whose catch is used for making *fishmeal (often with fish oil as a by-product), which is then used to feed animals, e.g., pigs, chicken or farm-raised salmon. An example of a fish that is mostly 'reduced' to fishmeal is the Peruvian anchoveta (*Engraulis ringens*) and in the Red Sea anchovies and sardines were reduced to fish meal in the 1950s and 1960s. Most fish used for fishmeal are perfectly edible, and thus reduction fisheries are usually competing with fisheries whose catch is used for direct human consumption.

Regional Fisheries Management Organization International governmental organization tasked with managing the fisheries of a region of the ocean (including the *High Sea areas) for the benefit of member states.

Resilience/resilient The capacity of a system to tolerate impacts without irreversible change in its outputs or structure. In a species or population, resilience is usually understood as the capacity to withstand exploitation.

RFMO *Regional Fisheries Management Organization.

Sambuk One of the commonly used wooden boats by artisanal fishers in the Red Sea. It is the largest of the artisanal boats with inboard engine, can take higher number of crew and can stay longer fishing.

Scientific names The unique names of marine fishes and invertebrates that are consistent with the Code of Zoological Nomenclature. Scientific names – for species – consist of a unique generic name (always capitalized) and a species epithet (never capitalized), with both always written in Roman characters and in agreement with the rule of Latin.

Sea Around Us (The) Title of a 1951 bestselling book by Rachel Carson, who inspired the project of the same name, launched in July 1999.

Selective Refers to a gear with enable fishers to choose or select the species and sizes that they are going to catch. A spear gun is highly selective; a trawl is unselective. Most other gears have selectivities ranging between these two extremes.

Sharm A local (Arabic) term for *lagoon used in the Middle East.

Shelf The sea (floor) between the coast and the 200 m *isobath around the continents ('continental shelf'), and less commonly, around islands. Shelves are the most productive parts of the oceans, and support their most important fisheries.

Shellfish An aquatic animal, such as a mollusc or crustacean, that has a shell or shell-like external skeleton.

Slope (continental) Region of the outer edge of a continent between the generally shallow continental shelf and the deep ocean floor, i.e., from 200 to 2,000 m; often steep.

Small-scale *Artisanal, *subsistence and *recreational.

Soviet Russian word for 'council'; usually pertaining to the former-USSR, i.e., the now defunct Union of Socialist Soviet Republics.

Spear fishing Fishing with devices functioning like crossbow or airguns. Spear fishing using SCUBA is widely forbidden and should be everywhere.

Sport fishing *Recreational fishing.

SST Sea surface temperature.

Standing stock Synonym for *biomass.

Stern trawlers A term used formerly to distinguish *trawlers in which the trawl is pulled on deck via a slipway at the back ('stern') from others, i.e., side trawlers. Not used much nowadays, as the overwhelming majority of trawlers are stern trawlers.

Stock The exploited part of an exploited fish population.

Stock assessment A set of mathematical procedures through which the current and probable future abundance or biomass of exploited fish stocks can be estimated, using data from life-history studies, environmental surveys and catch statistics. Generally forms the basis for setting *Total Allowable Catches.

Subsidies Government funds made available to a segment of the population of a country, or a sector of its economy. When given to a well-developed fishery, subsidies tend to encourage overfishing.

Subsistence fishing A form of *small-scale (inshore) fishing (or 'gleaning'), often practised by women and children, where the catch (often small fish and invertebrates, particularly bivalves) is mainly caught for self- or family-consumption, or bartered against other commodities. The primary aim or motivation of subsistence fishing is generally not commercial sale of the catch.

Surplus (1) In so-called *surplus-production models, the surplus is the biomass that is produced in excess of what is needed to maintain the stock at its current abundance level, and which can thus be taken by a fishery without the stock declining further. *MSY is the highest surplus level, and in the Schaefer model, it occurs when the biomass is reduced – by fishing – to half its unexploited biomass; (2) surplus is also a term used in the context of *UNCLOS, wherein the vessels of one or several distant-water fishing countries should be given access to the *EEZ of a coastal state (against payment of an 'access fee') if it is not itself fully exploiting the fishery resources therein.

Surplus-production model One of the simplest analytical *stock assessment models, which pools *recruitment, mortality and growth of a *stock into a single *production function.

Surveillance A key element of 'Monitoring, Control and Surveillance' (*MCS), required to support fisheries management in the context of extended jurisdiction, i.e., in larger *EEZs. Surveillance refers degree and types of observations required to maintain compliance with the regulatory controls imposed on fishing activities. Radar, including coastal, airborne, and space borne systems, even if intended for national security or law enforcement, can provide information to fisheries management and environmental protection authorities.

Sustainable An activity or process, whose properties are such that it could last indefinitely (or at least into the long-term foreseeable future). 'Sustainable growth', by this definition, is an oxymoron.

TAC *Total Allowable Catch.

Target This term has two meanings in fisheries. One refers to the fish (species or group) that are meant to be caught; this contrasts to *bycatch, which is caught because the gear targeting a certain resource type is not or insufficiently selective. The other meaning of target refers to fisheries management, which often defines *MSY as it associated fishing mortality (*F_{MSY}) as a limit, with the target being slightly more conservative.

Taxon (plural: taxa) According to the International Code of Zoological Nomenclature, any formal unit or category of organisms (species, genus, family, order, class, etc.). Derived terms are: taxonomist, taxonomic, taxonomically.

Territorial waters The area beyond the tidal base line of the open coasts of a country over which that country exercises full control except for innocent passage of foreign vessels. Set at a maximum of 12 nautical miles in breadth by the 1982 *UNCLOS. The United States and a few other countries claim territorial waters only three nautical miles in width (See also EEZ).

Thermocline The distinct interface between surface waters and cooler, deeper waters; region between the warm upper layer and the lower cold layer of the sea or lake, where temperature declines abruptly (1 °C m^{-1} or more) with increasing depth.

Threatened Species of animals, plants, fungi, etc., which are vulnerable to endangerment. The International Union for Conservation of Nature (IUCN) is the foremost authority on threatened species, and treats threatened species (in their 'Red List') as a group of three categories, depending on the degree to which they are threatened: Vulnerable species; Endangered species; Critically Endangered species. Less-than-threatened categories are Near Threatened, and Least Concern. Species which have not been evaluated (NE), or do not have sufficient data (Data Deficient) also are not considered 'threatened' by the IUCN.

Tidal mixing The mixing of water layers due to the action of alternating high and low tides.

Total allowable catch (TAC) The amount of fish (or *quota) that can be taken legally by a given fishery in a given period (usually a year, or a fishing season), as determined by a fishery *management agency.

Traditional An adjective misleadingly used in some countries to describe *artisanal fishers and fisheries.

Trammel net A set-net consisting of three layers of netting, designed so that a fish entering through one of the large-meshed outer sections will push part of the finer-meshed central section through the large meshes on the further side, forming a pocket in which the fish is trapped.

Transhipment (at sea) The transfer of goods (here: fish) from one boat to the other while at sea, often to avoid controls available in ports.

Trash fish The earlier and badly misleading name for the fraction of the *bycatch for which no market had been identified, and which was therefore mostly *discarded.

Trawl/Trawler/Trawling Fishing methods where a vessel – a trawler – tows a large bag-shaped trawl net. A wide range of *benthic (also called demersal or bottom) or *pelagic (mid-water) species of fish are taken by this fishing method. The trawl net usually features a buoyed head (top) rope, a weighted foot (bottom) rope and two heavy 'otter' doors to keep the net mouth open. Variation include beam trawls that use a horizontal beam instead of otter doors and foot rope to keep the net open, or pair-trawls in which two vessels are used to tow a single, often considerably larger net. Bottom trawling is not *selective and is destructive of habitats. Thus, it is gradually being banned from areas that people care about. All trawls are here considered *industrial gear, whatever the size of the vessel pulling them.

Trolling A fishing method where baited hooks or lures are towed behind a boat; not to be mistaken for *trawling.

Trophic level Numbers expressing the relative 'height' of an organism within the food web, with plants having a trophic level (TL) of 1 (by definition), herbivores 2, their predators 3, etc. Because fishes have mixed diets, they tend to have intermediate TL values, e.g., 2.5 for an omnivore feeding half on plants and half on herbivores.

Tropics/Tropical A climate zone ranging north and south from the Equator to the limits of the Subtropical Zone, and generally limited to *SST above 20 °C.

UNCLOS *United Nations Convention on the Law of the Sea.

United Nations Convention on the Law of the Sea (UNCLOS) Also called the Law of the Sea Convention or the Law of the Sea treaty, is the international agreement which came into force in 1994 and which defines the rights and responsibilities of nations with respect to their use of the world's oceans, establishing guidelines for businesses, the environment, and the management of marine natural resources. Among other things, UNCLOS enabled countries to declare an *EEZ out to a maximum of 200 nautical miles.

Upwelling An oceanographic phenomenon involving wind-driven rise of dense, cooler, and usually nutrient-rich water towards the ocean surface, where it replaces (and pushes offshore) warmer, usually nutrient-depleted surface water. The cold, but nutrient-rich upwelled water stimulates the growth of primary producers (mainly *phytoplankton), and *secondary producers (mainly *zooplankton), upon which the rest of the *ecosystem (*forage fish, other fishes, seabirds and marine mammals) depend.

Vessel Monitoring System Used by agencies tasked with monitoring the position, time at a position, and course and speed of fishing vessels. VMSs are a key part of *monitoring, control and surveillance (MCS) programs. VMS may be used to monitor vessels in the territorial waters of a country or a subdivision of a country, or in their EEZs.

Withdrawal(s) Synonym of *catch or catches when applied to fish; same as 'removals.

Working Group An ad hoc or permanent group of experts with a specific task, e.g., assessing the state of an exploited *stock of fish.

World Heritage Site Area identified by UNESCO as being of great cultural and/or historical importance.

Yield Catch in weight during a conventional period, e.g., a year; see also *Maximum Sustainable Yield or *MSY.

Index

D. Tesfamichael, D. Pauly (eds.), *The Red Sea Ecosystem and Fisheries*, Coral Reefs of the World 7,
DOI 10.1007/978-94-017-7435-2

MIX
Papier aus verantwortungsvollen Quellen
Paper from responsible sources
FSC® C105338

If you have any concerns about our products,
you can contact us on
ProductSafety@springernature.com

In case Publisher is established outside the EU,
the EU authorized representative is:
Springer Nature Customer Service Center GmbH
Europaplatz 3, 69115 Heidelberg, Germany

Printed by Libri Plureos GmbH
in Hamburg, Germany